DIFFERENTIAL CALCULUS

DIFFERENTIAL CALCULUS

by

R.K. Mohanty

ANMOL PUBLICATIONS PVT. LTD.

NEW DELHI - 110 002 (INDIA)

ANMOL PUBLICATIONS PVT. LTD.
4374/4B, Ansari Road, Daryaganj
New Delhi - 110 002
Ph.: 23261597, 23278000
Visit us at: www.anmolpublications.com

Differential Calculus

First Published, 2004

PRINTED IN INDIA

Published by J.L. Kumar for Anmol Publications Pvt. Ltd., New Delhi - 110 002 and Printed at Mehra Offset Press, Delhi.

Contents

Preface

Differential calculus is intended to the students appearing for **undergraduate examinations** conducted by different universities of our country. Much efforts have been made to present this subject matter in a manner as to enable a student of ordinary caliber to solve the problems without any external to the student of higher caliber in solving challenging problems.

To achieve this aim, the chapters are provided with examples, diagrams and basic concepts. The problems have been carefully selected and well graded. They are designed to test comprehension. A serious effort has been made to keep the book free from mistakes while preparing this book, I have not sacrificed the traditional approach of learning mathematics. On the contrary, care has been taken to provide an extensive foundation of knowledge required for the University Examination. A large number of exercises have been provided in order that mathematical skills may be developed, mastered and maintained.

I take this opportunity to thank Anmol Publication Pvt. Ltd, New Delhi, the publishers, for their help and co-operation during the production and Mr. R. K. Ahuja for his kind assistance during the preparation of the manuscript. I also express my loving thanks to Ms. Suchismita Mohanty and Er. Prashant Kumar Sahu for their valuable discussions. I had with them from time to time and to several friends and students for their valuable suggestions.

Suggestions for further improvement of the book are most welcome.

New Delhi **Er. R. K. Mohanty**

1

Real Numbers

1.1. INTRODUCTION

The real number system is the foundation on which the entire branch of Mathematics known as Real Analysis rests. It specially introduces and deals with the limit operation, the algebraic operations of Addition and Multiplication and their Inverses, Subtraction and Division. Thus, it is applied to Geometry, Mechanics and other branches of Theoretical Physics, Economics and Psychology.

Ancient Greek Mathematicians knew that a satisfactory theory of real numbers was not available until late in the 19th century. During the second half of the 19th century, three different theories for constructing the real numbers were put forth by three Greman Mathematicians- Karl Weierstrass (1815-1897), Richard Dedekind (1831-1916) and Georg Cantore (1845-1918). In 1889, the famous Italian Mathematician Guiseppe Peano (1858-1932) enunciated five axioms for the natural numbers which could be taken as the starting point for the construction of real numbers. The importance of real numbers for the study of the subject in hand being thus clear. We will, in some of the following articles, see how starting from the set of natural numbers, we arrive at the set of real numbers.

The real number system satisfies a deep philosophical need to base the theory part of Calculus on the notation of number alone to the entire exclusion of every physical basis.

1.2. RATIONAL NUMBER

1.2.1. Natural Numbers

The numbers 1,2,3,4,5 etc. is known as the set of Natural numbers. It is denoted as N. $N=\{1,2,3,\ldots\ldots\}$. Natural numbers are positive integers.

1.2.2. Fractional Numbers

The set of numbers like x/y, where x and y are natural numbers i.e., $x \in N$ and $y \in N$ and $y \neq 0$. Thus, fractional numbers are a new set of numbers of which natural numbers are subset. In this case $q = 1$.

1.2.3. Rational Numbers

The set of all positive fractions, negative fractions and zero is known as the set of Rational numbers. It is denoted as Q.

If we take gain as positive number, then loss will be negative number. Similarly, *rise* is taken as positive number where as *fall* is taken as negative number.

The natural numbers 1,2,3,4,5.... are called positive integers and the numbers -1,-2,-3,-4.... are called negative integers.

The set $\{0, \pm1, \pm2, \pm3, \pm4\ldots..\}$, called the set of integers. It is denoted as I.

1.3. FUNDAMENTAL OPERATIONS ON RATIONAL NUMBERS

There are two fundamental algebraic operations on the set of rational numbers.

1. ***Closure for Addition:***

   The set Q is closed with respect to addition if a and b be any two rational numbers, then $a + b$ is also a real number.

2. ***Commutative Law of Addition and Multiplication (C.L.A. & C.L.M.)***

   If a and b be any two rational numbers, then

$$a + b = b + a$$

$$ab = ba$$

3. ***Associative Law of Addition and Multiplication (A.L.A. & A.L.M.)***

   Addition and Multiplication of rational numbers is associative. If a, b and c be any three rational numbers, then

$$a + (b + c) = (a + b) + c$$
$$a(bc) = (ab)c$$

4. ***Identity element for Addition and Multiplication***

There exists a rational number, namely 0(zero), such that

$$a + 0 = 0 + a = a \text{ for all } a \in Q$$

Again, there exist a rational number, namely 1(one), such that

$$a.1 = 1.a = a \text{ for all } a \in Q$$

5. ***Existence of negatives and inverses.***

Corresponding to each real number *a*, there exist a real number *b* (called negative of *a* in case of addition and inverse of *a* in case of Multiplication) such that

$$a + b = b + a = 0$$
$$ab = ba = 1$$

6. ***Distributive law of Multiplication over addition (D.M.A.)***

If *a*, *b* and *c* be any three rational numbers, then according to distributive law of Multiplication over addition,

$$a (b + c) = ab + ac$$

1.4. MEANINGLESS OPERATION OF DIVISION BY ZERO.

Suppose we want to divide *a* by *b*.

Let $\quad a \div b = c$

$\therefore \quad bc = a$

The division will be intelligible, if and only if, the determination of c is possible.

Any number which when multiplied by zero produces zero. This means that there is, no number which when multiplied by zero produces a number other than zero.

$\therefore \quad a/0$ is no number when $a \neq 0$

Also any number when multiplied by zero produces zero so that 0/0 may be any number.

On account of this impossibility in one case and indefiniteness in the other, the operation of division by zero must be always avoided. This can be better understood by taking an example.

Let $x = 2$ or $x - 2 = 0$ ...(1)

$\Rightarrow$ $x^2 = 4$

$\Rightarrow$ $x^2 - 4 = 0$ ...(2)

Equating (1) & (2), we have

$x^2 - 4 = x - 2$

$\Rightarrow$ $(x+2)(x-2) = x - 2$

Dividing both sides by $(x - 2)$, we get

$x + 2 = 1$

Putting the value of $x = 2$, in the above equation, we have

$2 + 2 = 1$

$\Rightarrow$ $4 = 1$

which is clearly absurd.

Division by $(x-2)$, which is zero, is responsible for this absurd conclusion.

1.5. REPRESENTATION OF RATIONAL NUMBERS BY POINTS ON A STRAIGHT LINE.

Consider any straight line. Mark an arbitary point O, on the line and calling it the origin. The number zero will be represented by the point O. The point O divides the straight line into two parts. The right hand side of O is the positive part and the left hand side of O is the negative part. Take any point A on the positive part and call it the unit length. The number 1 is represented by the point A

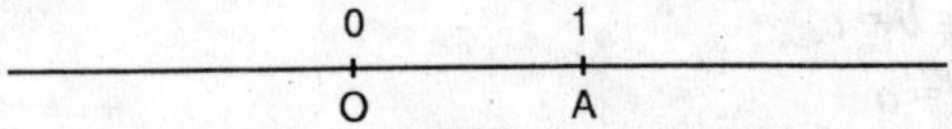

Figure 1.1

We are in a position to determine a point representing any given rational number as explained below:

Positive Integers. To represent a positive integer, m, we take a point on the positive part of O such that its distance from O is m times the unit length OA. This point represents the positive integer, m.

Negative Integers. To represent a negative integer, $-m$, we take a point on the negative part of O such that its distance from O is m time the unit length OA. This point represents the negative integer, $-m$.

Real Numbers. Let x/y be any rational number; y being a positive integer. Let OA be divided into y equal parts; OB being one of them. We take a point on the positive or negative side of O is x times (or, $-x$ times if x is negative) the distance OB. The point so obtained represents the rational number x/y.

1.6. IRRATIONAL NUMBERS

Construct a square with one of its sides as OA of unit length and take a point P on the line such that OP is equal in length to the diagonal of this square. It will now be shown that the length of OP cannot have a rational number as its measure. If possible, let its measure be a rational number x/y.

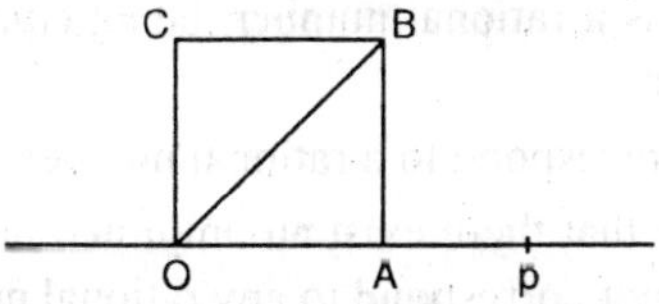

Figure 1.2

By the Pythagora's theorm, we have

$$(x/y)^2 = 1^2 + 1^2 = 2$$

$$\Rightarrow \quad x^2 = 2y^2 \qquad \ldots(i)$$

We suppose that the natural numbers x and y have no common factor, for such factors, if any, can be cancelled to begin with.

Firstly, we see that

$$(2n)^2 = 4n^2$$

$$\Rightarrow \quad (2n+1)^2 = (4n^2 + 4n) + 1$$

so that the square of an even number is even and that of an odd number is odd.

From eqn (i), we see, that p^2 is an even number.

Therefore p must be even.

Let, $p = 2n$, where n is an integer

We have $\quad 4n^2 = 2q^2$

$\Rightarrow \quad q^2 = 2n^2$

$\Rightarrow \quad q^2$ is even

$\Rightarrow \quad q$ is even

Hence, p and q have a common factor 2.

This conclusion contradicts the hypothesis that they have no common factor. Thus the measure $\sqrt{2}$ of OP is not a rational number.

Again, we take a point L on the line such that the length OL is any rational multiple say, x/y, of OP.

The length *OL* cannot have a rational measure. If possible, let *m/n* be the measure of *OL* so that,

$$\frac{x}{y}\sqrt{2} = \frac{m}{n}$$

$$\Rightarrow \quad \sqrt{2} = \frac{my}{nx}$$

which states that $\sqrt{2}$ is a rational number, being equal to *mq/np*. Thus, we arrive at a contradiction.

Hence, *L* cannot correspond to a rational number.

Thus we conclude that there exist an unlimited number of points on the number line which do not correspond to any rational number.

Real Number. *A number, rational or irrational is called a real number.* The set of real numbers is the set of rational and irrational numbers. This set is denoted by *R*.

1.6.1. Number and Point

If any real number *x* is represented by a point *p*, then we say that the point *p* is *x*. Thus the terms, number and point, are generally used in an indistinguishable manner.

1.6.2. Intervals

Let *a*, *b* be two given numbers such that $a < b$. Then the set of numbers *x* such that

$$a \le x \le b$$

is called a **closed** interval and denoted by $[a,b]$.

In symbols, $[a,b] = \{x : a \le x \le b\}$

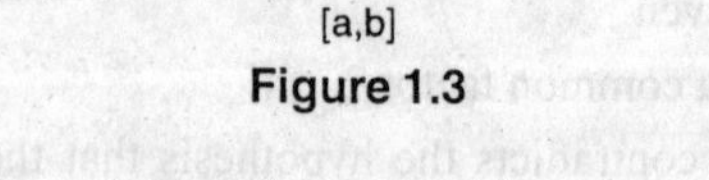

Figure 1.3

The set of numbers *x* such that $a < x < b$ is called an **open** interval and is denoted by $]a,b[$.

In symbols, $]a,b[= \{x : a < x < b\}$

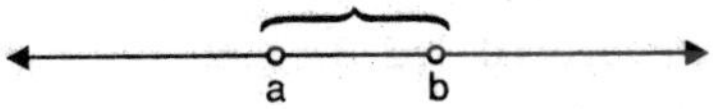

]a,b[

Figure 1.4

The set of numbers x is such that $a < x \leq b$ and $a \leq x < b$ are called **semi-closed** or **semi-open** intervals and are denoted by $]a,b]$, $[a,b[$.

In symbols, $]\,a,b\,] = \{x: a < x \leq b\}$

and $[a, b[\, = \{x: a \leq x < b\}$

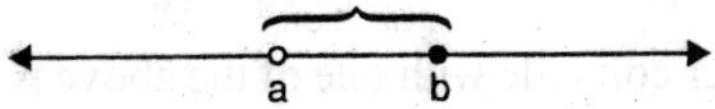

]a,b]

Figure 1.5

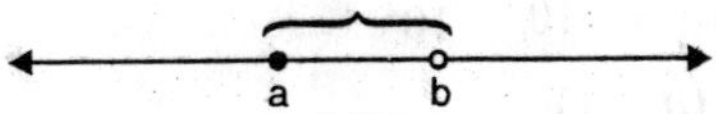

[a,b[

Figure 1.6.

1.7. DECIMAL REPRESENTATION OF REAL NUMBERS

Let P be any given point of the number line.

Suppose that the point p lies on the positive side of O. Let the points corresponding to integers be marked on the number line so that the whole line is divided into intervals of length one each.

Now if P coincides with some one of these points of divison, it corresponds to an integer and we need procced no further. In case P falls between two points of division, say, a, $a + 1$, we sub-divide the interval $[a, a + 1]$ into 10 equal parts so that the length of each part is 1/10. the points of division, now, are

$$a,\ a + \frac{1}{10}, a + \frac{2}{10}, a + \frac{3}{10}, ... a + \frac{9}{10}, a + 1$$

If P coincides with any of these points of division, then it corresponds to a rational number. In the alternative case, it falls between two points of division, say

$$a + \frac{a_1}{10},\ a + \frac{a_1 + 1}{10}$$

i.e., $a.a_1, a.(a_1+1)$,

where, a_1, is any of the integers 0, 1, 2, 3, ..., 9.

We again sub-divide the interval

$$\left[a+\frac{a_1}{10},\ a+\frac{a_1+1}{10}\right]$$

into 10 equal parts so that the length of each part is $1/10^2$. These points of divison, now, are

$$a+\frac{a_1}{10}, a+\frac{a_1}{10}+\frac{1}{10^2}, \ldots\ldots, a+\frac{a_1}{10}+\frac{9}{10^2}, a+\frac{a_1+1}{10}$$

The point P will either coincide with one of the above points of divison or will be between two points of divison, say

$$a+\frac{a_1}{10}+\frac{a_2}{10^2}, a+\frac{a_1}{10}+\frac{a_2+1}{10^2},$$

i.e., $a.a_1.a_2, a.a_1(a_2+1)$,

where a_2 is one of the integers 0, 1, 2, ..., 9.

After a number of steps, say n, the point will either be found to coincide with some point of division or lie between two points of the form

$$a+\frac{a_1}{10}+\frac{a_2}{10^2}+\ldots\ldots\ldots\frac{a_n}{10^n}, a+\frac{a_1}{10}+\frac{a_2}{10^2}+\ldots\ldots+\frac{a_{n+1}}{10^n},$$

i.e., $a.a_1a_2\ldots\ldots a_n, a.a_1a_2a_3\ldots\ldots(a_n+1)$,

the distance between which is $1/10^n$ and which gets smaller and smaller as n increases.

The successive intervals in which P lies go on shrinking in length and will clearly close up to the point P. This point P is then represented by the infinite decimal.

$$a.a_1.a_2.a_3\ldots\ldots\ldots$$

1.8. MODULUS OR ABSOLUTE VALUE OF A REAL NUMBER

Definition: If x be any given real number, then its absolute value is defined by the rule

$$|x| = \begin{cases} x \text{ if } x \geq 0 \\ -x \text{ if } x < 0 \end{cases}$$

It would be seen that for all $x \in R$, we have

$$|x| \geq 0$$

Also, $x_1 = x_2 \Rightarrow |x_1| = |x_2|$

Theorem 8.1: For every $x \in R$, $|x| = \text{Max.}\{-x, x\}$

Proof: According to the law of Trichotomy, one and exactly one of the following is true:

(i) $x > 0$, (ii) $x = 0$, (iii) $x < 0$

If $x \geq 0$, then $|x| = x$ and $x \geq -x$

If $x < 0$, then $|x| = -x$ and $-x > x$

Thus in either case, $|x|$ is greater of the two numbers x and $-x$, i.e.,

$|x| = \text{Max}\{x, -x\}$

Theorem 8.2: For every $x \in R$, $|x|^2 = x^2 = |-x|^2$

Proof: From definition,

$$|x| = \begin{cases} x \text{ if } x \geq 0 \\ -x \text{ if } x < 0 \end{cases}$$

In either case,

$|x|^2 = x^2$

and $|-x|^2 = (-x)^2 = x^2$

Therefore, $|x|^2 = x^2 = |-x|^2$

Theorem 8.3: For every $x \in R$, $|x| = |-x|$

Proof: We know that,

$$\begin{aligned} |-x| &= \text{Max.}\{-x, -(-x)\} \\ &= \text{Max.}\{-x, x\} \\ &= |x| \end{aligned}$$

Theorem 8.4: For $x, y \in R$, $|x.y| = |x|.|y|$

Proof: $$\begin{aligned} |x.y|^2 &= (xy)^2 \\ &= x^2y^2 \\ &= x^2.y^2 \\ &= |x|^2.|y|^2 \\ &= (|x|.|y|)^2 \end{aligned}$$

Since $|x.y|$ amd $|x|.|y|$ are both positive, therefore equating the positive square roots of both sides, we have

$$|x.y| = |x|.|y|$$

Theorem. 8.5: (Triangle Inequality). The modulus of the sum of two numbers is less than or equal to the sum of their moduli.

For all real numbers x and y,

$$|x+y| \leq |x| + |y|$$

Proof:

Case 1. $x + y \geq 0$

In this case,

$$|x+y| = x+y$$

Since $x \leq |x|$ and $y \leq |y|$,

therefore, $x + y \leq |x| + |y|$

$\Rightarrow \quad |x+y| \leq |x| + |y|$

Case 2. $x + y < 0$

In this case,

$$-(x+y) > 0$$

$$\Rightarrow \quad (-x) + (-y) > 0$$

Now, $\quad |x+y| = |-(x+y)|$

$$= |(-x) + (-y)|$$

$$\leq |-x| + |-y|$$

Since $|-x| = |x|$ and $|-y| = |y|$,

therefore, $|x+y| \leq |x| + |y|$

Theorem. 8.6: For all x and y, $|x-y| \geq ||x| - |y||$

Proof: From the triangle inequatity, we have

$$|x| = |(x-y) + y| \leq |x-y| + |y|$$

$\Rightarrow \quad |x| - |y| \leq |x-y| \qquad \ldots$(i)

Again, $\quad |y| = |(y-x) + x| \leq |y-x| + |x|$

$\Rightarrow \quad |y| - |x| \leq |y-x|$

$\Rightarrow \quad -(|x| - |y|) \leq |y-x|$

$\Rightarrow \quad -|x| - |y| \leq |x-y|$ since $|y-x| = |x-y| \qquad \ldots$(ii)

Now, $\quad |x| - |y| = \text{Max.}\{|x| - |y| - (|x| - |y|)\}$

$\leq |x-y| \qquad$ [by (i) & (ii)]

Examples

Ex.1: Calculate the cube root of 2 to three decimal places.

Sol: We have, $1^3 = 1 < 2$

and $2^3 = 8 > 2$

$\Rightarrow \quad 1 < \sqrt[3]{2} < 2$

The numbers considered are,

$1, 1.1, 1.2, \ldots\ldots\ldots, 1.9, 2$

These numbers divide the interval [1,2] into 10 equal parts and find two successive numbers such that the cube of the first is < 2 and that of the second is > 2.

Now, $(1.2)^2 = 1.728 < 2$

and $(1.3)^2 = 2.197 > 2$

$\Rightarrow 1.2 < \sqrt[3]{2} < 1.3$

Again, we consider the numbers

$1.2, 1.21, 1.22, \ldots\ldots\ldots, 1.29, 1.3$

which divide the interval, [1.2, 1.3] into 10 equal parts and find two successive numbers such that the cube of the first is < 2 and that of the second is > 2.

Now, $(1.25)^3 = 1.953125 < 2$

and $(1.26)^3 = 2.000376 > 2$

$\Rightarrow 1.25 < \sqrt[3]{2} < 1.26$

Again, we consider the numbers

$1.25, 1.251, 1.252, \ldots\ldots\ldots\ldots, 1.259, 1.26$

which divide the interval [1.25, 1.26] into 10 equal parts.

Now, $(1.259)^3 = 1.99561979 < 2$

and $(1.26)^3 = 2.000376 > 2$

$\Rightarrow 1.259 < \sqrt[3]{2} < 1.26$

Hence, $\sqrt[3]{2} = 1.259\ldots\ldots$

Thus, to three decimal places, we have $\sqrt[3]{2} = 1.259$.

Ex.2: If x,y be any real numbers, show that

$$|x+y|^2 + |x-y|^2 = 2(|x|^2 + |y|^2)$$

Sol: $|x+y|^2 + |x-y|^2$

$$= (x+y)^2 + (x-y)^2$$
$$= 2(x^2+y^2)$$
$$= 2\{(x^2)+(y^2)\}$$
$$= 2(|x|^2+|y|^2)$$

Ex.3: If a, b, c be real numbers, and $c > 0$, show that

$$|a-b| < c \Leftrightarrow b-c < a < b+c$$

Sol: $|a-b| < c$

$\Leftrightarrow$ Max $\{(a-b), -(a-b)\} < c$,

$\Leftrightarrow$ $a-b < c$ and $b-a < c$

$\Leftrightarrow$ $a < b+c$ and $b-c < a$

$\Leftrightarrow$ $b-c < a < b+c$

Exercise – 1

1. Prove:

   (i) $|x-y| \leq |x|+|y|$

   (ii) $|x+y+z| \leq |x|+|y|+|z|$

2. Show that:

   (i) $\left|\frac{1}{x}\right| = \frac{1}{|x|}$

   (ii) $\left|\frac{x}{y}\right| = \frac{|x|}{|y|}$

   (iii) $|x| < \in \Leftrightarrow -\in < x < \in$

3. If $y = |x| + |x-1|$, show that

$$y = \begin{cases} 1-2x & \text{for all } x \leq 0 \\ 1, & \text{for all } 0 < x < 1 \\ 2x-1 & \text{for all } x \geq 1 \end{cases}$$

4. If $x, y, \in$ be real numbers and $\in > 0$, then

$$|x-y| < \in \Leftrightarrow y-\in < x < y+\in$$

5. Show that:

$$\left.\begin{matrix} |a-b| < l \\ |b-c| < m \end{matrix}\right\} \Rightarrow |a-c| < l+m$$

6. Show that every rational number is expressible as a terminating or a recurring decimal.

7. Calculate the cube root of 5 to 2 decimal places.

   (*Ans:* 1.71)

8. For what values of x are the following equalities not valid:

   (i) $\frac{x}{x} = 1$ (*Ans:* 0)

   (ii) $\frac{x^2-a^2}{x-a} = x+a$ (*Ans:* a)

(iv) $\dfrac{1-x}{1-\sqrt{x}} = 1+\sqrt{x}$ (*Ans:* 1)

(iv) $\dfrac{1-\cos x}{\sin x} = \tan\dfrac{x}{2}$ (*Ans:* $\{n\pi : n \in I\}$)

9. Give the equilvalents of the following by doing away with the modulus notation:

(i) $|x+2|<3$ (*Ans:* $-5<x<1$)

(ii) $|x-1|\le 2$ (*Ans:* $1\le x<3$)

(iii) $0<|x-1|<2$ (*Ans:* $-1<x<3$)

10. Give the equivalents of the following inequalities in terms of the modulus notation:

(i) $-2\le x\le 8$

(ii) $3<x<6$

(iii) $1-\in<x<1+\in$

(iv) $-1\le x\le 3$

2

Functions

2.1. DOMAIN AND RANGE OF A FUNCTION

A function f from a set A to a set B associates to each element x in A a unique element in B which we denote by $f(x)$.

The set A is called the **Domain** of the function f.

If x denotes a member of the set A, then the number of the set B, which is the function f associates to $x \in A$, is denoted by $f(x)$ called the value of the function f at x. the function may be described as $x \to f(x)$ or $y = f(x)$ where $x \in$ A and y $\in$ B.

Range of a function: A function f from a set A to a set B so that the domain of f is the set A, then the set of all the function values is called range of f. Thus,

$$\text{Range of } f = \{f(x) : x \in A\}$$

Functions whose domain and codomain are both subsets of R, the set of all real numbes, are called real valued function of a real variable.

Usually the domain of a function is an interval; open, closed, semi-closed or semi-open. For finite real numbes a and b the closed interval with end points a and b, denoted by $[a, b]$.

$$[a, b] = \{x \in R : a < x \leq b\}$$

The open interval with end points a and b, denoted by

$$(a, b) = \{x \in R : a < x < b\}$$

We also have intervals which are closed at only one end point and are defined by

$$(a, b] = \{x \in R : a < x \leq b\}$$
$$[a, b) = \{x \in R : a \leq x < b\}$$

2.2. GRAPHS OF FUNCTIONS

Let f be a function with domain [a, b]

We have, $y = f(x), x \in [a, b]$

The function f associates to each $x \in [a, b]$ a number denoted by $f(x)$.

Generally, two straight lines are taken to represent the function graphically, called co-ordinate axes, at right angles to each other. We take O as origin for both the axes and select unit intervals on OX, OY of same length.

For a number $x \in [a, b]$ correspond a point M on x-axis such that OM = x. Again the function f associates to $x \in [a, b]$ a number $f(x)$. We write $f(x) = y$. We have a point N on Y-axis such that ON = $f(x) = y$.

When we complete the rectangle OMPN, we get a point P which is called to correspond to the pair of numbers $[x, f(x)]$ i.e., P(x, y).

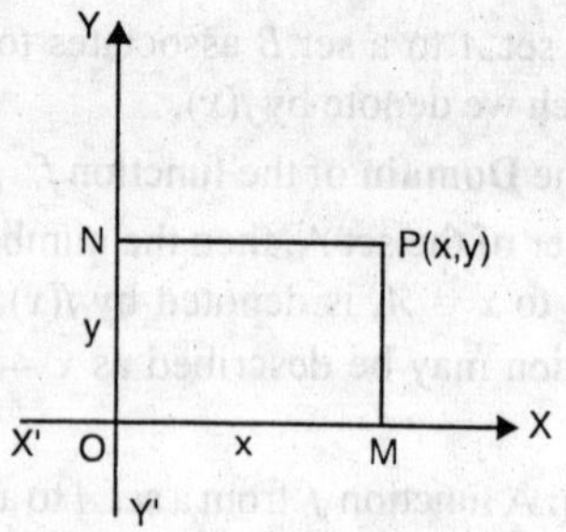

Figure 2.1

The set of points (x, y) obtained by giving different values to x, is said to be the graph of the function f. The graph of f is also said to be the graph of the equation $y = f(x)$.

Examples

Example 1: Find the domain and range of $f(x) = \sqrt{9 - x^2}$

Solution: We have, $f(x) = \sqrt{9 - x^2}$

Domain: Clearly, $f(x)$ is not defined for $9 - x^2 < 0$.

So, $f(x)$ is defined only when $9 - x^2 \geq 0$

Now, $9 - x^2 \geq 0$

$\Rightarrow \quad x^2 - 9 \leq 0$

$\Rightarrow \quad (x+3)(x-3) \leq 0$

$\Rightarrow \quad -3 \leq x \leq 3$

$\Rightarrow \quad x \in [-3, 3]$

Therefore, domain $f(x) = [-3, 3]$

Range: Let $f(x) = y \Rightarrow \sqrt{9 - x^2} = y$

$$\Rightarrow \quad x = \pm\sqrt{9 - y^2}$$

Clearly, x is defined when $9 - y^2 \geq 0$

Now, $9 - y^2 \geq 0$

$\Rightarrow \quad y^2 - 9 \leq 0$

$\Rightarrow \quad (y+3)(y-3) \leq 0$

$\Rightarrow \quad -3 \leq y \leq 3$

But $f(x)$ attains only non-negative values.

Therefore, $y \in [0, 3]$

Hence, range $f(x) = [0, 3]$

Example 2: Find the range of the function f given by $f(x) = 1 + 3\cos 2x$

Solution: Recall that

$-1 \leq \cos 2x \leq 1$

$\Rightarrow \quad -3 \leq 3\cos 2x \leq 3$

$\Rightarrow \quad 1 - 3 \leq 1 + 3\cos 2x \leq 1 + 3$

$\Rightarrow \quad -2 \leq 1 + 3\cos 2x \leq 4$

$\Rightarrow \quad -2 \leq f(x) \leq 4$

Therefore, the range of the given function is the interval $[-2, 4]$

Example 3: Find the domain and range of the function $f(x) = \dfrac{x}{1+x^2}$.

Solution: We have, $f(x) = \dfrac{x}{1+x^2}$

Domain: clearly, $f(x)$ is defined for all real values of x. so, domain$(f) = r$

Range: Let $f(x) = y$

$$\Rightarrow \quad \frac{x}{1+x^2} = y$$

$$\Rightarrow \quad x^2y - x + y = 0$$

$$\Rightarrow \quad x = \frac{1 \pm \sqrt{1-4y^2}}{2y}$$

Now, $\dfrac{1 \pm \sqrt{1-4y^2}}{2y}$ is a real number if

$$1 - 4y^2 \geq 0 \text{ and } y \neq 0$$

$$\Rightarrow \quad 4y^2 - 1 \leq 0 \text{ and } y \neq 0$$

$$\Rightarrow \quad y^2 - \frac{1}{4} \leq 0 \text{ and } y \neq 0$$

$$\Rightarrow \quad \left(y+\frac{1}{2}\right)\left(y-\frac{1}{2}\right) \leq 0 \text{ and } y \neq 0$$

$$\Rightarrow \quad -\frac{1}{2} \leq y \leq \frac{1}{2} \text{ and } y \neq 0$$

$$\Rightarrow \quad y \in \left[-\frac{1}{2}, 0\right) \cup \left(0, \frac{1}{2}\right]$$

hence, range $(f) = \left[-\frac{1}{2}, 0\right] \cup \left[0, \frac{1}{2}\right]$

Example 4: Find the domain of the following functions:

(i) $\dfrac{\sin^{-1} x}{x}$ (ii) $\dfrac{\cos^{-1} x}{[x]}$

Solution:

(i) Let $f(x) = \dfrac{\sin^{-1} x}{[x]}$ $g(x) = \sin^{-1}x$ and $h(x) = x$

Then, $f(x) = \dfrac{g(x)}{h(x)}$

$\Rightarrow$ Domain (f) = Domain(g) $\cap$ Domain (h) – $\{x \mid h(x) = 0\}$

Clearly, domain $(g) = [-1, 1]$, domain (h) = R and $\{x \mid h(x) = 0\} = \{0\}$

$\therefore$ Domain $(f) = [-1, 1] \cap R - \{0\} = [-1, 0] \cup \{0, 1\}$

(ii) Let $f(x) = \dfrac{\cos^{-1} x}{[x]}$, $g(x) = \cos^{-1}x$ and $h(x) = [x]$

Then, $f(x) = \dfrac{g(x)}{h(x)}$

$\Rightarrow$ Domain (f) = Domain(g) $\cap$ Domain (h) – $\{x \mid h(x) = 0\}$

Clearly, domain $(g) = [-1, 1]$,

domain (h) = R and $\{x \mid h(x) = 0\} = x \mid [x] = 0 = [0, 1]$

$\therefore$ Domain $(f) = [-1, 1] \cap R - [0, 1]$

$= [-1, 0] \cup \{1\}$

Example 5: Draw the graph of the function f given by

$$f(x) = \begin{cases} -1, & \text{if } x < 0 \\ x^2, & \text{if } 0 \le x \le 2 \\ 4, & \text{if } x > 2 \end{cases}$$

Solution: The given function for $x < 0$ is a constant function given by $f(x) = -1$ for $x > 2$, it is again a constant function given by $f(x) = 4$.

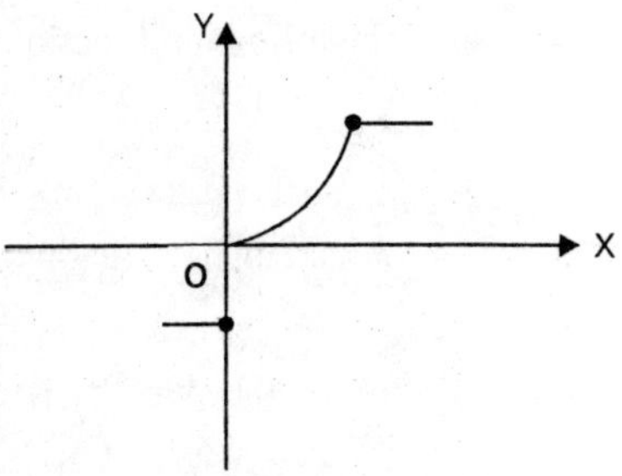

Figure 2.2

However, for $0 \leq x \leq 2$ the graph is parabola.

Example 6: What are the domain and the range of the function f given by $f(x) = e^{-2x}$. Also draw its graph.

Solution: The given function is well defined for each real number. Therefore, the domain of the function is R. Also, for each real number, positive or negative, e^{-2x} is positive. Hence the range of the given function is $(0, \infty)$

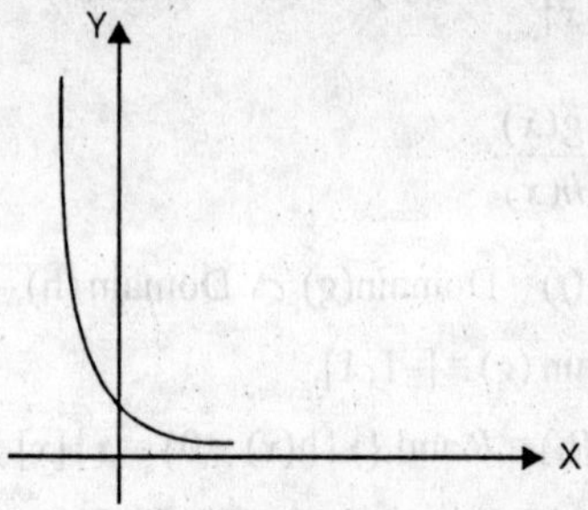

Figure 2.3

Exercise – 2.1

1. *Find the domain of each of the following functions:*

(i) $f(x) = x\,|x|$ (ii) $f(x) = \dfrac{1}{\sqrt{x+|x|}}$

(iii) $f(x) = \dfrac{x^3 - 5x + 3}{x^2 - 1}$ (iv) $f(x) = e^{x + \sin x}$

(v) $f(x) = [x] + x$ (vi) $f(x) = \dfrac{\sin^{-1} x}{x}$

(vii) $f(x) = \sqrt{\left(x - \dfrac{x}{1-x}\right)}$ (viii) $f(x) = \sqrt{1 + 2\sin x}$

(ix) $f(x) = \sqrt{\dfrac{1-|x|}{2-|x|}}$

2. *Find the differences between the domains of the following pairs of functions.*

(i) $y = \sqrt{x^2(x+1)}$ and $y = x\sqrt{(x+1)}$

(ii) $y = \sqrt{(1-x)}$ and $y = \sqrt{1-x^2}$

(iii) $y = \sqrt{(1-x^2)}$ and $y = \sqrt{(x^2-1)}$

(iv) $y = x - 5$ and $y = \sqrt{(x-5)^2}$

3. *Find the range of each of the following functions given by*

(i) $f(x) = \dfrac{3}{2 - x^2}$ (ii) $f(x) = \dfrac{x}{1 + x^2}$

(iii) $f(x) = 1 - |x - 2|$ (iv) $f(x) = \dfrac{|x-4|}{x-4}$

(v) $f(x) = \sqrt{16 - x^2}$ (vi) $f(x) = \dfrac{1}{\sqrt{x-5}}$

4. *Find the domain and the range of the following functions given by*

(i) $f(x) = x!$ (ii) $f(x) = \dfrac{1}{\sqrt{x-[x]}}$

(iii) $f(x) = 1 - |x-3|$ (iv) $f(x) = \sin^2(x^3) + \cos^2(x^3)$

(v) $f(x) = \dfrac{1}{\sqrt{4+3\sin x}}$

5. *Draw the graph of each of the following functions given by*

(i) $f(x) = x - [x]$ (ii) $f(x) = 3^x$

(iii) $f(x) = \begin{cases} \dfrac{x}{|x|}, \text{if } x \neq 0 \\ 0 \ , \text{ if } x = 0 \end{cases}$ (iv) $f(x) \begin{cases} 1 \ , \text{if } x \leq 0 \\ x^2 + 1, \text{if } 0 \leq x \leq 2 \\ 5 \ \ , \text{if } x \geq 2 \end{cases}$

(v) $f(x) = x + \dfrac{\sqrt{(x-1)^2}}{x-1}$

(vi) $f(x) = x + \dfrac{\sqrt{(x-1)^2}}{x-1} + \dfrac{\sqrt{(x-2)^2}}{x-2} + \dfrac{\sqrt{(x-3)^2}}{x-3}$

(vii) $f(x) = 3^{[3x]}$ (viii) $f(x) = x^2 + [x]^2$

(ix) $f(x) = [|x|]$ (x) $f(x) = |[x]|$

2.3. OPERATION ON REAL FUNCTIONS

2.3.1. *Addition of two real functions.*

Let f: X $\rightarrow$ R and g : X $\rightarrow$ R be any two*r*eal functions where X $\subset$ R. then, we define $f+g$: X $\rightarrow$ R by

$$(f+g)(x)-f(x)+g(x), \text{ for all } x \in X.$$

2.3.2. *Subtraction of a real function from another*

Let f: X $\rightarrow$ R and g : X $\rightarrow$ R be any two real functions whe*R*e X $\subset$ R then, we define $f-g$: X $\rightarrow$ R by

$$(f-g)(x)-f(x)-g(x), \text{ for all } x \in X.$$

2.3.3. *Multiplication by a scalar*

Let f : x be a real valued function and a be any scalar. Then, the product af is a function from x to R defined by

$$(a\alpha)(x) = \alpha f(x), \text{ for all } x \in X.$$

2.3.4. *Multiplication of two real functions*

The multiplication of two real functions f: X$\rightarrow$ R and g : X $\rightarrow$ R is *a* function f, then fg: X $\rightarrow$ R defined by

$$(fg)(x) = f(x).g(x), \text{ for all } x \in X.$$

2.3.5. *Quotient of two real functions*

Let f and g be two real functions. Then, the quotient of f by g denoted by f/g is *a* fraction from x – {x: $g(x) = 0$} to R defined by

$$\left(\frac{f}{g}\right)(x) = \frac{f(x)}{g(x)}$$

2.4. POLYNOMIAL FUNCTIONS AND RATIONAL FUNCTIONS

A function of the form

$$f(x) = a_0x^n + a_1x^{n-1} + \ldots\ldots\ldots + a_{n-1}x + a_n$$

where $a_0, a_1, a_2, \ldots\ldots\ldots\ldots, a_n$ are given real numbers and $a_0 \neq 0$ is called a **polynomial function** of degree *n*. The domain of every polynomial function is *r*.

A function of the form

$$g(x) = \frac{a_0x^n + a_1x^{n-1} + \ldots\ldots\ldots\ldots + a_{n-1}x + a_n}{b_0x^m + b_1x^{m-1} + \ldots\ldots\ldots\ldots + b_{m-1}x + b_m}$$

where $a_0x^n + a_1x^{n-1} + \ldots\ldots\ldots + a_n, b_0x^m + b_1x^{m-1} + b_m$ are polynomials is called a **Rational function.**

The domain of a rational function is the set of all those real numbers for which the value of polynomial in the denominator is not zero.

2.5. CONSTANT FUNCTIONS AND IDENTITY FUNCTIONS

A function f: A $\rightarrow$ B, A, B $\subset$ *R*, is said to be a **constant function** if there exist a real number k such that $f(x)$ = k, for all $x \in$ A.

Domain: A

Range: {k}

A function f: *R* $\rightarrow$ *R* is said to be an **identity function** if for each *x* in *R*, $f(x) = x$ and is usually denoted by I.

Domain: R

Range: R

2.6. MODULUS FUNCTION

The function f: *R* $\rightarrow$ *R* defined by

$$f(x) = |x| = \begin{cases} x \;, & \text{if } x \geq 0 \\ -x, & \text{if } x < 0 \end{cases}$$

is called the **Modulus function.**

It is also called **Absolute value function.**

Domain: R

Range: R_0 = {*x*:*x* is *a* non-negative numbe*R*}

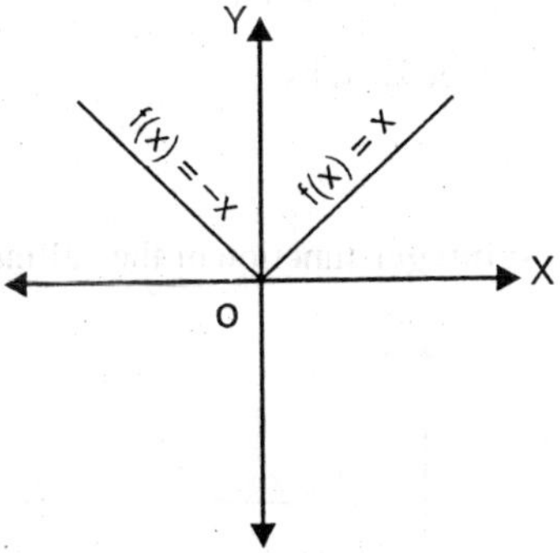

Figure 2.4

2.7. THE GREATEST AND THE SMALLEST INTEGER FUNCTION

For any real number x, we denote [x], the greatest integer less than or equal to x. For example, [2.45] = 2, [-2.11] = -3, [0.32] = 0 etc.

The function $f: R \to R$ defined by

$$f(x) = [x] \text{ or } \lfloor x \rfloor,\ x \in R$$

is called the **greatest integer function** or **the floor function**. It is also called a **step function**.

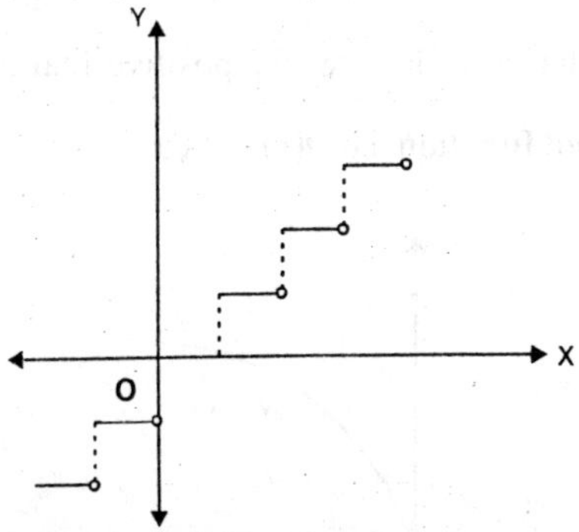

Figure 2.5

For a real number x, we denote by $\lceil x \rceil$, the smallest integer greater than or equal to x.

For example $\lceil 5.2 \rceil = 6, \lceil -5.2 \rceil = -5$ etc.

The function $f: R \to R$ defined by

$$f(x) = \lceil x \rceil, x \in R$$

is called the **smallest integer** function or the **calling function.**

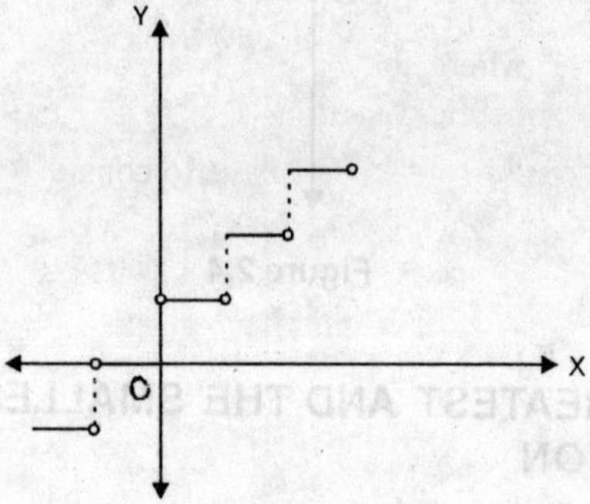

Figure 2.6

It is also called a **step function.**

Domain: R

Range: Z

2.8. SQUARE ROOT FUNCTION

The function that associates every positive real number x to $+\sqrt{x}$ is called the **square root function,** i.e., $f(x) = +\sqrt{x}$.

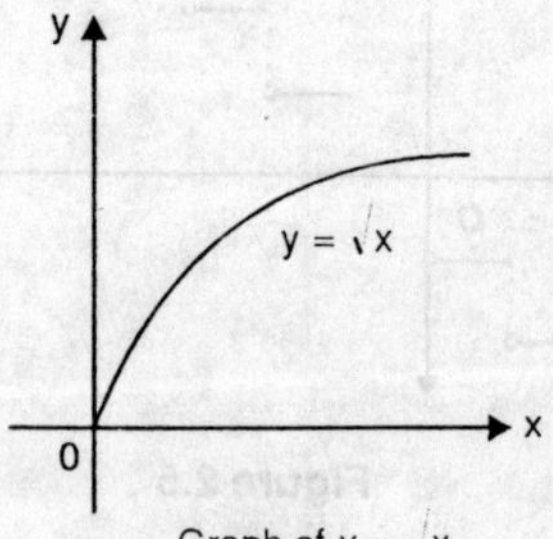

Graph of y = √x

Figure 2.7

Since negative real numbers do not have real square roots. So, $f(x)$ is not defined when x is a negative real number. Therefore, domain of f is the set of all non-negative numbers.

Domain: $[0, \infty]$

Range : $[0, \infty]$

2.9. EXPONENTIAL FUNCTION

The function $y = a^x$ where the base a is a constant and the index, x a variable is called an exponential function.

To consider the graph of $y = a^x$, we have to consider two cases.

(i) $a > 1$

(ii) $0 < a < 1$

Let us take $a = 2$ to have an indication of the nature of variation of the function $y = a^x$, we consider the case when x varies over the sequence

$$\ldots\ldots\ldots -3, -2, -1, 0, 1, 2, 3, \ldots\ldots\ldots$$

of integers and the corresponding sequence of the value of a^x is

$$\ldots\ldots 1/8, 1/4, 1/2, 1, 2, 4, 8, 16, \ldots\ldots\ldots$$

Every member of this sequence is positive.

Now take $a = 1/2 < 1$

The sequence of values of a^x now obtained is

$$\ldots\ldots\ldots, 16, 8, 4, 2, 1, 1/2, 1/4, 1/8, \ldots\ldots$$

In this case, the number on the left can be made as large as we like and those on the right as near 0 as we like.

1. $y = a^x; a > 1$

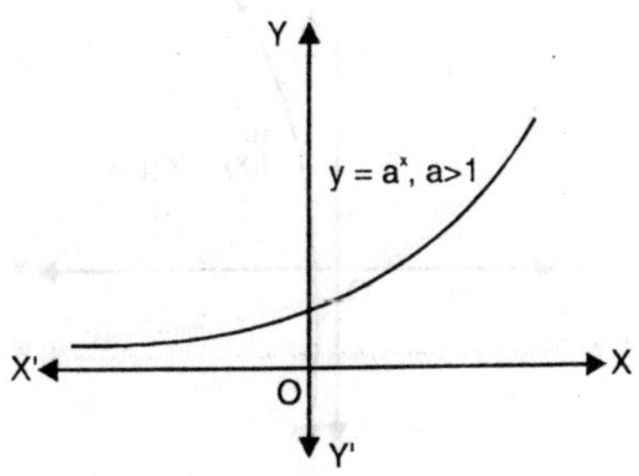

Figure 2.8

(i) *The domain is* $]-\infty, \infty[$
(ii) *The range is* $]0, \infty[$
(iii) *The function is strictly increasing.*

2. $y = a^x;\ 0 < a < 1$

(i) *The domain is* $]-\infty, \infty[$
(ii) *The range is* $]0, \infty[$
(iii) *The function is strictly decreasing.*

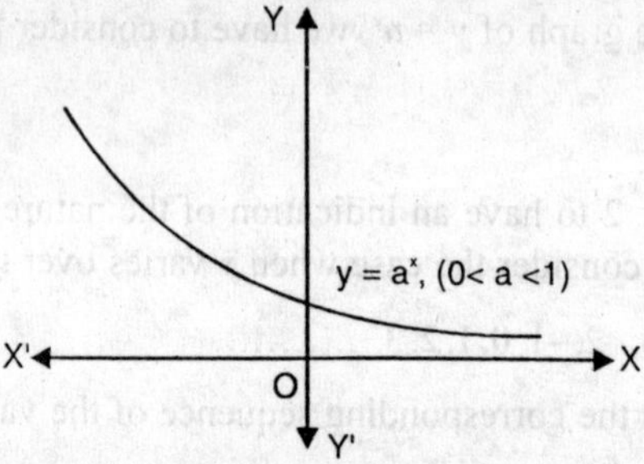

Figure 2.9

2.10. LOGARITHMIC FUNCTION

If $a > 0$ and $a \neq 1$, then the function $\log_a : R+ \to R$ is given by $\log_a x = y$ if and only if $a^y = x$ is called the **logarithmic function**.

Domain: $R+$, the set of all positive real numbers.

Range: R

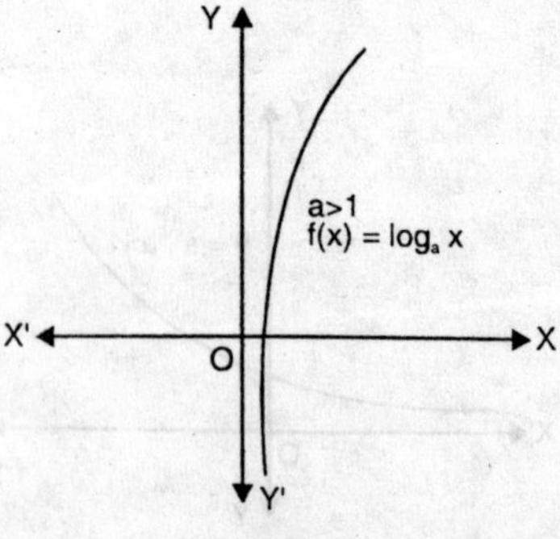

Figure 2.10

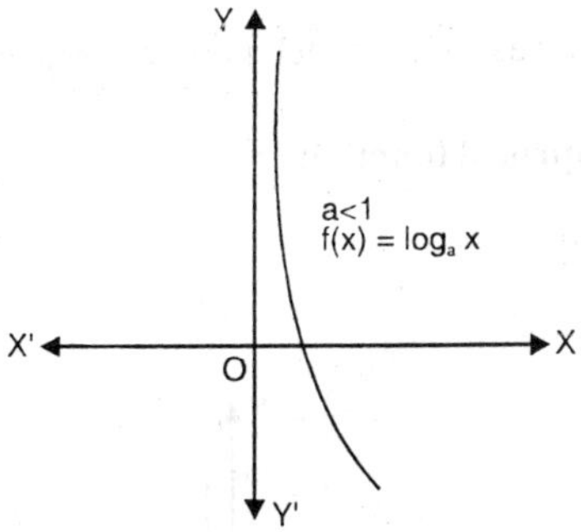

Figure 2.11

2.11. SIGNUM FUNCTION

The function f defined by

$$f(x) = \begin{cases} 1 & , \text{ if } x > 0 \\ 0 & , \text{ if } x = 0 \\ -1 & , \text{ if } x > 0 \end{cases}$$

is called the **signum function**.

Domain: R
Range: $\{-1, 0, 1\}$

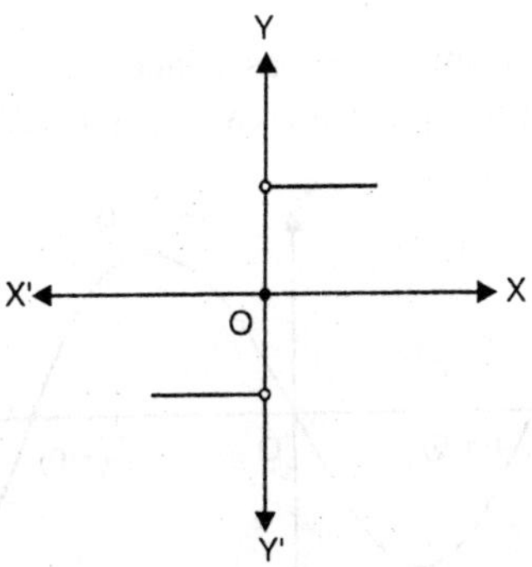

Figure 2.12

2.12. RECIPROCAL FUNCTION

The function that associates each non-zero real number x to its reciprocal $\frac{1}{x}$ is called the **reciprocal function**.

Domain: $R - \{0\}$

Range: $R - \{0\}$

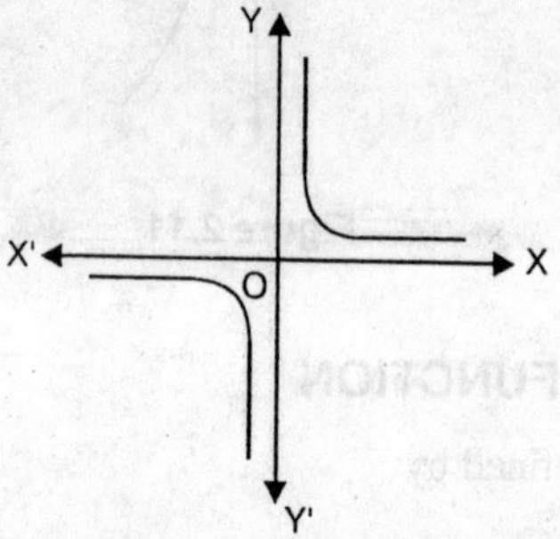

Figure 2.13

2.13. TRIGONOMETRIC FUNCTION

(i) Sine function. The function that associates to each real number x to $\sin x$ is called the *sine function*. Here x is the radian measure of the angle.

Domain: R

Range: $[-1, 1]$

It increases strictly from -1 to 1 as x increases from $-\pi/2$ to $\pi/2$, and decreases strictly from 1 to -1 as x increases from $\pi/2$ to $3\pi/2$ and so on.

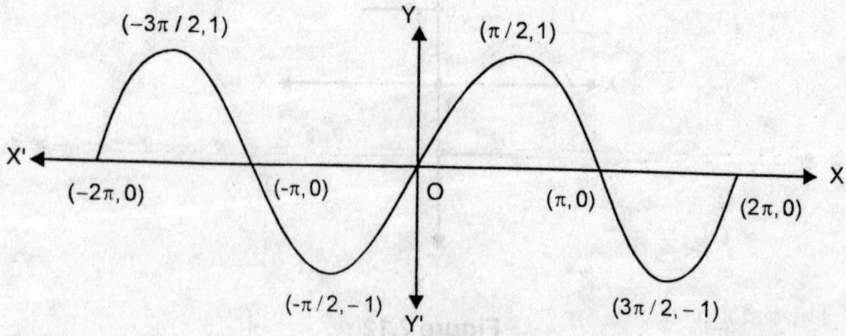

Figure 2.14

(ii) Cosine function. The function that associate to each real number x to $\cos x$ is called the *cosine function*. He*R*e x is the *R*adian measu*R*e of the angle.

Domain: R

Range: $[-1, 1]$

The function decreases strictly from 1 to -1 as x increase form 0 to π and decreases strictly from -1 to 1 as x increases form π to 2π and so on.

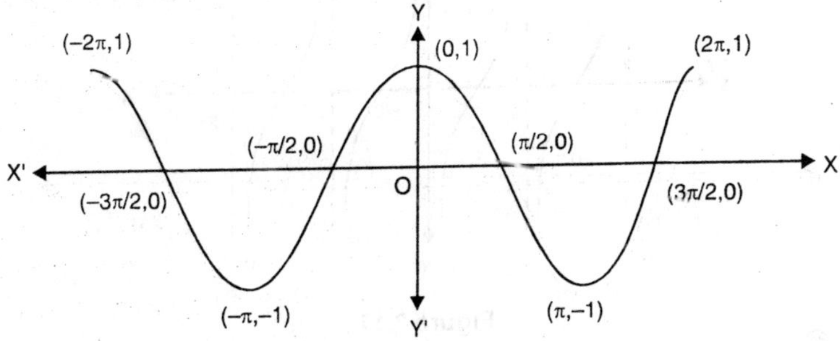

Figure 2.15

(iii) Tangent function. The function that associates a real number x to $\tan x$ is called the *tangent function.*

Clearly, the tangent function is not defined at odd multiples of $\pi/2$, i.e., $\pm \pi/2, \pm 3\pi/2$ etc.

Domain: $R - \{(2n+1)\pi/2\}$

Range: *R*

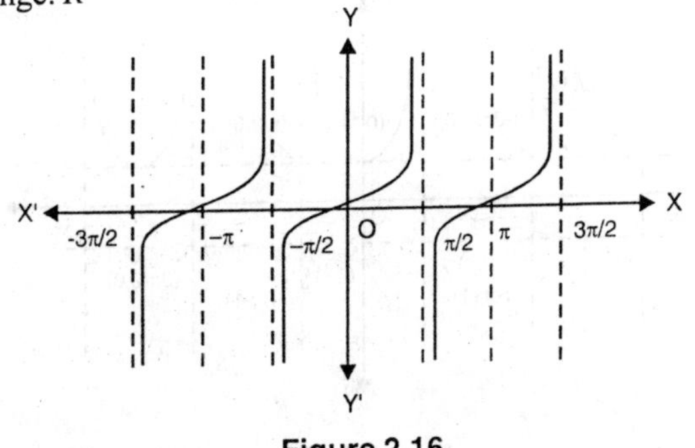

Figure 2.16

The function, $y = \tan x$ increases strictly from $-\infty$ to ∞ as x increases in $]\pi/2, \pi/2[$

(iv) Cotangent function. The function that associates a real number x to $\cot x$ is called the *cotangent function.*

Clearly, cot x is not defined at $x = n\pi$, $n \in Z$

Domain: $R - \{n\pi \mid n \in Z\}$

Range: R

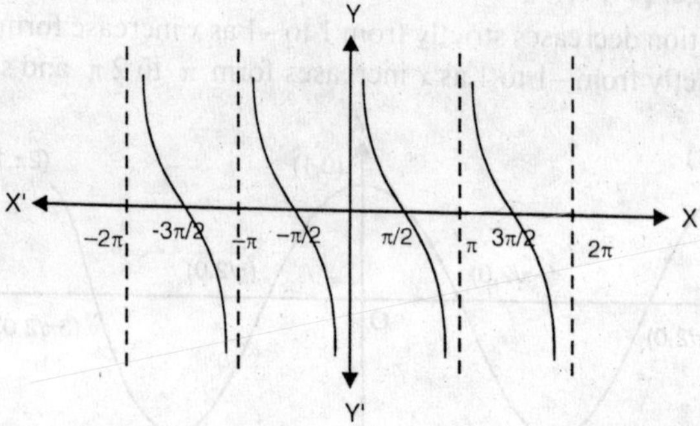

Figure 2.17

(v) Secant function. The function that associates a real number x to sec x is called the *secant function.*

Clearly, sec x is not defined at odd multiples of $\pi/2$ i.e., $(2n + 1)\pi/2$, where $n \in Z$. So,

Domain: $R - \{(2n + 1)\pi/2 \mid n \in Z\}$

Range: $(-\infty, -1] \cup [1, \infty)$

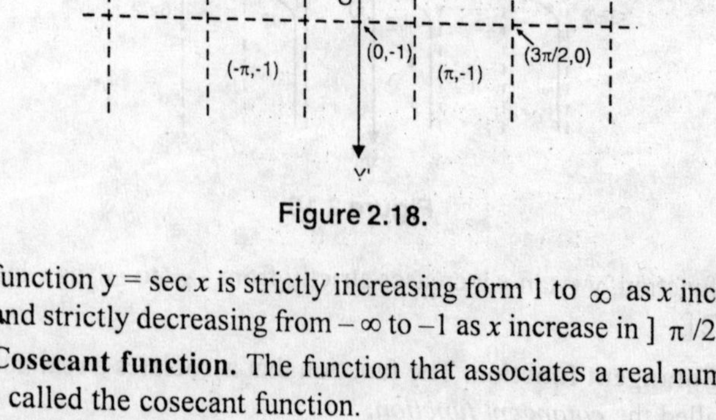

Figure 2.18.

The function y = sec x is strictly increasing form 1 to ∞ as x increases in $[0, \pi/2[$ and strictly decreasing from $-\infty$ to -1 as x increase in $]\pi/2, \pi]$

(vi) Cosecant function. The function that associates a real number x to cosec x is called the cosecant function.

Clearly, cosec x is not defined at $x = n\pi$, $n \in Z$ i.e., $0, \pm\pi, \pm 2\pi$, etc.

Domain: $R - \{n\pi \mid n \in Z\}$

Range: $(-\infty, -1] \cup [1, \infty)$

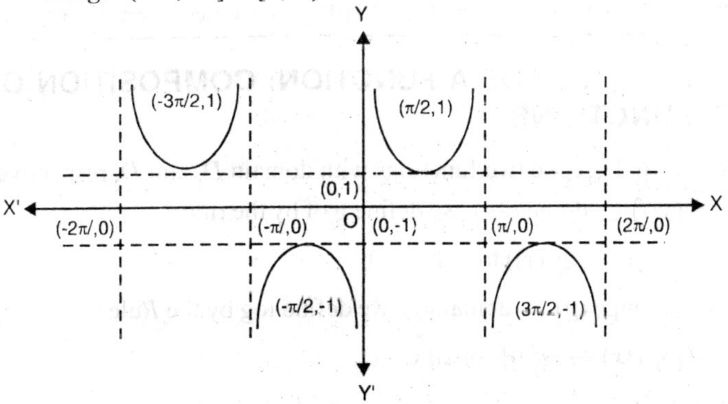

Figure 2.19

The function, $y = \text{cosec}\, x$ decreases strictly from -1 to $-\infty$ as x increases in $[-\pi/2, 0[$ and decreases from $+\infty$ to 1 as x increases in $]0, \pi/2]$.

2.14. INVERSE TRIGONOMETRICAL FUNCTIONS

The inverse trigonometrical functions $\sin^{-1}x$, $\cos^{-1}x$ etc. are *defined as the inverse of the corresponding trigonometrical functions.* For example, $\sin^{-1}x$ is defined as the angle whose sine is x. But since for each value of $\sin^{-1}x$, there corresponding an unlimited number of values of the angles whose sine is x and for values of x with $|x| > 1$, $\sin^{-1}x$ does not exist. Thus, we cannot define the inverse function of sinx unless we modify the definition of sinx in such a way that it becomes a bisection. Consequently, $\sin^{-1}x$ exits and it is a function from $[-1, 1]$ to $[-\pi/2, \pi/2]$.

Similarly, the other inverse trigonometrical functions can be defined. The domain and range of the inverse trigonometrical functions are as stated below:

Function	*Domain*	*Range*	*Definition of the function*
$\sin^{-1}x$	$[-1, 1]$	$[-\pi/2, \pi/2]$	$y = \sin^{-1}x \Leftrightarrow x = \sin y$
$\cos^{-1}x$	$[-1, 1]$	$[0, \pi]$	$y = \cos^{-1}x \Leftrightarrow x = \cos y$
$\tan^{-1}x$	R	$]-\pi/2, \pi/2[$	$y = \tan^{-1}x \Leftrightarrow x = \tan y$
$\cot^{-1}x$	R	$]0, \pi[$	$y = \cot^{-1}x \Leftrightarrow x = \cot y$

Function	Domain	Range	Definition of the function
$\sec^{-1}x$	R -]-1, 1[	[0, π]- { π /2}	$y = \sec^{-1}x \Leftrightarrow x = \sec y$
$\text{cosec}^{-1}x$	R -]-1, 1[	[- π /2, π /2]-{0}	$y = \text{cosec}^{-1}x \Leftrightarrow x = \text{cosec}y$

2.15. FUNCTION OF A FUNCTION: COMPOSITION OF FUNCTIONS

Let $f(x)$ and $g(x)$ be two functions with domain D_1 and D_2 respectively. If range(f) $\subset$ domain(g), we define g of by the rule

$$(\text{gof})\,(x) = g(f(x)) \text{ for all } x \in D_1$$

Also, if range (g) $\subset$ domain(f), we define fog by the *R*ule

$$(\text{fog})\,(x) = f\{g(x)\} \text{ for all } x \in D_2$$

2.16. INVERTIBLE FUNCTIONS

Consider a one-one function with domain A and range B. Let $y \in B$. The function f being one-one, the member $y \in B$ arises form one and only one member $x \in A$ such that $f(x) = y$.

Thus, we define a new function, say g, such that

$$g(y) = x \Leftrightarrow f(x) = y$$

Also the domain of the function g is the range of the given function f and vice-versa.

The function g is said to be an inve*R*se of the function f. We also say that f is an invertible function.

Clearly if g is the inverse of f, then f is also the inverse of g.

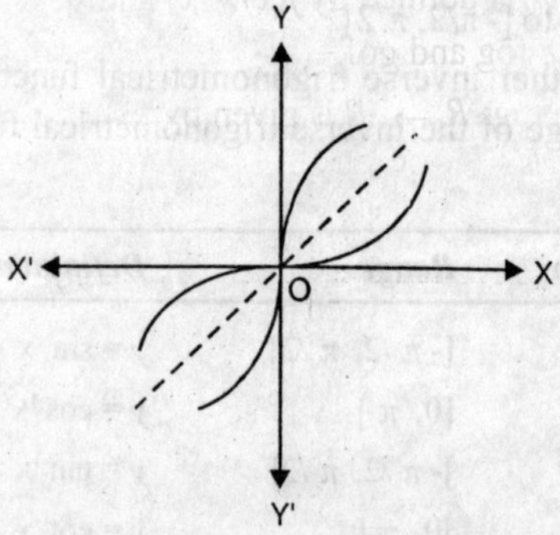

Figure 2.20

Examples

Example 1: What is the sum and difference of the identity function and the modulus function.

Solution: Let $f(x) = x, x \in R$ be the identity function and

$g(x) = |x|, x \in R$ be the modulus function. Then

$(f+g)(x) = f(x) + g(x) = x + |x|$

$$= \begin{cases} 2x, if x \geq 0 \\ 0 \ , if x < 0 \end{cases}$$

and $(f-g)(x) = f(x) - g(x) = x - |x|$

$$= \begin{cases} 0 \ , if x \geq 0 \\ 2x, if x < 0 \end{cases}$$

Example 2: Let f be the sine function and let g be the function $2x$.
Find (i) fog, (ii) gof, (iii) $f$of and (iv) gog

Solution: We have,

$f(x) = \sin x$

and $g(x) = 2x$

(i) $(fog)(x) = f(g(x)) = f(2x) = \sin 2x$

(ii) $(gof)(x) = g(f(x)) = g(\sin x) = 2\sin x$

(iii) $(fof)(x) = f(f(x)) = f(\sin x) = \sin(\sin x)$

(iv) $(gog)(x) = g(g(x)) = g(2x) = 2(2x) = 4x$

Example3: If $f: R \rightarrow R$ is defined by $f(x) = x^2$ and $g: R \rightarrow R$ is defined by $g(x) = \sin x$ then what is fog and gof.

Solution: The function fog: $R \rightarrow R$ is given by

$(fog)(x) = f(g(x)) = f(\sin x) = \sin^2 x$, for all $x \in R$

the function $gof(x) = g(f(x))$

$= g(x^2) = \sin(x^2)$, for all $x \in R$

Example 4: Let f be the exponential function and g be the logarithm function. Find

(i) $(f+g)(1)$ (ii) $(f.g)(1)$

(iii) $(3f)(1)$ (iv) $(5g)(1)$

(v) $(fog)(1)$ (vi) $(gof)(1)$

Solution: We have,

$f(x) = e^x$ and $g(x) = \ln(x)$

(i) $(f+g)(1) = f(1) + g(1)$

$= e^1 + \ln(1)$

$= e$

(ii) $(f.g)(1) = f(1).g(1)$

$= e^1 . \ln(1) = 0$

(iii) $(3f)(1) = 3f(1)$

$= 3 \times e^1 = 3e$

(iv) $(5g)(1) = 5g(1)$

$= 5 \times \ln(1) = 0$

(v) $(fog)(1) = f(g(1))$

$= f(\ln(1)) = f(0) = e^0 = 1$

(vi) $(gof)(1) = g(f(1))$

$= g(e^1) = \ln e = 1$

Example 5: Find the inverse function of the function f given by $f(x) = (x + 1)$, $x \in R$.

Solution. Let g be the inverse of f.

Then, for each $x \in R$, we have

$g(f(x)) = x$

$\Rightarrow g(x+1) = x$

$\Rightarrow g(x) = x - 1$

Thus, $g(x) = x - 1$ is the inverse of the function f given by $f(x) = x + 1$.

Exercise – 2.2

1. Show that the function $f(x) = (2x + 3)/ (x - 3)$ is invertible. Find the inverse g, of f and verify that gof is the identity function.
2. If $f(x)$ be defined on [-2, 2] and is given by

$$f(x) = \begin{cases} -1 \ , & -2 \le x \le 0 \\ x-1, & 0 < x \le 2 \end{cases}$$

and $g(x) = f(x) + |f(x)|$. Find $g(x)$.

3. Let $f(x) = \begin{cases} 1+x \ , & 0 \le x \le 2 \\ 3-x, & 2 < x \le 3 \end{cases}$

Find fof.

4. Let $f(x) = \sqrt{x+1}$ and $g(x) = \sqrt{1-x}$. Describe (i) f+ g (ii) f - g (iii) f.g (iv) f/g.
5. Express the following functions as the composites of appropriate functions:

(i) $y = \sqrt{\sin x}$ (ii) $y = \tan(\tan x)$

(iv) $y = \sin\sqrt{(x^2+1)}$ (iv) $y = \sqrt{(x^2-3x+2)}$

6. Consider the three functions:

$f(x) = x^2$, $g(x) = \sin x$, $h(x) = \sqrt{x}$

Compute:

(i) [(gof)oh] (x),
(ii) [go(foh)] (x),
(iii) [(fog)oh] (x),
(iv) [fo(goh)] (x).

3

Limits and Continuity

3.1. INTRODUCTION

In previous chapter, we have discussed functions, types of functions, etc. In this chapter, we intend to study the theory of limits in order to clarify the concept of continuity which would lay the foundation for the study of differentiability and differentiation of a function.

If x_1, x_2 are any two members of the domain of a function f so that $f(x_1)$, $f(x_2)$ are the corresponding values of the function, the $|f(x_2) - f(x_1)|$ may be large even though $|x_2 - x_1|$ is small. We now purpose to study the change $|f(x_2) - f(x_1)|$ relative to the change $|x_2 - x_1|$ by introducing the notation of continuity and discontinuity of a function.

3.2. LIMIT OF A FUNCTION

Let f be a function defined in a domain which we take for the sake of definiteness to be an interval I. We shall introduce and study the concept of limit of a function f at a point $C \in I$. Though the concept of limit is an abstract one, we shall explain the meaning of it or to have a feel of it, with the help of examples. First let us understand what do we mean by real value of x approaches (or tends to) a real number c.

Consider the graph of a straight line $y = x$. Let $P(x, y)$ be any point on the line. Draw PQ perpendicular on x-axis. Then $PQ = x(\because x = y)$. Now observe that as the point P approaches the origin O along the graph, the length x of the perpendicular keeps on decreasing.

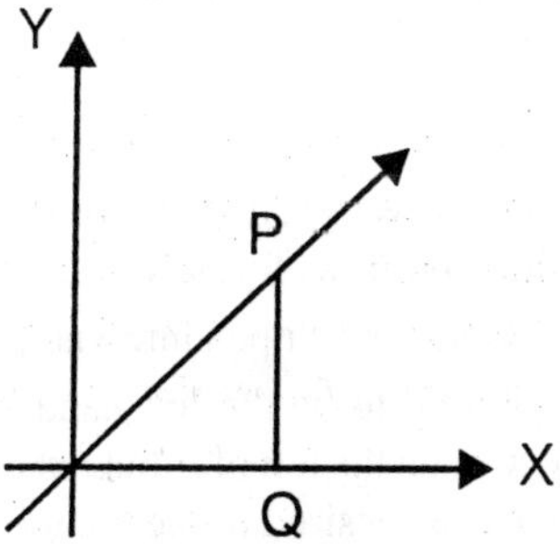

Figure 3.1

In fact at origin O, the length x of the perpendicular becomes zero. In other words, we say that x approaches (or tends to) zero and we write $x \to 0$.

For example, consider the function $f(x) = x^2, x \in R$.

Table 3.1: $x \to -2$ from left

x	$f(x)$
–2.5	6.25
–2.4	5.76
–2.3	5.29
–2.2	4.84
–2.1	4.41
–2.05	4.20
–2.04	4.16
–2.02	4.08
–2.001	4.004

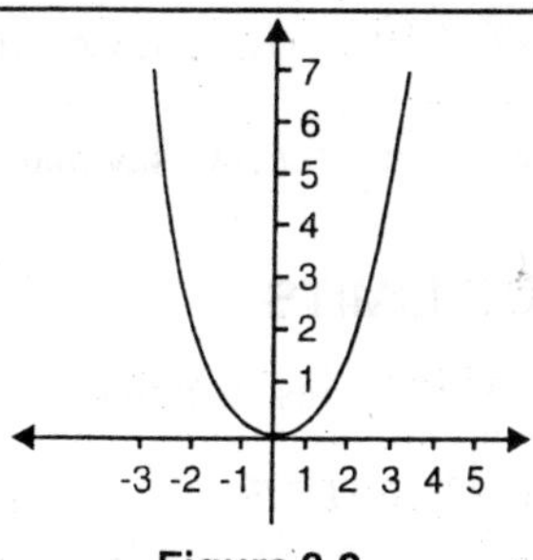

Figure 3.2

Table 3.2: $x \to -2$ from right

x	$f(x)$
–1.5	2.25
–1.6	2.56
–1.7	2.89
–1.8	3.24
–1.9	3.61
–1.93	3.80
–1.96	3.84
–1.99	3.96
–1.996	3.996

Here we observe that:

(i) as x approaches –2 form the left of –2, the graph of $f(x) = x^2$ approaches the point (–2, 4) and the value of $f(x)$ approaches 4. In this case, we say that left hand limit of f at $x = -2$ exists and is equal to 4. We use the notation

$$\lim_{x \to (-2)^-} f(x) = 4$$

(ii) as x approaches –2 from the right of –2, the graph of $f(x) = x^2$ approaches the point (–2, 4) and the value of $f(x)$ approaches 4. In this case, we say that right hand limit of f at $x = -2$ exists and is equal to 4. We use the notation

$$\lim_{x \to (-2)^+} f(x) = 4$$

On the basis of the above discussion, we may informally explain the concept of one-sided limits as follows:

Let f be a function defined on an open interval $(c-h, c+h)$, $(h > 0)$. Then,

(i) A real number l, is said to be the left hand limit of f at c, if for all x sufficiently close to c, on the left of c, but not equal to c, the value of $f(x)$ can be made as close as desired to the number l. we write

$$\text{L.H.L.} = \lim_{x \to c^-} f(x) = l$$

(ii) A real number l' is said to be the right hand limit of f at c, if for all x sufficiently close to c, on the right of c, but not equal to c, the value of $f(x)$ can be made as close as desired to the number l', we write,

$$\text{R.H.L.} = \lim_{x \to c} f(x) = l'$$

(iii) If $l = l'$, i.e., the left hand limit and the right hand limit of f at c are equal, we say that the limit of f at c exists and we write $l = \lim_{x \to c} f(x)$ exists and equals l $(= l')$. However, if $l = l'$, we say that $\lim_{x \to c} f(x)$ does not exist.

3.3. ALGEBRA OF LIMITS

Let f and g be two real functions such that

$$\lim_{x \to a} f(x) = l \quad \text{and} \quad \lim_{x \to a} g(x) = m$$

then,

(i) $\lim_{x \to a} (f+g)(x) = \lim_{x \to a} f(x) + \lim_{x \to a} g(x) = l + m$

i.e., the limit of the sum of two functions is equal to the sum of their limits.

(ii) $\lim_{x \to a} (f-g)(x) = \lim_{x \to a} f(x) - \lim_{x \to a} g(x) = l - m$

i.e., the limit of the difference of two functions is equal to the difference of their limits.

(iii) $\lim_{x\to a}(fg)(x) = \lim_{x\to a} f(x).\lim_{x\to a} g(x) = l.m$

i.e., the limit of the product of two functions is equal to the product of their limits.

(iv) $\lim_{x\to a}(f \div g)(x) = \lim_{x\to a} f(x) \div \lim_{x\to a} g(x) = l \div m, (m \neq 0)$

i.e., the limit of the quotient of two functions is equal to the quotient of their limits provided the limit of the division is not zero.

(v) $\lim_{x\to a} Kf(x) = K.\lim_{x\to a} f(x)$, where K is constant.

(vi) $\lim_{x\to a}|f(x)| = |\lim_{x\to a} f(x)| = |l|$

(vii) $\lim_{x\to a}(f(x))^{g(x)} = l^m$

(viii) $\lim_{x\to a} fog(x) = f\left(\lim_{x\to a} g(x)\right) = f(m)$

(ix) If $\lim_{x\to a} f(x) = +\infty$ or $-\infty$, then $\lim_{x\to a}\frac{1}{f(x)} = 0$

3.4. INFINITE LIMITS AND VARIABLES TENDING TO INFINITY

(i) $\lim_{x\to c} f(x) = \infty$

A function f is said to be tend to ∞ as x tends to c if to any positive number G, however large, there corresponds to a positive number δ such that for all values of $x \in]c-\delta\ [\cup]c, c+\delta\ [$,

$f(x) > G$

(ii) $\lim_{x\to c} f(x) = -\infty$

A function f is said to be tend to $-\infty$ as x tends to c if to any positive number G, however large, there corresponds a positive number δ such that for all values of $x \in]c-\delta, c[\cup]c, c+\delta\ [$,

$f(x) < -G$

(iii) $\lim_{x\to\infty} f(x) = l$

A function is said to be tend to l as x tends to ∞ if to any given number $\in > 0$ there corresponds $G > 0$ such that

$|f(x) - l| < \in$ for all $x > G$

(iv) $\lim_{x \to -\infty} f(x) = l$

A function is said to be tend to l as x tends to $-\infty$ if to any given number $\in > 0$ there corresponds $G > 0$ such that

$|f(x) - l| < \in$ for all $x < -G$

(v) $\lim_{x \to \infty} f(x) = \infty$

A function f is said to be tend to ∞, as x tends to ∞, if to any given number $\Delta > 0$ there corresponds $G > 0$ such that

$f(x) > \Delta$ for all $x > G$

(vi) $\lim_{x \to -\infty} f(x) = \infty$

A function is said to be tend to ∞ as x tends to $-\infty$ if to given $\Delta > 0$ there corresponds $G > 0$ such that

$f(x) > \Delta$ for all $x < -G$

3.5 EXTENSION OF OPERATIONS ON LIMITS

3.5.1

Lim f	*Lim g*	*Lim(f + g)*
l	$+\infty$	$+\infty$
l	$-\infty$	$-\infty$
$+\infty$	$+\infty$	$+\infty$
$-\infty$	$-\infty$	$-\infty$
$+\infty$	$-\infty$	No general conclusion

3.5.2

Lim |f|	*Lim |g|*	*Lim |fg|*
$l \neq 0$	$+\infty$	$+\infty$
0	$+\infty$	No general conclusion
$+\infty$	$+\infty$	$+\infty$

3.5.3

Lim |g|	*Lim l/|g|*
0	$+\infty$
$+\infty$	0

3.5.4

Lim |f|	Lim |g|	Lim |f| / |g|
$l \neq 0$	0	$+\infty$
0	0	No general conclusion
l	$+\infty$	0
$+\infty$	l	$+\infty$
$+\infty$	$+\infty$	No general conclusion

3.6 LIMITS OF TRIGONOMETRIC FUNCTIONS

In order to prove

$$\lim_{x \to 0} \frac{\sin x}{x} = 1$$

We first discuss without proof the behaviour of limit with respect to comparison of functions.

If f and g be two real functions, within the same domain, we say that $f \leq g$ if $f(x) \leq g(x)$, for every x in there domain.

Theorem 1: Let f and g be real functions defined on an open interval containing c such that $\lim_{x \to c} f(x)$ and $\lim_{x \to c} g(x)$ both exits. Then

$$f \leq g \Rightarrow \lim_{x \to c} f(x) \leq \lim_{x \to c} g(x)$$

Theorem 2: If f is a function defined on an interval containing c, then $\lim_{x \to c} f(x) = 0$ if and only if $\lim_{x \to c} |f(x)| = 0$.

Theorem 3: (Sandwich Theorem)

Let f, g and h be real functions defined on an open interval containing c such that $f \leq g \leq h$ and

$$\lim_{x \to c} f(x) = l = \lim_{x \to c} h(x), \quad l \in R$$

Then $\lim_{x \to c} g(x)$ exists and is equal to zero.

Assuming these results, we shall now prove some trigonometric inequalities, which are useful to prove the main result stated in the beginning.

R-I: $\cos x < \dfrac{\sin x}{x} < 1$ for all $0 < |x| < \dfrac{\pi}{2}$

Proof: Since

$$\frac{\sin(-\theta)}{-\theta} = \frac{\sin\theta}{\theta} \text{ and } \cos(-\theta) = \cos\theta$$

It is enough to prove the result for positive values of x, i.e., for $0 < |x| < \dfrac{\pi}{2}$.

In fig 3.1., OAB is a triangle with $\angle AOB = x$ radian and $\angle OAB = \dfrac{\pi}{2}$ radian (=90°). Take a point C on the hypotenuse OB such that $OC = OA$. Draw a line CD parallele to AB meeting OA at D. draw the circle with centre O and radius OA and note that it passes through C, because OA=OC. Now, we have,

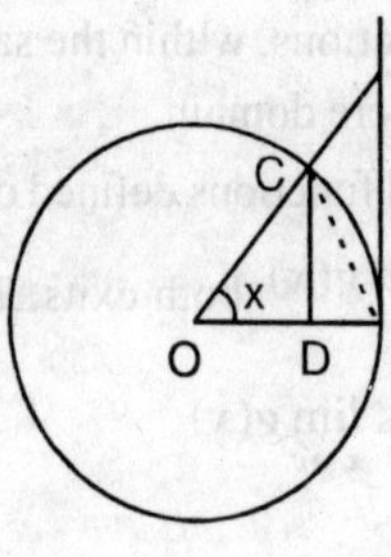

Figure 3.3

Area of ΔOAC < area of sector OAC < area of ΔAOB

$$\Rightarrow \frac{1}{2} OA.CD < \frac{x}{2\pi}.\pi.(OA)^2 < \frac{1}{2} OA.\,AB$$

$$\Rightarrow CD < x.OA < AB \qquad \ldots(1)$$

From ΔODC, we have

$$CD = OC \sin x = OA \sin x \qquad (\because OC = OA)$$

Also, from ΔOAB, we have $AB = OA \tan x$

Therefore, (1) reduces to

$$OA \sin x < OA.x < OA \tan x$$

$$\Rightarrow \sin x < x < \tan x \qquad \ldots(2)$$

Dividing throughout by the positive quantity $\sin x$, we get

$$1 < \frac{x}{\sin x} < \frac{1}{\cos x}$$

Taking reciprocals, we have

$$\cos x < \frac{\sin x}{x} < 1$$

R-II: $|\sin x| \le |x|$, holds for all x.

Proof: If $0 \le x \le 1 < \frac{\pi}{2}$, then by (2), we get

$$|\sin x| = \sin x < x \le |x|$$

If $-1 \le x \le 0$, then $0 \le -x \le 1$ and so by using (2) again, we get

$$|\sin x| = -\sin x = \sin(-x) \le -x \le |x|$$

If neither of the two cases hold, i.e., if $|x| \ge 1$, then

$$|\sin x| \le 1 \le |x|$$

Thus in all cases, we have $|\sin x| \le |x|$.

R-III: $\lim_{x \to 0} \sin x = 0$

Proof: In view of R-II, we have

$$0 \le |\sin x| \le |x|$$

Since $\lim_{x \to 0} |x| = 0$, it follows from the Sandwich theorem that $\lim_{x \to 0} |\sin x| = 0$ and so $\lim_{x \to 0} \sin x = 0$.

R-IV: $1 - \frac{x^2}{2} \le \cos x \le 1$, holds for all x.

Proof: By R-II, we have

$$|\sin x| \le |x|, \text{ for all } x$$

Squaring, we get

$$|\sin x|^2 \le |x|^2, \text{ for all } x$$

$\Rightarrow \sin^2 x \le x^2$, for all x.

In particular, changing x to $\frac{x}{2}$, we get

$$\sin^2 \frac{x}{2} \le \frac{x^2}{4}, \text{ for all } x$$

$$\Rightarrow \frac{1-\cos x}{2} \le \frac{x^2}{4}, \text{ for all } x$$

$$\Rightarrow 1-\frac{x^2}{2} \le \cos x, \text{ for all } x.$$

But $\cos x \le 1$ for all x.

Hence $1-\frac{x^2}{2} \le \cos x \le 1$, for all x.

R-V: $\lim_{x\to 0} \cos x = 1$

Proof: By R-IV, we have

$$1-\frac{x^2}{2} \le \cos x \le 1, \text{ for all } x.$$

Therefore, by Sandwich theorem, we get

$$\lim_{x\to 0} \cos x = 1 \quad (\text{since } \lim_{x\to 0} 1-\frac{x^2}{2} = 1)$$

Theorem-VI: Let x be measured in radians. Then

$$\lim_{x\to 0} \frac{\sin x}{x} = 1$$

Proof: By R-I, we have

$$\cos x < \frac{\sin x}{x} < 1, \text{ for all } 0 < |x| < \frac{\pi}{2}$$

But by R-V, $\lim_{x \to 0} \cos x = 1$

Therefore, by Sandwich theorem., we get

$$\lim_{x \to 0} \frac{\sin x}{x} = 1$$

Examples

Example 1: Evaluate

(i) $\lim_{x \to 2}\left[\frac{1}{x-2} - \frac{2(2x-3)}{x^3 - 3x^2 + 2x}\right]$

(ii) $\lim_{x \to 0} \frac{\sqrt{1+x} - \sqrt{1-x}}{x}$

(iii) $\lim_{x \to 1} \frac{(2x-3)(\sqrt{x}-1)}{2x^2 + x - 3}$

Solution:

(i) We have,

$$\lim_{x \to 2}\left[\frac{1}{x-2} - \frac{2(2x-3)}{x^3 - 3x^2 + 2x}\right] = \lim_{x \to 2}\left[\frac{1}{x-2} - \frac{2(2x-3)}{x(x-1)(x-2)}\right]$$

$$= \lim_{x \to 2}\left[\frac{x(x-1) - 2(2x-1)}{x(x-1)(x-2)}\right]$$

$$= \lim_{x \to 2}\left[\frac{x^2 - 5x + 6}{x(x-1)(x-2)}\right]$$

$$= \lim_{x \to 2}\left[\frac{x-3}{x(x-1)}\right] = \frac{-1}{2}$$

(ii) $$\lim_{x \to 0} \frac{\sqrt{1+x} - \sqrt{1-x}}{x} = \lim_{x \to 0}\left[\frac{\sqrt{1+x} - \sqrt{1-x}}{x} \times \frac{\sqrt{1+x} + \sqrt{1-x}}{\sqrt{1+x} + \sqrt{1-x}}\right]$$

$$= \lim_{x\to 0} \frac{(1+x)-(1-x)}{x(\sqrt{1+x}+\sqrt{1-x})}$$

$$= \lim_{x\to 0} \frac{2x}{x(\sqrt{1+x}+\sqrt{1-x})}$$

$$= \lim_{x\to 0} \frac{2}{\sqrt{1+x}+\sqrt{1-x}}$$

$$= \frac{2}{2} = 1$$

(iii) $$\lim_{x\to 1} \frac{(2x-3)(\sqrt{x}-1)}{2x^2+x-3} = \lim_{x\to 1} \frac{(2x-3)(\sqrt{x}-1)}{(2x+3)(x-1)}$$

$$= \lim_{x\to 1} \frac{(2x-3)(\sqrt{x}-1)}{(2x+3)(\sqrt{x}-1)(\sqrt{x}+1)}$$

$$= \lim_{x\to 1} \frac{(2x-3)}{(2x+3)(\sqrt{x}+1)}$$

$$= \frac{-1}{5\times 2} = \frac{-1}{10}$$

Example 2: Show that $\lim_{x\to 0} \frac{\log(1+x)}{x} = 1$

Solution: Let $\frac{\log(1+x)}{x} = y$, then

$$\log(1+x) = xy$$

$$\Rightarrow \quad 1+x = e^{xy}$$

$$\Rightarrow \quad \frac{e^{xy}-1}{x} = 1$$

$$\Rightarrow \quad \frac{e^{xy}-1}{xy}.y = 1$$

$$\Rightarrow \quad \lim_{xy\to 0} \frac{e^{xy}-1}{xy} \cdot \lim_{x\to 0} y = 1 \qquad (\because x \to 0 \text{ gives } xy \to 0)$$

$$\Rightarrow \quad \lim_{x\to 0} y = 1$$

$$\Rightarrow \quad \lim_{x\to 0} \frac{\log(1+x)}{x} = 1$$

Example 3: Find each of the following limits, if they exist.

(i) $\lim_{x\to+\infty} \frac{2x-1}{x+2}$ (ii) $\lim_{x\to-\infty} \frac{x^2+3}{1-x}$

Solution: (i) x is the highest power present in the expression $\lim_{x\to+\infty} \frac{2x-1}{x+2}$. So we divide each term in the numerator and denominator by x to obtain

$$\lim_{x\to+\infty} \frac{2x-1}{x+2} = \lim_{x\to+\infty} \left[\frac{\frac{2x}{x} - \frac{1}{x}}{\frac{x}{x} + \frac{2}{x}} \right]$$

$$= \lim_{x\to+\infty} \left[\frac{2 - \frac{1}{x}}{1 + \frac{2}{x}} \right]$$

$$= \frac{2-0}{1+0} \quad \left[\text{Since } \lim_{x\to+\infty} \frac{1}{x} = 0 \right]$$

$$= 2$$

(ii) x^2 is the highest power present in the expression $\lim_{x\to-\infty} \frac{x^2+3}{1-x}$. So we divide each term in the numerator and denominator by x^2 to obtain

$$\lim_{x \to -\infty} \frac{x^2+3}{1-x} = \lim_{x \to -\infty} \left[\frac{\frac{x^2}{x^2}+\frac{3}{x^2}}{\frac{1}{x}-\frac{x}{x^2}} \right]$$

$$= \lim_{x \to -\infty} \left[\frac{1+\frac{3}{x^2}}{\frac{1}{x^2}-\frac{1}{x}} \right]$$

Since the numerator tends to a finite non-zero limit $x \to -\infty$ while the denominator tends to zero as x tends to $x \to -\infty$

so, $\lim\limits_{x \to -\infty} \frac{x^2+3}{1-x}$ does not exist.

Example 4: Evaluate $\lim\limits_{x \to -\infty} e^x$

Solution: Put $y = -x$

As $x \to -\infty$, then $y \to +\infty$

So, $\lim\limits_{x \to -\infty} e^x = \lim\limits_{y \to \infty} e^{-y}$

Now, for all positive y, we have

$$e^y \geq 1+y$$

$$\Rightarrow \quad e^{-y} = \frac{1}{e^y} \leq \frac{1}{1+y}$$

$$\Rightarrow \quad 0 \leq e^{-y} \leq \frac{1}{1+y} \quad \text{(since } e^y \text{ is always positive)}$$

But $\lim\limits_{x \to +\infty} \frac{1}{1+y} = \lim\limits_{y \to +\infty} \left[\frac{1/y}{1/y+y} \right] = \frac{0}{0+1} = 0$

Example-5: Examine $\lim_{x\to 0}(x\sin x)$

Solution:

Case 1: Let $x > 0$

Now, $0 < x < \pi/2 \Rightarrow \sin x > 0$

For $0 < x < \pi/2$, we have

$$\sin x < x \Rightarrow x\sin x < x^2$$

Let $\in$ be an arbitrarily assigned positive number.

For values of x which are positive and less than $\sqrt{\in}$, we take $x^2 < \in$. Thus

$$0 < x\sin x < \in \text{ when } 0 < x < \sqrt{\in}$$

It follows that

$$\lim_{x\to(0+0)} x\sin x = 0$$

Case 2: Let $x < 0$

Now, $-\pi/2 < x < 0 \Rightarrow \sin x < 0$

The values of the function for two values of x which are equal in magnitude but opposite in signs are equal. Hence, as in case1, we see that for any value of x in the interval $\left]-\sqrt{\in},0\right[$, the numerical value of the difference between $x\sin x$ and 0 is less than $\in$. Thus $\lim_{x\to 0-0} x\sin x = 0$

Case 3: Combining the conclusions arrived at in the last two cases, we see that corresponding to any positive number $\in$ arbitrarily assigned, there exits an interval $\left]-\sqrt{\in},\sqrt{\in}\right[$ around 0, such that for any x belonging to this interval, the numerical value of the difference between $x\sin x$ and $0 < \in$ i.e.,

$$|x\sin x - 0| < \in$$

Thus, $$\lim_{x\to 0} x\sin x = 0$$

Example 6: Show that $\lim_{x\to 0+0}\frac{1}{x} = \infty$, $\lim_{x\to(0-0)}\frac{1}{x} = -\infty$ and $\lim_{x\to 0}\frac{1}{x}$ does not exist.

Solution: The function $f(x) = 1/x$ is not defined for $x = 0$. We write

$$y = \frac{1}{x}$$

Case 1: Let $x>0$ so that y is positive. If G be any positive number taken arbitrarily, then

$$1/x > G, \text{ if } 0 < x < 1/G$$

$$\Rightarrow \quad \lim(1/x) = \infty \text{ when } x \to (0+0)$$

Case 2: Let $x < 0$ so that $y = 1/x$ is negative.

If G be any positive number taken arbitrarily, then

$$1/x < -G \text{ if } -1/G < x < 0$$

$$\Rightarrow \quad \lim(1/x) = -\infty \text{ when } x \to (0-0)$$

Case 3: Clearly when $x \to 0$, $\lim(1/x)$ does not exist.

Example 7: Find $\lim_{x \to -\infty} (7x^3 + 8x^2 + 5x - 7)$

Solution: We have $7x^3 + 8x^2 + 5x - 7$

$$= x^3\left(7 + \frac{8}{x} + \frac{5}{x^2} - \frac{7}{x^3}\right)$$

Now, $\lim_{x \to -\infty} \frac{8}{x} = \lim_{x \to -\infty} \frac{-7}{x^3} = 0$

Also, $\lim_{x \to -\infty} x^3 = -\infty$

It follows that $\lim_{x \to -\infty} (7x^3 + 8x^2 + 5x - 7) = -\infty$

Exercise – 3.1

1. Show that $\lim\limits_{x\to(1+0)} \dfrac{\sqrt{x}-1}{x-1} = \dfrac{1}{2}$

2. Show that $\lim\limits_{x\to 0} \cos(1/x)$ does not exist.

3. Show that $\lim\limits_{x\to(1-0)} [x] = 0$, $\lim\limits_{x\to(1+0)} [x] = 1$.

4. Show that $\lim\limits_{x\to(3-0)} \dfrac{5x+2}{x-3} = -\infty$, $\lim\limits_{x\to(3+0)} \dfrac{(5x+2)}{x-3} = +\infty$.

5. Show that $\lim\limits_{x\to+\infty} x^3 = +\infty$, $\lim\limits_{x\to\infty} (x^3 - x^2) = +\infty$.

6. Find $\lim \dfrac{x^2+1}{x^2-4}$ when x tends to 2,-2, ∞, $-\infty$.

7. Given that $f(x) = \dfrac{x^2}{(x-1)(x-2)}$, show that

   $$\lim_{x\to+\infty} f(x) = 1 = \lim_{x\to-\infty} f(x)$$

8. If $f(x) = \dfrac{1}{x^2} + 3$, $g(x) = \dfrac{1}{x^2} + 1$ show that

   $$\lim_{x\to 0} f(x) = \infty = \lim_{x\to 0} g(x)$$

9. Show that:

   $$\lim_{x\to+\infty} \frac{\sqrt{(x^2+1)}}{x+1} = 1 =; \lim_{x\to-\infty} \frac{\sqrt{(x^2+1)}}{x+1} = -1$$

10. Show that

    $$\lim_{x\to 0} \frac{\operatorname{cosec}^2 x}{\cot^2 x} = 1$$

11. Show that $\lim\limits_{x\to 0} (\operatorname{cosec} x - \cot x) = 0$

12. Evaluate

(i) $\lim\limits_{x \to 1} \frac{1}{|1-x|}$ (ii) $\lim\limits_{x \to \infty} \frac{x^3 + x^2 - 6x + 8}{4x^3 + 5x - 8}$

(iii) $\lim\limits_{x \to \infty} \frac{(3x-1)(4x-2)}{(x+8)(x-1)}$ (iv) $\lim\limits_{x \to \infty} \sqrt{x^2 + x - 1} - x$

(v) $\lim\limits_{x \to 0} \frac{e^x - e^{-x}}{x}$ (vi) $\lim\limits_{x \to 0} \frac{e^{3+x} - \sin x - e^3}{x}$

(vii) $\lim\limits_{x \to a} \frac{\sin x - \sin a}{\sqrt{x} - \sqrt{a}}$ (viii) $\lim\limits_{x \to 0} \frac{\sin 2x + 3x}{2x + \sin 3x}$

(ix) $\lim\limits_{x \to \frac{\pi}{6}} \frac{\sqrt{3}\sin x - \cos x}{x - \frac{\pi}{6}}$ (x) $\lim\limits_{x \to \pi} \frac{1 + \cos x}{\tan^2 x}$

(xi) $\lim\limits_{x \to 0} \frac{1 - \cos mx}{1 - \cos nx}$, where m and n are fixed non-zero real numbers.

(xii) $\lim\limits_{x \to 0} \frac{\sqrt{1+x^2} - \sqrt{1+x}}{\sqrt{1+x^3} - \sqrt{1+x}}$ (xiii) $\lim\limits_{x \to 2} \frac{x^4 - 16}{x - 2}$

(xiv) $\lim\limits_{x \to 1/\sqrt{2}} \frac{x - \cos(\sin^{-1} x)}{1 - \tan(\sin^{-1} x)}$ (xv) $\lim\limits_{x \to \infty} \left(\sqrt{x + \sqrt{x + \sqrt{x}}} - \sqrt{x} \right)$

3.7. CONTINUITY

Definition 1. Let f be a real function and let c be any point in the domain of f. Then f is said to be continuous at c if $\lim_{x \to c} f(x)$ exists and $\lim_{x \to c} f(x) = f(c)$

if a function f is not continuous at a point c, we say that f is discontinuous at c and c is called a point of discontinuity of f.

Definition 2. A real function is said to be continuous in an interval (open or closed), if it is continuous at every point of the interval.

When a function f is considered on a closed interval $[a, b]$, then f is continuous on $[a, b]$ means that it is continuous at every point of $[a, b]$ including the end points a and b. By the continuity at the end points a and b of the interval $[a, b]$, we mean the following:

The function f is said to be continuous at a if

$$\lim_{x \to a+} f(x) = f(a)$$

Similarly, the function f is said to be continuous at b if

$$\lim_{x \to b-} f(x) - f(b)$$

3.8. TYPES OF DISCONTINUITY

*A function is said to be **discontinuous** at a point if it is not continuous at that point.*

Let f be a function defined on an interval I. If f be discontinuous at a point $P \in I$, then we say that

(i) Removable Discontinuity

f has a discontinuity of the first kind from the left at p if $\lim_{x \to p} f(x)$ exists but is not equal to $f(p)$.

(ii) Discontinuity of the first kind:

f has a discontinuity of the first kind from the left at p if $\lim_{x \to (p-0)} f(x)$ exists but is not equal to $f(p)$.

f has a discontinuity of the first kind from the right at p if $\lim_{x \to (p+0)} f(x)$ exists but is not equal to $f(p)$.

f has a discontinuity of the first kind at p if $\lim\limits_{x \to (p-0)} f(x)$ and $\lim\limits_{x \to (p+0)} f(x)$ exists but are unequal.

(iii) Discontinuity of the second kind:

f has a discontinuity of the second kind from the left at p if $\lim\limits_{x \to (p-0)} f(x)$ does not exist.

f has a discontinuity of the second kind from the right at p if $\lim\limits_{x \to (p+0)} f(x)$ does not exist.

f has a discontinuity of the second kind at p if neither of $\lim\limits_{x \to (p-0)} f(x)$ and $\lim\limits_{x \to (p+0)} f(x)$ exists.

3.9. ALGEBRA OF CONTINUOUS FUNCTIONS

If f and g are two continuous real functions, then

(i) $f+g$ is continuous

(ii) $f-g$ is continuous.

(iii) αf is continuous, where α is any real number.

(iv) fg is continuous.

(v) $\dfrac{1}{f}$ is continuous, for all x such that $f(x) \neq 0$

(vi) $\dfrac{f}{g}$ is continuous, for all x such that $g(x) \neq 0$

3.10. CONTINUITY OF ELEMENTARY FUNCTIONS

Consider the polynomial function

$$f(x) = a_0 x^n + a_1 x^{n-1} + \ldots\ldots.. + a_{n-1} x + a_n$$

We know

$$f = a_0 I^n + a_1 I^{n-1} + \ldots\ldots.. + a_{n-1} I + a_n \,,$$

where $a_0 .... a_n$ denote constant functions and I is the identity function.

Every constant function and the identity function are continuous over R.

A rational expression

$$g(x) = \frac{a_0 x^n + a_1 x^{n-1} + ... + a_n}{b_0 x^n + b_1 x^{n-1} + ... + b_n}$$

is continuous for every value of x except for those for which the denominator becomes zero.

Thus, the domain of continuity of each of the functions with function values sin$x$, cos$x$, tan$x$, cot$x$, sec$x$ *and* cosec$x$ *coincides with the corresponding domain of definition.*

3.11. CONTINUITY OF THE INVERSE OF A CONTINUOUS INVERTIBLE FUNCTION

If f is an invertible continuous function and g denotes its inverse, then g is also continuous.

It now follows that the functions

$$y = \sin^{-1} x, \; y = \cos^{-1} x, \; y = \tan^{-1} x$$

$$y = \cot^{-1} x, \; y = \sec^{-1} x, \; y = \operatorname{cosec}^{-1} x$$

are continuous in their domains.

Examples

Example 1: The modulus function given by $f(x) = |x|, x \in R$ is continuous at every point of R.

Solution: We have,

$$f(x) = \begin{cases} x\;, & \text{if } x > 0 \\ 0\;, & \text{if } x = 0 \\ -x\;, & \text{if } x < 0 \end{cases}$$

Let c be any real number.

If $c = 0$, then

$$\lim_{x \to 0+} f(x) = 0 = \lim_{x \to 0-} f(x)$$

$$\Rightarrow \lim_{x\to 0} f(x) = 0 = f(0)$$

$\Rightarrow$ f is continuous at $c=0$

If $c > 0$, then

$$\lim_{x\to c+} f(x) = c = \lim_{x\to c-} f(x)$$

$$\Rightarrow \lim_{x\to c} f(x) = c = f(c) \quad (\because c > 0)$$

$\Rightarrow$ f is continuous at $c > 0$

If $c < 0$, then

$$\lim_{x\to c+} f(x) = -c = \lim_{x\to c-} f(x)$$

$$\Rightarrow \lim_{x\to c} f(x) = -c = f(c) \quad (\because c < 0)$$

$\Rightarrow$ f is continuous at $c < 0$

Hence, f is continuous at any real number c.

Example 2: Show that the logarithmic function given by $f(x) = \log x, x \in (0,\infty)$ is continuous at every point of its domain.

Solution: Let c be any real point in the domain $(0,\infty)$. Then

$$\lim_{x\to c} f(x) = \lim_{x\to c} \log x$$

$$= \lim_{h\to 0} \log(c+h)$$

$$= \lim_{h\to 0} \log\left[c\left(1+\frac{h}{c}\right)\right] \qquad (c>0)$$

$$= \lim_{h\to 0}\left[\log c + \log\left(1+\frac{h}{c}\right)\right]$$

$$= \log c + \lim_{h\to 0} \log\left(1+\frac{h}{c}\right)$$

$$= \log c + \lim_{\frac{h}{c}\to 0} \frac{\log(1+h/c)}{h/c} . \lim_{h\to 0} \frac{h}{c}$$

$$= \log c + 1.0\left(as \ \lim_{x \to 0} \frac{\log(1+x)}{x} = 1 \right)$$

$$= \log c$$

Also, $f(c) = \log c$. Therefore, $\lim_{x \to c} f(x) = f(c)$. Hence, f is continuous at every point of $(0, \infty)$.

Example 3: Show that the greatest integer function is discontinuous at all integral values.

Solution: Let $f(x) = [x], \ x \in R$

Let n be any integer.

$$\lim_{x \to n+} f(x) = \lim_{x \to n+} [x]$$

$$= \lim_{x \to n+} n = n$$

$$\lim_{x \to n-} f(x) = \lim_{x \to n-} [x]$$

$$= \lim_{x \to n-} (n-1) = n-1$$

Since $n \neq n-1$ for any n, we conclude that the given function is not continuous at n. Hence it is discontinuous at every integral value.

Example 4: If $f(x) = \begin{cases} 1 & , \ \text{if } x \leq 3 \\ ax+b, & \text{if } 3 < x < 5 \\ 7 & , \ \text{if } 5 \leq x \end{cases}$

Determine the values of a and b so that $f(x)$ is continuous.

Solution: The given function is a constant function for all $x < 3$ and for all $x > 5$ so it is continuous for all $x < 3$ and for all $x > 5$.

For $x = 3$, we have

$$\lim_{x \to 3-} f(x) = \lim_{x \to 3} 1 = 1$$

$$\lim_{x \to 3+} f(x) = \lim_{x \to 3} ax + b = 3a + b$$

$$f(3) = 1$$

so, for $f(x)$ be continuous at $x=3$, we must have

$$\lim_{x\to 3+} f(x) = \lim_{x\to 3+} f(x) = f(3)$$

$\Rightarrow \quad 1 = 3a + b$ ...(i)

For $x = 5$, we have

$$\lim_{x\to 5-} f(x) = \lim_{x\to 5} ax + b = 5a + b$$

$$\lim_{x\to 5+} f(x) = \lim_{x\to 5} 7 = 7$$

and $f(5) = 7$

so, for $f(x)$ to be continuous at $x=5$, we must have

$$\lim_{x\to 5-} f(x) = \lim_{x\to 5+} f(x)$$

$\Rightarrow \quad 5a + b = 7$ ...(ii)

Solving (i) and (ii) we get

$a = 3$ and $b = -8$.

Exmaple 5: Examine the continuity of the function defined by

$$f(x) = \begin{cases} -x^2 & , x \le 0 \\ 5x - 4 & , 0 < x \le 1 \\ 4x^2 - 3x, & 1 < x < 2 \\ 3x + 4 & , x \ge 2 \end{cases}$$

at the points $x = 0, 1, 2$.

Solution: We have

$$\lim_{x\to 0+0} f(x) = \lim_{x\to 0+0} (5x - 4) = -4$$

$$\lim_{x\to 0-0} f(x) = \lim_{x\to 0-0} (-x^2) = 0$$

$$\lim_{x\to 0+0} f(x) \ne \lim_{x\to 0-0} f(x)$$

$\Rightarrow \quad f$ is not continuous at $x = 0$.

Again $\lim_{x\to 1+0} f(x) = \lim_{x\to 1+0} (4x^2 - 3x) = 1$

and $\lim_{x \to 1-0} f(x) = \lim_{x \to 1-0} (5x - 4) = 1$

Thus, $\lim_{x \to 1+0} f(x) = \lim_{x \to 1-0} f(x) = 1 = f(1)$

$\Rightarrow$ f is continuous at $x = 1$.

Also, $\lim_{x \to 2+0} f(x) = \lim_{x \to 2+0} (3x + 4) = 10$

and $\lim_{x \to 2-0} f(x) = \lim_{x \to 2-0} (4x^2 - 3x) = 10$

$\therefore \lim_{x \to 2+0} f(x) = \lim_{x \to 2-0} f(x) = f(2)$

Thus f is continuous at $x = 2$.

Exercise – 3.2

1. Examine the continuity of the following.

(i) $f(x)=\begin{cases} x \ , & \text{if } x<0 \\ x^2 \ , & \text{if } x \geq 0 \end{cases}$ at x = 0

(ii) $f(x)=\begin{cases} 5x-4 \ , & \text{if } x \leq 1 \\ 4x^2-3x \ , & \text{if } x>1 \end{cases}$ at x = 1

(iii) $f(x)=\begin{cases} \dfrac{\sin x}{x} \ , & \text{if } x<0 \\ x+1 \ , & \text{if } x \geq 0 \end{cases}$ at x = 0

(iv) $f(x)=\begin{cases} \dfrac{x}{|x|} \ , & \text{if } x \neq 0 \\ 0 \ , & \text{if } x=0 \end{cases}$ at x = 0

(v) $f(x)=\begin{cases} x^2 \sin\dfrac{1}{x} , & \text{if } x \neq 0 \\ 0 \ , & \text{if } x=0 \end{cases}$ at x = 0

(vi) $f(x)=\begin{cases} \dfrac{x^2-1}{x+1} , & \text{if } x \neq -1 \\ -2 \ , & \text{if } x=-1 \end{cases}$ at x = –1

(vii) $f(x)=\begin{cases} \dfrac{x^2-x-6}{x^2-2x-3} , & \text{if } x \neq 3 \\ \dfrac{5}{3} \ , & \text{if } x=3 \end{cases}$ at x = 3

(viii) $f(x)=\begin{cases} \dfrac{\sin x}{x}+\cos x , & \text{if } x \neq 0 \\ 5 \ , & \text{if } x=0 \end{cases}$ at x = 0

(ix) $f(x)=\begin{cases}\dfrac{e^x-1}{\log_e(1+2x)}, & \text{if } x\neq 0\\ 7 & , \text{ if } x=0\end{cases}$ at $x=0$

(x) $f(x)=\begin{cases}\dfrac{e^{x^2}}{e^{1/x^2}-1}, & \text{if } x\neq 0\\ 1 & , \text{ if } x=0\end{cases}$ at $x=0$

2. Examine the limit of the function $f(x)$ as $x\to 2$, where

$$f(x)=\begin{cases}\dfrac{|x-2|}{x-2}, & x\neq 2\\ 0 & , x=2\end{cases}$$

3. Examine the continuity of the function

$$f(x)=\begin{cases}2x-[x]+\dfrac{1}{\sin x}, & x\neq 0\\ 0 & , x=0\end{cases}$$

4. A function f is defined on [0, 1] as follows:

$$f(x)=\frac{1}{2^n} \text{ when } \frac{1}{2^{n+1}}<x\le\frac{1}{2^n} \text{ (n=0, 1, 2,......)}$$

Show that f is discontinuous at the points $\frac{1}{2},\left(\frac{1}{2}\right)^2,\left(\frac{1}{2}\right)^3,$.......... and examine the nature of discontinuity.

5. In each of the following, find the value of the constant k so that the given function is continuous at the indicated point.

(i) $f(x)=\begin{cases}\dfrac{k\cos x}{\pi-2x}, & \text{if } x\neq\dfrac{\pi}{2}\\ 3 & , \text{ if } x=\dfrac{\pi}{2}\end{cases}$ at $x=\dfrac{\pi}{2}$

(ii) $f(x)=\begin{cases}k(x^2+3x), & \text{if } x<0\\ \cos 2x & , \text{ if } x\ge 0\end{cases}$

(iii) $f(x)=\begin{cases}\dfrac{\sqrt{1+kx}-\sqrt{1-kx}}{x}, & \text{if } -1\le x<0\\[2ex] \dfrac{2x+1}{x-2}, & \text{if } 0\le x\le 1\end{cases}$

6. Find the values of a and b so that the function $f(x)$ defined by

$$f(x)=\begin{cases} x+a\sqrt{2}\sin x, & \text{if } 0\le x<\dfrac{\pi}{4}\\[2ex] 2x\cot x+b, & \text{if } \dfrac{\pi}{4}\le x\le\dfrac{\pi}{2}\\[2ex] a\cos 2x-b\sin x, & \text{if } \dfrac{\pi}{2}\le x\le\pi\end{cases}$$

becomes continuous on $[0,\pi]$

7. Show that the function f given by

$f(x)=|x|+|x-1|\cdot\ x\in R$ is continuous both at $x=0$ and $x=1$.

4

Derivatives

4.1. INTRODUCTION

To find the equation of tangent to a curve at a specific point was one of the most important problems in mathematics. It was quite easy for circles because the tangent is just a perpendicular to the raius at that point. Even for parabolas, ellipses and hyperbolas, the standard equations of tangents were available. So in this case the problem was not that difficult. But the real difficulty was seen with curves having complicated graphs like $y = x^3$, $y^2 = x^3$ etc.

There was one more problem that of the path in which planets were revolving round the sun. After many years of watching the sky. Kelper concluded that the plantes are moving round the sun in elliptic orbits. Though it was quite a major breakthrough, Kelper could not give a logical reasoning for his claim.

Newton settled the problem of orbits of planets mathematically while Leibnitz settled the problem of tangents. Both of them independently invented 'Differential Calculus', one using physical approach and the other using geometric approach. Today, anyone who is well familiar with the methods of calculus can very easily explain it convincingly.

4.2. DERIVABILITY AND DERIVATIVE

Consider a function f with some domain D. Let 'c' be an interior point of D.

Take another member $(c + h)$ of the domain which lies to the right or left of c (i.e, $c + h > c$ or, $c + h < c$) according as h is positive or negative.

The values of the function correspond to c and $c + h$ are $f(c)$ and $f(c+h)$ respectively. Now, h, is the change in x and $f(c + h) - f(c)$ is the corresponding change in $f(x)$.

Definition: A function f is said to be derivable at c if,

$$\lim_{h\to 0} \frac{f(c+h)-f(c)}{h}$$

exists and the limit is called the **derivative** of the function at c and is denoted by $f'(c)$ the function f is said to be finitely **derivable** at c if the derivative at c is finite.

Right Hand Derivative:

If $\lim_{h\to(0+0)} \frac{f(c+h)-f(c)}{h}$ exists, is called the **right hand derivative** at c and is denoted by $f'(c+o)$ or $Rf'(c)$.

Left Hand Derivative:

If $\lim_{h\to(0-0)} \frac{f(c+h)-f(c)}{h}$ exists, is called the **left hand derivative** at c and is denoted by $f'(c-o)$ or $Lf'(c)$.

So, f is derivable at c if $f'(c+o)$ and $f'(c-o)$ both exists and are equal.

Examples

Example 1: Show that the function $f(x) = x^2$ is derivable for $x = 1$.

Solution: Now, $f(x) = x^2$

$$\Rightarrow f(1) = 1^2 = 1$$

$$\Rightarrow f'(x) = \frac{f(1+h)-f(1)}{h}$$

$$= \frac{1+2h+h^2-1}{h}$$

$$= 2+h \quad (\because h \neq 0)$$

$$\Rightarrow \lim_{h\to 0} \frac{f(1+h)-f(1)}{h} = 2$$

$$\Rightarrow f'(1) = 2$$

Therefore, f is derivable at 1 and the derivative $f'(1)$ is 2.

Example 2: Show that $f(x) = |x|$ is not derivable at 0.

Solution:

$$f(x) = |x|$$

$$\Rightarrow f(h) = h \quad \text{if } h > 0$$

$$= -h \quad \text{if } h < 0$$

$$\therefore \frac{f(0+h) - f(0)}{h} = \frac{f(h)}{h} = \frac{h}{h} = 1 \quad \text{if } h > 0$$

$$= \frac{-h}{h} = -1 \quad \text{if } h < 0$$

Thus, $f'(0+0) = 1$
and $f'(0-0) = 1$
Now, $f'(0+0)$ and $f'(0-0)$ are not equal.
So $f'(0)$ does not exits and the function is not derivable at 0.

Example 3: Let f be defined by setting

$f(x) = |x-1| + |x+1|$, for all $x \in R$.

Show that f is not derivable at the points $x = -1$ and $x = 1$ and is derivable at every other point.

Solution.

We know,

$$|x-1| = \begin{cases} x-1 &, \text{ if } x-1 \geq 0 \\ -(x-1), & \text{if } x-1 < 0 \end{cases}$$

$$|x+1| - \begin{cases} x+1 &, \text{ if } x+1 \geq 0 \\ -(x+1), & \text{if } x+1 < 0 \end{cases}$$

If $x+1 < 0$ and $x-1 < 0$,

then, $f(x) = |x-1| + |x+1|$

$$= -(x-1)+[-(x+1)]$$

$$= -2x$$

If $x+1 \geq 0$ and $x-1 < 0$

then, $f(x) = -(x-1)+(x+1)$

$$=2$$

If $x+1 \geq 0$ and $x+1 > 0$

then, $f(x) = (x-1)+(x+1)$

$$=2x$$

Thus,

now, the right hand and left hand derivatives at $x = -1$ and $x = 1$ is

$$Rf'(-1) = \lim_{h \to (0+0)} \frac{f(-1+h)-f(1)}{h}$$

$$= \lim_{h \to (0+0)} \frac{2-2}{h} = 0$$

$$Lf'(-1) = \lim_{h \to (0-0)} \frac{f(-1-h)-f(-1)}{h}$$

$$= \lim_{h \to (0+0)} \frac{-2(-1-h)-(2)}{-h}$$

$$= -2$$

Here, $Rf'(-1) \neq Lf'(-1)$.

Therefore, f is not derivable at $x = -1$

Again,

$$Rf'(1) = \lim_{h \to (0+0)} \frac{f(1+h)-f(1)}{h}$$

$$= \lim_{h \to (0+0)} \frac{2(1+h)-2}{h}$$

$$=2$$

$$Lf'(1) = \lim_{h \to (0-0)} \frac{f(1-h)-f(1)}{-h}$$

$$= \lim_{h \to (0-0)} \frac{2-2}{-h}$$

$$= 0$$

Since $Rf'(1) \neq Lf'(1)$, therefore f is not derivable at $x = 1$.

Exercise – 4.1

1. Let f be defined on R by setting

   $$f(x) = x \frac{e^{1/x} - e^{-1/x}}{e^{1/x} + e^{-1/x}}, \text{ if } x \neq 0$$

   $$f(0) = 0$$

   Show that f is not derivable at $x = 0$.

2. Find the derivatives for the given values of x if

   (i) $f(x) = 4x^2 - 3x + 5$ for $x = 3/2$

   (ii) $f(x) = 1/x$ for $x = 3$

   (iii) $f(x) = \sqrt[3]{x}$ for $x = 1$

   (iv) $f(x) = 1/\sqrt{x}$ for $x = 1$

   (v) $f(x) = x^3 - 3$ for $x = 2$

4.3. DERIVED FUNCTION

Let f be a function with domain $[a, b]$. Suppose f is a finitely derivable at every point of $]a, b[$ and also the right-hand and left-hand derivatives exist finitely at a and b respectively.

Thus, we have another function by f' such that $f'(x)$ denotes the derivative of f for $x \in [a,b]$. The function f' is called **derived function**. The derived function f is given as follows:

$$\frac{dy}{dx} = f'(x), \quad x \in [a,b]$$

Another Notation of derived function. Consider a derivable function f. Let $f(x) = y$

Let x be a member of the domain of f.

Usually, Δx denotes the change in x in palce of h.

$$y = f(x)$$

$$\Rightarrow y + \Delta y = f(x + \Delta x)$$

$$\Rightarrow \Delta y = f(x + \Delta x) - f(x)$$

$$\Rightarrow \frac{\Delta y}{\Delta x} = \frac{f(x + \Delta x) - f(x)}{\Delta x}$$

Taking, limits of both sides as $\Delta x \to 0$, we get

$$\lim_{\Delta x \to 0} \frac{\Delta y}{\Delta x} = \lim_{\Delta x \to 0} \frac{f(x + \Delta x) - f(x)}{\Delta x}$$

$$\Rightarrow \lim_{\Delta x \to 0} \frac{\Delta y}{\Delta x} = f'(x)$$

$$\Rightarrow \frac{dy}{dx} = f'(x)$$

The derivatives at c is in terms of this notation, is denoted by

$$\left(\frac{dy}{dx}\right)_{x=c} \quad \text{so that we have} \quad f'(c) = \left(\frac{dy}{dx}\right)_{x=c}$$

4.4. DERIVABILITY AND CONTINUITY

Theorem. If f is finitely derivable at c, then f is also continuous at c.

Proof. Since f is derivable at c, therefore,

$$\lim_{h\to 0}\frac{f(c+h)-f(c)}{h} \text{ exists and equals } f'(c).$$

Now,

$$f'(c+h)-f(c)=\frac{f(c+h)-f(c)}{h}\times h$$

$$\Rightarrow \lim_{h\to 0}[f'(c+h)-f(c)]=\lim_{h\to 0}\left[\frac{f(c+h)-f(c)}{h}\times h\right]$$

$$=\lim_{h\to 0}\frac{f(c+h)-f(c)}{h}\times\lim_{h\to 0}(h)$$

$$=f'(c)\times 0=0$$

$$\Rightarrow \lim_{h\to 0}f(c+h)=f(c)$$

$$\Rightarrow \lim_{x\to c}f(x)=f(c)$$

$\Rightarrow f$ is continuous at c.

Examples

Example 1: Show that $f(x)=x^2|x|$ is derivable at the origin.

Solution:

We have, $f(x)=x^2|x|$

$$\Rightarrow f(x)=\begin{cases}x^3, & if\, x\geq 0\\ -x^3, & if\, x<0\end{cases}$$

$$Lf'(0)=\lim_{x\to(0-0)}\frac{f(x)-f(0)}{x-0}$$

$$= \lim_{x \to (0-0)} \frac{-x^3}{x}$$

$$= 0$$

$$Rf'(0) = \lim_{x \to (0+0)} \frac{f(x) - f(0)}{x - 0}$$

$$= \lim_{x \to (0+0)} \frac{x^3}{x}$$

$$= 0$$

$\therefore Lf'(0) = Rf'(0)$

$\Rightarrow f$ is derivable at the origin.

Example 2: Show that the function defined by

$$f(x) = \begin{cases} \dfrac{x}{1+e^{1/x}}, & x \neq 0 \\ 0 & , x = 0 \end{cases}$$

is continuous at $x = 0$ but not derivable at that point.

Solution.

We have,

$$f(0+0) = \lim_{x \to 0+0} \frac{x}{1+e^{1/x}}$$

$$= 0$$

$$f(0-0) = \lim_{x \to 0-0} \frac{x}{1+e^{1/x}}$$

$$= 0$$

Also $f(0) = 0$

$\therefore f(0+0) = f(0-0) = f(0)$

$\Rightarrow f$ is continuous at $x = 0$.

Again,

$$f'(0+0) = \lim_{x \to 0+0} \frac{f(x) - f(0)}{x - 0}$$

$$= \lim_{x \to 0+0} \frac{\dfrac{x}{1+e^{1/x}} - 0}{x}$$

$$= \lim_{x \to 0+0} \frac{1}{1+e^{1/x}}$$

$$= 1$$

$$f'(0-0) = \lim_{x \to 0-0} \frac{f(x) - f(0)}{x - 0}$$

$$= \lim_{x \to 0-0} \frac{\dfrac{x}{1+e^{1/x}} - 0}{x}$$

$$= \lim_{x \to 0-0} \frac{1}{1+e^{1/x}}$$

$$= 0$$

Since $f'(0+0) \neq f'(0-0)$, the derivable of $f(x)$ at $x = 0$ does not exist.

Example 3: Show that the function f defined by

$$f(x) = |x-2| + |x| + |x+2|$$

is not derivable at x = –1, 0 and 1.

Solution. We have,

$$f(x) = \begin{cases} -(x-1) - x - (x+1) = -3x \quad, & x < -1 \\ -(x-1) - x + (x+1) = -x + 2, & -1 \le x \le 0 \\ -(x-1) + x + (x+1) = x + 2 \quad, & 0 \le x < 1 \\ (x-1) + x + (x+1) \quad = 3x \quad, & x \ge 1 \end{cases}$$

$$f'(-1+0) = \lim_{x \to -1+0} \frac{f(x) - f(-1)}{x+1}$$

$$= \lim_{x \to -1+0} \frac{-x+2-3}{x+1}$$

$$= \lim_{x \to -1+0} \frac{-(x+1)}{x+1}$$

$$= -1$$

$$f'(-1-0) = \lim_{x\to -1-0} \frac{f(x)-f(1)}{x+1}$$

$$= \lim_{x\to -1-0} \frac{-3x-3}{x+1}$$

$$= -3$$

$\because f'(-1+0) \neq f'(-1-0)$

$\therefore f$ is not derivable at $x = -1$.

Again,

$$f'(0+0) = \lim_{x\to 0+0} \frac{f(x)-f(0)}{x-0}$$

$$= \lim_{x\to 0+0} \frac{x+2-2}{x}$$

$$= 1$$

$$f'(0-0) = \lim_{x\to 0-0} \frac{f(x)-f(0)}{x-0}$$

$$= \lim_{x\to 0-0} \frac{-x+2-2}{x}$$

$$= -1$$

Since $f'(0+0) \neq f'(0-0)$,

therefore f is not derivable at $x = 0$.

Also,

$$f'(1+0) = \lim_{x\to 1+0} \frac{f(x)-f(1)}{x-1}$$

$$= \lim_{x\to 1+0} \frac{3x-3}{x-1}$$

$$= 3$$

$$f'(1-0) = \lim_{x\to 1-0} \frac{f(x)-f(1)}{x-1}$$

$$= \lim_{x\to 1-0} \frac{x+2-3}{x-1}$$

$$= 1$$

$$\therefore \quad f'(1+0) \neq f'(1-0)$$

$\Rightarrow f$ is not derivable at $x = 1$.

Example 4: Discuss the derivability of the function

$$f(x) = \begin{cases} 1+x & , \quad x \leq 1 \\ x & , \quad 0 < x < 1 \\ 2-x & , \quad 1 \leq x \leq 2 \\ 3x-x^2, & \quad x > 2 \end{cases}$$

at $x = 0$, 1 and 2.

Solution.

We have,

$$f'(0+0) = \lim_{x \to 0+0} \frac{f(x) - f(0)}{x - 0}$$

$$= \lim_{x \to 0+0} \frac{x-1}{x}$$

$$= -\infty$$

$$f'(0-0) = \lim_{x \to 0-0} \frac{f(x) - f(0)}{x - 0}$$

$$= \lim_{x \to 0-0} \frac{1+x-1}{x}$$

$$= 1$$

Since $f'(0+0) \neq f'(0-0)$,

therefore, f is not derivable at $x = 0$.

Again,

$$f'(1+0) = \lim_{x \to 1+0} \frac{f(x) - f(1)}{x - 1}$$

$$= \lim_{x \to 1+0} \frac{2-x-1}{x-1}$$

$$= -1$$

$$f'(1-0) = \lim_{x\to 1-0} \frac{f(x)-f(1)}{x-1}$$

$$= \lim_{x\to 1-0} \frac{x-1}{x-1}$$

$$= 1$$

Since $f'(1+0) \neq f'(1-0)$,

therefore, f is not derivable at $x = 1$.

Also,

$$f'(2+0) = \lim_{x\to 2+0} \frac{f(x)-f(2)}{x-2}$$

$$= \lim_{x\to 2+0} \frac{3x-x^2-0}{x-2}$$

$$= \infty$$

$$f'(2-0) = \lim_{x\to 2-0} \frac{f(x)-f(2)}{x-2}$$

$$= \lim_{x\to 2-0} \frac{2-x-0}{x-2}$$

$$= -1$$

Since, $f'(2+0) \neq f'(2-0)$,

therefore, f is not derivable at $x = 2$.

Example 5: If $f(x) = x^2 \sin(1/x)$ when $x \neq 0$ and $f(0) = 0$, show that f is derivable for every value of x but the derivative is not continuous for zero.

Solution.

When $x \neq 0$, $f'(x) = 2x\sin\frac{1}{x} + x^2\cos\left(\frac{1}{x}\right)\left(-\frac{1}{x^2}\right)$

$$= 2x\sin\frac{1}{x} - \cos\frac{1}{x}$$

When $x = 0$, we have

$$\frac{f(x)-f(0)}{x-0} = \frac{x^2 \sin\frac{1}{x}}{x}$$

$$= x \sin\frac{1}{x}$$

$$\Rightarrow f'(0) = 0$$

Thus the function possesses a derivative for every value of x given by

$$f'(x) = 2x\sin\frac{1}{x} - \cos\frac{1}{x}, \text{ when } x \neq 0,\ f'(0) = 0.$$

Now, we show that f is not continuous for $x = 0$

We write

$$\cos\frac{1}{x} = 2x\sin\frac{1}{x} - \left(2x\sin\frac{1}{x} - \cos\frac{1}{x}\right)$$

We have,

$$\lim_{x\to 0}\left(2x\sin\frac{1}{x}\right) = 0$$

In case $\lim_{x\to 0} f'(x)$, had existed, it would follow from(i) that $\lim_{x\to 0}\left(\cos\frac{1}{x}\right)$ would also exist.

But this is not the case. Hence $\lim_{x\to 0} f'(x)$, does not exist so that f is not continuous for $x = 0$.

Exercise – 4.2

1. Show that the function f defined by

$$f(x) = |x-2| + |x| + |x+2|$$

is not derivable at $x = -2$, 0 and 2.

2. Examine the function f where

$$f(x) = \frac{x(e^{-1/x} - e^{1/x})}{e^{-1/x} + e^{1/x}}, \quad x \neq 0$$
$$= 0,$$

as regards continuity and derivability at the origin.

3. If $f(x) = x^{2n+1}\sin(1/x)$, when $x \neq 0$ and $f(0) = 0$, show that f^{n} is continuous at $x = 0$ but not derivable at $x = 0$.

4. Show that if $f(x) = |x| + |x-1|$, the function f is continuous for every value of x but not derivable for $x = 0$ and $x = 1$.

5. Discuss the continuity and derivability of the function.

$$f(x) = \begin{cases} x, & x < 1 \\ 2 - x, & 1 \le x \le 2 \\ -3 + 3x - x^2, & x > 2 \end{cases}$$

at $x = 1, 2$.

6. Show that the function

$$f(x) = x\left\{1 + \frac{1}{3}\sin(\log x^2)\right\}; \text{ for } \; x \neq 0,\; f(0) = 0$$

is everywhere continuous but has no derivable at $x = 0$.

7. Examine for continuity at $x = a$, the function f where

$$f(x) = \begin{cases} \dfrac{x^2}{a} - a, & 0 < x < a \\ 0, & x = a \\ a - \dfrac{a^3}{x^2}, & a < x \end{cases}$$

Also examine if the function is derivable at a.

8. Find the derivative of f where $f(x)=\frac{\sin x^2}{x}$, when $x \neq 0$ and $f(0)=0$ and show that the derivative is continuous at $x = 0$.

9. Examine the continuity and derivability in the interval $]-\infty,\infty[$ of the function defined as follows:

$$f(x) = 1 \text{ in }]-\infty,0[$$

$$= 1+\sin x \qquad \text{in } \left]0,\frac{\pi}{2}\right[$$

$$= 2+\left(x-\frac{\pi}{2}\right)^2 \text{ in } \left]\frac{\pi}{2},\infty\right[$$

10. Show that the function $f(x)=2|x-2|+5|x-3|$ for all $x \in R$ is not derivable at $x=2$ and $x=3$.

4.5. GEOMETRICAL INTERACTION OF A DERIVATIVE

Consider two points $P[c, f(c)]$ and $2[c+h, f(c+h)]$ on the curve $y = f(x)$.

Draw the ordinates PL, QM and draw $PN \perp MQ$.

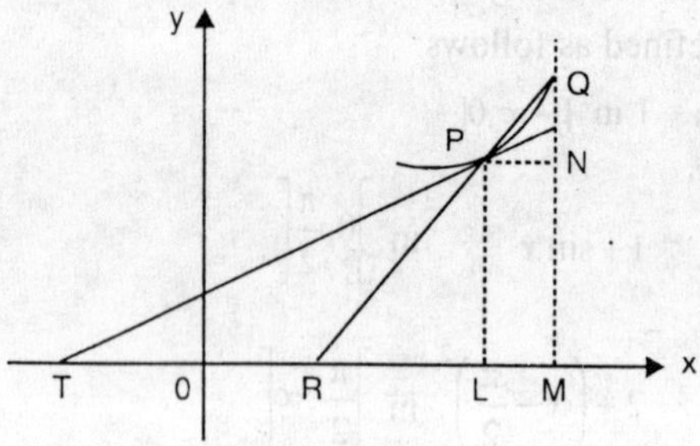

Figure 4.1

We have, $PN = LM = h$

and $NQ = MQ - LP$

$= f(c+h) - f(c)$

$\therefore \tan \angle XRQ = \tan \angle NPQ$

$$= \frac{NQ}{PN}$$

$$= \frac{f(c+h) - f(c)}{h} \qquad \ldots(1)$$

Here, $\angle XRQ$ is the angle which the chord PQ of the curve makes with x-axis.

As h approaches 0, the point Q moving along the curve approaches the point P, the chord PQ approaches the tangent line TP and $\angle XRQ$ which we denote by ψ. On taking limits, the equation(i) gives

$\tan \psi = f'(c)$

Thus $f'(c)$ is the slope of the tangent to the curve $y = f(x)$ at the point $P[c, f(c)]$.

The slope of the tangent at a point of a curve is also known as the ***Gradient*** *of the curve at the point.*

4.6. DIFFERENTIAL AND DIFFERENTIAL CO-EFFICIENT

Let a function f be differentiable in $[a, b]$.

Let $x \in [a,b]$

We write, $y = f(x)$

Now, when $\Delta x \to 0$, $\lim \frac{\Delta y}{\Delta x} = f'(x)$

So that, $\frac{\Delta y}{\Delta x}$ differs from $f'(x)$ by a variable which tends to zero when $\Delta x \to 0$

We write,

$$\frac{\Delta y}{\Delta x} = f'(x) + \alpha$$

$$\Rightarrow \Delta y = f'(x)\Delta x + \alpha \Delta x$$

$f'(x)\Delta x$ is more significant than that of $\alpha\Delta x$ in that $\alpha\Delta x$ is the product of two variables α and Δx each of which tends to zero. The part $f'(x)\Delta x$ is called the **differential** of y and is denoted by ***dy***.

$$\therefore dy = f'(x)\Delta x$$

When $f(x) = x, f'(x) = 1$

$$\therefore dx = 1 \times \Delta x = \Delta x$$

Thus, we have

$$dy = f'(x)dx$$

While Δy denotes the increment in y, dy stands for the differential of y.

The derivative $f'(x)$ being the co-efficient of the differential dx is also known as **differential co-efficient.**

4.7. KINETIC INTERPRETATION OF A DERIVATIVE

In practice, velocity at an instant is calculated by measuring the distance traveled in some short interval of time subsequent to the instant under consideration. This manner of calculating the velocity cannot clearly be considered precise, for different measuring agents may employ differential intervals for the purpose. In fact, this is only an approximate value of the actual velocity. The smaller the interval, the better is, of course, the approximation to the actual velocity.

The precise meaning of the velocity of a moving particle at any instant can only be given by employing the notation of derivative which we do in the following.

Expression for velocity. The motion of a particle along a straight line is analytically represented by an equation of the form

$$s = f(t)$$

where 's' represents the distance of the particle measured from some fixed point O on the line at time t.

Let P be the position of the particle at any time 't'.

Let Q be it's position after some interval Δt.

Let $PQ = \Delta s$

$$\text{Velocity at time, t} = \lim_{\Delta t \to 0} \frac{f(t+\Delta t) - f(t)}{\Delta t} = \lim_{\Delta t \to 0} \frac{\Delta s}{\Delta t} = \frac{ds}{dt}$$

Expression for acceleration. Let 'v' be the velocity at time 't' and Δv be the velocity after some short interval of time Δt.

The ratio $\dfrac{\Delta v}{\Delta t}$ is the average acceleration during this interval Δt and is an approximation to the actual acceleration at time t. Thus, we can define the measure of acceleration at time t as

$$\lim_{\Delta t \to 0} \frac{\Delta v}{\Delta t} = \frac{dv}{dt}$$

4.8. DERIVATIVE OF A CONSTANT FUNCTION

Consider a constant function

$y = c$ for all $x \in R$

where c is a given number.

We have, $\dfrac{\Delta y}{\Delta x} = \dfrac{0}{\Delta x} = 0$

$$\therefore \frac{dy}{dx} = \lim_{\Delta x \to 0} \frac{\Delta y}{\Delta x}$$

$$= 0 \quad \text{for all } x \in R$$

Thus, the derivative of a constant function is the zero function which is also a constant function.

4.9. ALGEBRA OF DERIVATIVES

4.9.1. Derivative of y = cu, where c is a constant and u is any derivable function of x.

Now, $y = c(u)$

$$\Rightarrow y + \Delta y = c(u + \Delta u)$$

$$\Rightarrow \Delta y = c(u + \Delta u) - cu$$

$$= c.\Delta u$$

$$\Rightarrow \frac{\Delta y}{\Delta x} = c.\frac{\Delta u}{\Delta x}$$

Taking limit of both sides as $\Delta x \to 0$, we have

$$\lim_{\Delta x \to 0} \frac{\Delta y}{\Delta x} = \lim_{\Delta x \to 0} c.\frac{\Delta u}{\Delta x}$$

$$\Rightarrow \frac{dy}{dx} = c.\frac{du}{dx}$$

$$\Rightarrow \frac{d(cu)}{dx} = c.\frac{du}{dx}$$

4.9.2. Derivative of the sum or difference of two functions:

Let u and v be two derivable function of x.

We write, $y = u + v$

Then $\frac{dy}{dx} = \lim_{\Delta x \to 0} \frac{[(u + \Delta u) + (v + \Delta v)] - (u + v)}{\Delta x}$

$$= \lim_{\Delta x \to 0} \left[\frac{\Delta u}{\Delta x} + \frac{\Delta v}{\Delta x} \right]$$

$$= \lim_{\Delta x \to 0} \frac{\Delta u}{\Delta x} + \lim_{\Delta x \to 0} \frac{\Delta v}{\Delta x}$$

$$= \frac{du}{dx} + \frac{dv}{dx}$$

Thus, $\frac{d}{dx}(u + v) = \frac{du}{dx} + \frac{dv}{dx}$

Similarly, we can write

$$\frac{d}{dx}(u - v) = \frac{du}{dx} - \frac{dv}{dx}$$

4.9.3. Derivative of the product of two functions:

Let $y = u.v$, where both u and v are derivable function of x.

Then $y + \Delta y = (u + \Delta u)(v + \Delta v)$

Therefore, $\Delta y = (u + \Delta u)(v + \Delta v) - u.v$

$$= u.\Delta v + v.\Delta u + \Delta u.\Delta v$$

Thus, $\frac{dy}{dx} = \lim_{\Delta x \to 0} \frac{\Delta y}{\Delta x} = \lim_{\Delta x \to 0} \left[\frac{u.\Delta v + v.\Delta u + \Delta u.\Delta v}{\Delta x} \right]$

$$= \lim_{\Delta x \to 0} \left[u.\frac{\Delta v}{\Delta x} + v.\frac{\Delta u}{\Delta x} + \frac{\Delta u}{\Delta x}.\Delta v \right]$$

$$= \lim_{\Delta x \to 0} u.\frac{\Delta v}{\Delta x} + \lim_{\Delta x \to 0} v.\frac{\Delta v}{\Delta x} + \lim_{\Delta x \to 0} \frac{\Delta u}{\Delta x}.\lim_{\Delta x \to 0} \Delta v$$

$$= u.\frac{dv}{dx} + v.\frac{du}{dx} + \frac{du}{dx}.0$$

Hence, $\frac{d}{dx}(u.v) = u.\frac{dv}{dx} + v.\frac{du}{dx}$

Note: *Derivative of the product of two functions = First function × derivative of the second function + second function × derivative of the first function.*

4.9.4. *Derivative of a Quotient:*

Let u and v be two derivable functions of x and $v(x) \neq 0$.

We write, $y = \frac{u}{v}$

Therefore, $y + \Delta y = \frac{u + \Delta u}{v + \Delta v}$

$$\Rightarrow \Delta y = \frac{u + \Delta u}{v + \Delta v} - \frac{u}{v}$$

$$= \frac{v.\Delta u - u.\Delta v}{v(v + \Delta v)}$$

$$\Rightarrow \lim_{\Delta x \to 0} \frac{\Delta y}{\Delta x} = \frac{\lim\limits_{\Delta x \to 0} v.\Delta u - \lim\limits_{\Delta x \to 0} u.\Delta v}{v. \lim\limits_{\Delta x \to 0} (v + \Delta v)}$$

$$\Rightarrow \frac{dy}{dx} = \frac{v.\frac{du}{dx} - u.\frac{dv}{dx}}{v^2}$$

Thus, $\frac{d}{dx}\left(\frac{u}{v}\right) = \frac{v.\frac{du}{dx} - u.\frac{dv}{dx}}{v^2}$

4.10. DERIVATIVE OF $y = x^n$

Let $y = x^n$, where x is any integer.

Therefore, $y + \Delta y = (x + \Delta x)^n$

$$\Rightarrow \Delta y = (x + \Delta x)^n - x^n$$

$$\Rightarrow \frac{\Delta y}{\Delta x} = \frac{(x + \Delta x)^n - x^n}{(x + \Delta x) - x}$$

$$\Rightarrow \lim_{\Delta x \to 0} \frac{\Delta y}{\Delta x} = \lim_{\Delta x \to 0} \left[\frac{(x + \Delta x)^n - x^n}{(x + \Delta x) - x} \right]$$

$$\Rightarrow \frac{dy}{dx} = nx^{n-1} \quad \left[\text{Using } \lim_{x \to a} \frac{x^n - a^n x}{x - a} = na^{n-1} \right]$$

Thus, $\dfrac{d}{dx}(x^n) = nx^{n-1}$

for all $x \in R$ and for all $x \in N$.

4.11. DERIVATIVE OF $y = e^x$

Let $y = e^x$

Therefore, $y + \Delta y = e^{x+\Delta x}$

$$\Rightarrow \Delta y = e^{x+\Delta x} - e^x$$

$$\Rightarrow \frac{\Delta y}{\Delta x} = \frac{e^{x+\Delta x} - e^x}{\Delta x}$$

$$\Rightarrow \lim_{\Delta x \to 0} \frac{\Delta y}{\Delta x} = \lim_{\Delta x \to 0} \frac{e^{x+\Delta x} - e^x}{\Delta x}$$

$$\Rightarrow \frac{dy}{dx} = \lim_{\Delta x \to 0} \frac{e^x (e^{\Delta x} - 1)}{\Delta x}$$

$$= e^x \lim_{\Delta x \to 0} \frac{e^{\Delta x} - 1}{\Delta x}$$

$$= e^x .1 = e^x$$

Thus, $\dfrac{d}{dx}(e^x) = e^x$

4.12. DERIVATIVE OF $y = a^x$

Let $y = a^x$, where $a > 0, a \neq 1$

$$\Rightarrow y + \Delta y = a^{x+\Delta x}$$

$$\Rightarrow \Delta y = a^{x+\Delta x} - a^x$$

$$\Rightarrow \frac{\Delta y}{\Delta x} = \frac{a^{x+\Delta x} - a^x}{\Delta x}$$

$$\Rightarrow \lim_{\Delta x \to 0} \frac{\Delta y}{\Delta x} = \lim_{\Delta x \to 0} \frac{a^{x+\Delta x} - a^x}{\Delta x}$$

$$\Rightarrow \frac{dy}{dx} = \lim_{\Delta x \to 0} \frac{a^x (a^{\Delta x} - 1)}{\Delta x}$$

$$= a^x \lim_{\Delta x \to 0} \frac{a^{\Delta x} - 1}{\Delta x}$$

$$= a^x \log_e a \quad \left[\because \lim_{x \to 0} \frac{a^x - 1}{x} = \log_e a\right]$$

Thus, $\boldsymbol{\frac{d}{dx}(a^x) = a^x \log_e a}$

4.13. DERIVATIVE OF $y = \log_e x$

Let $y = \log_e x$,

$$\Rightarrow y + \Delta y = \log_e (x + \Delta x)$$

$$\Rightarrow \Delta y = \log_e (x + \Delta x) - \log_e x$$

$$\Rightarrow \frac{\Delta y}{\Delta x} = \frac{\log_e \left(\frac{x + \Delta x}{x}\right)}{\Delta x}$$

$$\Rightarrow \lim_{\Delta x \to 0} \frac{\Delta y}{\Delta x} = \lim_{\Delta x \to 0} \frac{\log_e \left(1 + \frac{\Delta x}{x}\right)}{\Delta x}$$

$$\Rightarrow \frac{dy}{dx} = \lim_{\Delta x \to 0} \frac{(1 + \Delta x / x)}{\Delta x / x} \times \frac{1}{x}$$

$$= \frac{1}{x} \quad \left[\because \lim_{x \to 0} \frac{\log_e (1 + x)}{x} = 1\right]$$

Thus, $\frac{d}{dx}(\log_e x) = \frac{1}{x}$

4.14. DERIVATIVE OF TRIGONOMETRIC FUNCTIONS

(i) Derivative of sin*x*

Let $y = \sin x$,

$$\Rightarrow y + \Delta y = \sin(x + \Delta x)$$

$$\Rightarrow \Delta y = \sin(x + \Delta x) - \sin x$$

$$\Rightarrow \frac{\Delta y}{\Delta x} = \frac{\sin(x + \Delta x) - \sin x}{\Delta x}$$

$$\Rightarrow \lim_{\Delta x \to 0} \frac{\Delta y}{\Delta x} = \lim_{\Delta x \to 0} \frac{2\cos\left(\frac{2x + \Delta x}{2}\right)\sin\frac{\Delta x}{2}}{\Delta x}$$

$$\Rightarrow \frac{dy}{dx} = \lim_{\Delta x \to 0} \cos\left(\frac{2x + \Delta x}{2}\right) . \lim_{\Delta x \to 0} \frac{\sin \Delta x / 2}{\Delta x / 2}$$

$$= (\cos x).1 \quad \left[\because \lim_{\Delta x \to 0} \frac{\sin \Delta x / 2}{\Delta x / 2} = 1\right]$$

$$= \cos x$$

Thus, $\frac{d}{dx}(\sin x) = \cos x$

for all $x \in R$

(ii) Derivative of cos*x*

Let $y = \cos x$,

$$\Rightarrow y + \Delta y = \cos(x + \Delta x)$$

$$\Rightarrow \Delta y = \cos(x + \Delta x) - \cos x$$

$$\Rightarrow \frac{\Delta y}{\Delta x} = \frac{-2\sin\left(\frac{2x + \Delta x}{2}\right) . \sin\frac{\Delta x}{2}}{\Delta x}$$

$$\Rightarrow \lim_{\Delta x\to 0}\frac{\Delta y}{\Delta x}=\lim_{\Delta x\to 0}-\sin\left(\frac{2x+\Delta x}{2}\right).\frac{\sin\Delta x/2}{\Delta x/2}$$

$$=\lim_{\Delta x\to 0}-\sin\left(\frac{2x+\Delta x}{2}\right).\lim_{\Delta x\to 0}\frac{\sin\Delta x/2}{\Delta x/2}$$

$$\Rightarrow \frac{dy}{dx}=(-\sin x).1$$

$$=-\sin x$$

Thus, $\dfrac{d}{dx}(\cos x)=-\sin x$

for all $x\in R$

(iii) Derivative of tanx

Let $y=\tan x$,

$$\Rightarrow y+\Delta y=\tan(x+\Delta x)$$

$$\Rightarrow \Delta y=\tan(x+\Delta x)-\tan x$$

$$\Rightarrow \frac{\Delta y}{\Delta x}=\frac{\dfrac{\sin(x+\Delta x)}{\cos(x+\Delta x)}-\dfrac{\sin x}{\cos x}}{\Delta x}$$

$$=\frac{\sin(x+\Delta x)\cos x-\cos(x+\Delta x)\sin x}{\Delta x.\cos(x+\Delta x).\cos x}$$

$$\Rightarrow \lim_{\Delta x\to 0}\frac{\Delta y}{\Delta x}=\lim_{\Delta x\to 0}\frac{\sin(x+\Delta x-x)}{\Delta x.\cos(x+\Delta x).\cos x}$$

$$\Rightarrow \frac{dy}{dx}=\lim_{\Delta x\to 0}\frac{\sin\Delta x}{\Delta x}.\lim_{\Delta x\to 0}\frac{1}{\cos(x+\Delta x).\cos x}$$

$$=\frac{1}{\cos x.\cos x}$$

$$=\frac{1}{\cos^2 x}$$

$$=\sec^2 x$$

Thus, $\frac{d}{dx}(\tan x) = \sec^2 x$

for all $x \in R \sim \left\{(2n+1)\frac{\pi}{2}; n \in I\right\}$

(iv) Derivative of cotx

Let $y = \cot x$,

$$\therefore y + \Delta y = \cot(x + \Delta x)$$

$$\Rightarrow \Delta y = \cot(x + \Delta x) - \cot x$$

$$\Rightarrow \frac{\Delta y}{\Delta x} = \frac{1}{\Delta x}\left[\frac{\cos(x+\Delta x)}{\sin(x+\Delta x)} - \frac{\cos x}{\sin x}\right]$$

$$\Rightarrow \frac{\Delta y}{\Delta x} = \frac{1}{\Delta x}\left[\frac{\sin x.\cos(x+\Delta x) - \cos x.\sin(x+\Delta x)}{\sin(x+\Delta x).\sin x}\right]$$

$$\Rightarrow \frac{\Delta y}{\Delta x} = \frac{\sin x(x - x - \Delta x)}{\Delta x.\sin(x+\Delta x).\sin x}$$

$$\Rightarrow \lim_{\Delta x \to 0} \frac{\Delta y}{\Delta x} = \lim_{\Delta x \to 0} \frac{\sin(-\Delta x)}{\Delta x}. \lim_{\Delta x \to 0} \frac{1}{\sin(x+\Delta x).\sin x}$$

$$= \frac{-1}{\sin x.\sin x} = -\cos ec^2 x$$

Thus, $\frac{d}{dx}(\cot x) = -\mathrm{cosec}^2 x$

for all $x \in R \sim \{n\pi; n \in I\}$

(v) Derivative of secx

Let $y = \sec x$,

$$\therefore y + \Delta y = \sec(x + \Delta x)$$

$$\Rightarrow \Delta y = \sec(x + \Delta x) - \sec x$$

$$\Rightarrow \frac{\Delta y}{\Delta x} = \frac{1}{\Delta x}\left[\frac{1}{\cos(x+\Delta x)} - \frac{1}{\cos x}\right]$$

$$\Rightarrow \frac{\Delta y}{\Delta x} = \frac{1}{\Delta x}\left[\frac{\cos x - \cos(x+\Delta x)}{\cos(x+\Delta x).\cos x}\right]$$

$$\Rightarrow \lim_{\Delta x \to 0} \frac{\Delta y}{\Delta x} = \lim_{\Delta x \to 0} \frac{2\sin\left(\frac{2x+\Delta x}{2}\right)\sin\frac{\Delta x}{2}}{\Delta x.\cos(x+\Delta x).\cos x}$$

$$\Rightarrow \frac{dy}{dx} = \lim_{\Delta x \to 0} \frac{\sin\left(\frac{2x+\Delta x}{2}\right)}{\cos(x+\Delta x).\cos x}.\lim_{\Delta x \to 0}\frac{\sin \Delta x/2}{\Delta x/2}$$

$$= \frac{\sin x}{\cos x} \times \frac{1}{\cos x} \times 1$$

$$= \sec x.\tan x$$

Thus, $\frac{d}{dx}(\sec x) = \sec x.\tan x$

for all $x \in R \sim \left\{(2n+1)\frac{\pi}{2}; n \in I\right\}$

(vi) Derivative of cosecx

Let $y = \operatorname{cosec} x$,

$\therefore y + \Delta y = \operatorname{cosec}(x+\Delta x)$

$$\Rightarrow \Delta y = \operatorname{cosec}(x+\Delta x) - \operatorname{cosec} x$$

$$\Rightarrow \frac{\Delta y}{\Delta x} = \frac{1}{\Delta x}\left[\frac{1}{\sin(x+\Delta x)} - \frac{1}{\sin x}\right]$$

$$\Rightarrow \frac{\Delta y}{\Delta x} = \frac{1}{\Delta x}\left[\frac{\sin x - \sin(x+\Delta x)}{\sin(x+\Delta x).\sin x}\right]$$

$$\Rightarrow \lim_{\Delta x \to 0} \frac{\Delta y}{\Delta x} = \lim_{\Delta x \to 0} \frac{-2\cos\left(x+\frac{\Delta x}{2}\right).\sin\frac{\Delta x}{2}}{\Delta x.\sin(x+\Delta x).\sin x}$$

$$\Rightarrow \frac{dy}{dx} = \lim_{\Delta x \to 0} \frac{-\cos\left(x + \frac{\Delta x}{2}\right)}{\sin(x+\Delta x).\sin x} . \lim_{\Delta x \to 0} \frac{\sin \Delta x / 2}{\Delta x / 2}$$

$$= \frac{-\cos x}{\sin x . \sin x}.1$$

Thus, $\frac{d}{dx}(\text{cosec}x) = -\text{cosec}x.\cot x$

4.15. DERIVATIVES OF INVERSE TRIGONOMETRIC FUNCTIONS

(1) Derivative of $\sin^{-1}x$

Let $y = \sin^{-1} x$,

Then $-1 < x < 1$

$\Rightarrow$ the principal value of $y = \sin^{-1} x$ lies between $\frac{-\pi}{2}$ and $\frac{\pi}{2}$.

$$\Rightarrow y \in \left(-\frac{\pi}{2}, \frac{\pi}{2}\right)$$

Now, $y = \sin^{-1} x$

$\Rightarrow x = \sin y$

Let Δx be a small change in x and let Δy be the corresponding change in y. Then,

$$x + \Delta x = \sin(y + \Delta y)$$

$$\Rightarrow \Delta x = \sin(y+\Delta y) - x$$

$$\Rightarrow \Delta x = \sin(y+\Delta y) - \sin y \qquad [\because x = \sin y]$$

$$\Rightarrow \frac{\Delta x}{\Delta y} = \frac{\sin(y+\Delta y) - \sin y}{\Delta y}$$

$$\Rightarrow \frac{\Delta y}{\Delta x} = \frac{\Delta y}{\sin(y+\Delta y) - \sin y}$$

$$\Rightarrow \lim_{\Delta x \to 0} \frac{\Delta y}{\Delta x} = \lim_{\Delta y \to 0} \frac{\Delta y}{\sin(y + \Delta y) - \sin y} \quad [\because \Delta x \to 0 \Rightarrow \Delta y \to 0]$$

$$\Rightarrow \frac{dy}{dx} = \lim_{\Delta y \to 0} \frac{\Delta y}{2\sin\left(\frac{\Delta y}{2}\right)\cos\left(y + \frac{\Delta y}{2}\right)}$$

$$\Rightarrow \frac{dy}{dx} = \lim_{\Delta y \to 0} \frac{\Delta y / 2}{\sin(\Delta y / 2)} \cdot \lim_{\Delta y \to 0} \frac{1}{\cos\left(y + \frac{\Delta y}{2}\right)}$$

$$\Rightarrow \frac{dy}{dx} = \frac{1}{\cos y}$$

Now, $y \in \left(-\frac{\pi}{2}, \frac{\pi}{2}\right)$

$$\Rightarrow \cos y > 0$$

$$\Rightarrow \cos y = +\sqrt{1 - \sin^2 y}$$

$$\Rightarrow \frac{dy}{dx} = \frac{1}{\cos y} = \frac{1}{\sqrt{1 - \sin^2 y}} = \frac{1}{\sqrt{1 - x^2}}$$

Thus, $\frac{d}{dx}(\sin^{-1} x) = \frac{1}{\sqrt{1 - x^2}}$, where $-1 < x < 1$

(2) Derivative of $\cos^{-1} x$

Let $y = \cos^{-1} x$,

Then $-1 < x < 1$

The principal value of $y = \cos^{-1} x$ lies between 0 and π.

$$\Rightarrow y \in (0, \pi)$$

Now, $y = \cos^{-1} x$,

$$\Rightarrow x = \cos y$$

Let Δx be a small change in x and let Δy be the corresponding change in y. Then,

$$x + \Delta x = \cos(y + \Delta y)$$

$$\Rightarrow \Delta x = \cos(y + \Delta y) - \cos y \qquad [\because x = \cos y]$$

$$\Rightarrow \frac{\Delta x}{\Delta y} = \frac{\cos(y + \Delta y) - \cos y}{\Delta y}$$

$$\Rightarrow \frac{\Delta y}{\Delta x} = \frac{\Delta y}{\cos(y + \Delta y) - \cos y}$$

$$\Rightarrow \lim_{\Delta x \to 0} \frac{\Delta y}{\Delta x} = \lim_{\Delta y \to 0} \frac{\Delta y}{\cos(y + \Delta y) - \cos y}$$

$$\Rightarrow \frac{dy}{dx} = \lim_{\Delta y \to 0} \frac{\Delta y}{-2\sin\left(y + \frac{\Delta y}{2}\right)\sin\frac{\Delta y}{2}}$$

$$\Rightarrow \frac{dy}{dx} = -\lim_{\Delta y \to 0} \frac{\Delta y / 2}{\sin\left(\frac{\Delta y}{2}\right)} . \lim_{\Delta y \to 0} \frac{1}{\sin(y + \Delta y / 2)}$$

$$\Rightarrow \frac{dy}{dx} = -\frac{1}{\sin y}$$

Now, $y \in (0, \pi)$

$$\Rightarrow \sin y > 0$$

$$\Rightarrow \sin y = +\sqrt{1 - \cos^2 y}$$

$$\therefore \frac{dy}{dx} = -\frac{1}{\sin y} = -\frac{1}{\sqrt{1 - \cos^2 y}} = -\frac{1}{\sqrt{1 - x^2}}$$

Thus, $\boldsymbol{\frac{d}{dx}(\cos^{-1} x) = -\frac{1}{\sqrt{1 - x^2}}}$

(3) Derivative of $\tan^{-1} x$

Let $y = \tan^{-1} x$,

Then $x = \tan y$

Let Δx be a small change in x and let Δy be the corresponding change in y. Then,

$$x + \Delta x = \tan(y + \Delta y)$$

$$\Rightarrow \Delta x = \tan(y + \Delta y) - \tan y \qquad [\because x = \tan y]$$

$$\Rightarrow \frac{\Delta x}{\Delta y} = \frac{\tan(y + \Delta y) - \tan y}{\Delta y}$$

$$\Rightarrow \frac{\Delta y}{\Delta x} = \frac{\Delta y}{\tan(y + \Delta y) - \tan y}$$

$$\Rightarrow \lim_{\Delta x \to 0} \frac{\Delta y}{\Delta x} = \lim_{\Delta y \to 0} \frac{\Delta y}{\tan(y + \Delta y) - \tan y}$$

$$\Rightarrow \frac{dy}{dx} = \lim_{\Delta y \to 0} \frac{\Delta y}{\frac{\sin(y + \Delta y)}{\cos(y + \Delta y)} - \frac{\sin y}{\cos y}}$$

$$\Rightarrow \frac{dy}{dx} = \lim_{\Delta y \to 0} \frac{\Delta y . \cos y . \cos(y + \Delta y)}{\sin(y + \Delta y)\cos y - \cos(y + \Delta y)\sin y}$$

$$= \lim_{\Delta y \to 0} \frac{\Delta y}{\sin(y + \Delta y - y)} . \cos y . \cos(y + \Delta y)$$

$$= \lim_{\Delta y \to 0} \frac{\Delta y}{\sin \Delta y} . \lim_{\Delta y \to 0} [\cos y . \cos(y + \Delta y)]$$

$$= \cos^2 y$$

$$= \frac{1}{\sec^2 y} = \frac{1}{1 + \tan^2 y} = \frac{1}{1 + x^2}$$

Thus, $\frac{d}{dx}(\tan^{-1} x) = -\frac{1}{1 + x^2}$

(4) Derivative of $\cot^{-1} x$

Let $y = \cot^{-1} x$,

Then $x = \cot y$

Let Δx be a small change in x and let Δy be the corresponding change in y. Then,

$$x + \Delta x = \cot(y + \Delta y)$$

$$\Rightarrow \Delta x = \cot(y + \Delta y) - \cot y$$

$$\Rightarrow \frac{\Delta x}{\Delta y} = \frac{\cot(y + \Delta y) - \cot y}{\Delta y}$$

$$\Rightarrow \frac{\Delta y}{\Delta x} = \frac{\Delta y}{\cot(y + \Delta y) - \cot y}$$

$$\Rightarrow \lim_{\Delta x \to 0} \frac{\Delta y}{\Delta x} = \lim_{\Delta y \to 0} \frac{\Delta y}{\cot(y + \Delta y) - \cot y} \quad [\because \Delta x \to 0 \Rightarrow \Delta y \to 0]$$

$$\Rightarrow \lim_{\Delta x \to 0} \frac{\Delta y}{\Delta x} = \lim_{\Delta y \to 0} \frac{\Delta y}{\dfrac{\cos(y + \Delta y)}{\sin(y + \Delta y)} - \dfrac{\cos y}{\sin y}}$$

$$\Rightarrow \frac{dy}{dx} = \lim_{\Delta y \to 0} \frac{\Delta y \sin y \sin(y + \Delta y)}{\sin y \cos(y + \Delta y) - \cos y \sin(y + \Delta y)}$$

$$\Rightarrow \frac{dy}{dx} = \lim_{\Delta y \to 0} \frac{\Delta y \sin y \sin(y + \Delta y)}{-\sin(y + \Delta y - y)}$$

$$\Rightarrow \frac{dy}{dx} = -\lim_{\Delta y \to 0} \frac{\Delta y}{\sin \Delta y} . \lim_{\Delta y \to 0} [\sin y \sin(y + \Delta y)]$$

$$\Rightarrow \frac{dy}{dx} = -\sin^2 y = \frac{-1}{\operatorname{cosec}^2 y} = \frac{-1}{1 + \cot^2 y} = \frac{-1}{1 + x^2}$$

Thus, $\boldsymbol{\dfrac{d}{dx}(\cot^{-1} x) = \dfrac{-1}{1 + x^2}}$

(5) Derivative of $\sec^{-1}x$

Let $y = \sec^{-1} x$,

Then $x \le -1$ or $x > 1$

$\Rightarrow$ the principal value of $y = \sec^{-1} x$ lies between 0 and π but $y \neq \pi/2$.

$$\Rightarrow y \in \left(0, \frac{\pi}{2}\right) \cup \left(\frac{\pi}{2}, \pi\right)$$

Now, $y = \sec^{-1} x$

$\Rightarrow x = \sec y$

Let Δx be a small change in x and let Δy be the corresponding change in y. Then,

$$x + \Delta x = \sec(y + \Delta y)$$

$$\Rightarrow \Delta x = \sec(y + \Delta y) - \sec y$$

$$\Rightarrow \frac{\Delta y}{\Delta x} = \frac{\Delta y}{\sec(y + \Delta y) - \sec y}$$

$$\Rightarrow \lim_{\Delta x \to 0} \frac{\Delta y}{\Delta x} = \lim_{\Delta y \to 0} \frac{\Delta y}{\sec(y + \Delta y) - \sec y} \quad [\because \Delta x \to 0 \Rightarrow \Delta y \to 0]$$

$$\Rightarrow \frac{dy}{dx} = \lim_{\Delta y \to 0} \frac{\Delta y}{\dfrac{1}{\cos(y + \Delta y)} - \dfrac{1}{\cos y}}$$

$$\Rightarrow \frac{dy}{dx} = \lim_{\Delta y \to 0} \frac{\cos y \cos(y + \Delta y).\Delta y}{\cos y - \cos(y + \Delta y)}$$

$$\Rightarrow \frac{dy}{dx} = \lim_{\Delta y \to 0} \frac{\Delta y}{2 \sin\left(y + \dfrac{\Delta y}{2}\right) \sin \dfrac{\Delta y}{2}} . \cos y \cos(y + \Delta y)$$

$$\Rightarrow \frac{dy}{dx} = \lim_{\Delta y \to 0} \frac{(\Delta y / 2)}{\sin\left(\dfrac{\Delta y}{2}\right) \sin\left(y + \dfrac{\Delta y}{2}\right)} . \cos y \cos(y + \Delta y)$$

$$= \lim_{\Delta y \to 0} \frac{(\Delta y/2)}{\sin\left(\frac{\Delta y}{2}\right)} \cdot \lim_{\Delta y \to 0} \frac{\cos y \cos(y+\Delta y)}{\sin\left(y+\frac{\Delta y}{2}\right)}$$

$$= \frac{\cos y \cos y}{\sin y} = \frac{1}{\tan y \sec y}$$

$$= \frac{1}{\sec y\sqrt{\sec^2 y - 1}}$$

Now, $y \in \left(0, \frac{\pi}{2}\right) \cup \left(\frac{\pi}{2}, \pi\right)$

$$\Rightarrow \sec y \tan y > 0$$

$$\Rightarrow \frac{1}{\sec y \tan y} > 0$$

$$\therefore \frac{dy}{dx} = \frac{1}{\sec y \tan y} > 0$$

$$\Rightarrow \frac{dy}{dx} = \frac{1}{\sec y\sqrt{\sec^2 y - 1}} > 0$$

$$\Rightarrow \frac{dy}{dx} = \frac{1}{|x|\sqrt{x^2 - 1}}$$

Thus, $\frac{d}{dx}(\sec^{-1} x) = \frac{1}{|x|\sqrt{x^2 - 1}}$

(6) Derivative of cosec^{-1}x

Let $y = \operatorname{cosec}^{-1} x$,

Then $x \le -1$ or $x \ge 1$

$\Rightarrow$ the principal value of $y = \operatorname{cosec}^{-1} x$ lies between $-\frac{\pi}{2}$ and $\frac{\pi}{2}$ but $y \neq 0$.

Now, $y = \operatorname{cosec}^{-1} x$

$$\Rightarrow x = \operatorname{cosec} y$$

Let Δx be a small change in x and let Δy be the corresponding change in y. Then,

$$x + \Delta x = \operatorname{cosec}(y + \Delta y)$$

$$\Rightarrow \Delta x = \operatorname{cosec}(y + \Delta y) - \operatorname{cosec} y$$

$$\Rightarrow \frac{\Delta x}{\Delta y} = \frac{\operatorname{cosec}(y + \Delta y) - \operatorname{cosec} y}{\Delta y}$$

$$\Rightarrow \frac{\Delta y}{\Delta x} = \frac{\Delta y}{\operatorname{cosec}(y + \Delta y) - \operatorname{cosec} y}$$

$$\Rightarrow \lim_{\Delta x \to 0} \frac{\Delta y}{\Delta x} = \lim_{\Delta y \to 0} \frac{\Delta y}{\frac{1}{\sin(y + \Delta y)} - \frac{1}{\sin y}}$$

$$\Rightarrow \frac{dy}{dx} = \lim_{\Delta y \to 0} \frac{\Delta y \sin y \sin(y + \Delta y)}{\sin y - \sin(y + \Delta y)}$$

$$\Rightarrow \frac{dy}{dx} = \lim_{\Delta y \to 0} \frac{\Delta y}{2\cos\left(y + \frac{\Delta y}{2}\right)\sin\left(\frac{-\Delta y}{2}\right)} . \sin y \sin(y + \Delta y)$$

$$\Rightarrow \frac{dy}{dx} = -\lim_{\Delta y \to 0} \frac{(\Delta y / 2)}{\sin\left(\frac{\Delta y}{2}\right)} . \lim_{\Delta y \to 0} \frac{\sin y \sin(y + \Delta y)}{\cos\left(y + \frac{\Delta y}{2}\right)}$$

$$\Rightarrow \frac{dy}{dx} = -\frac{\sin y \sin y}{\cos y} = -\frac{1}{\cot y \operatorname{cosec} y}$$

Now, $y \in \left(-\frac{\pi}{2}, 0\right) \cup \left(0, \frac{\pi}{2}\right)$

$$\Rightarrow \frac{\sin^2 y}{\cos y} = \frac{1}{\cot y \operatorname{cosec} y} > 0$$

$$\Rightarrow -\frac{\sin^2 y}{\cos y} = -\frac{1}{\cot y \cos ecy} < 0$$

$$\Rightarrow \frac{dy}{dx} = -\frac{1}{\cos ecy \cot y} < 0$$

$$\Rightarrow \frac{dy}{dx} = -\frac{1}{\cos ecy\sqrt{\cos ec^2 y - 1}}$$

$$\Rightarrow \frac{dy}{dx} = -\frac{1}{|x|\sqrt{x^2 - 1}}$$

Thus, $\boldsymbol{\frac{d}{dx}(\operatorname{cosec}^{-1} x) = -\frac{1}{|x|\sqrt{x^2 - 1}}}$

Examples

Example 1: Find the derivative of the functions defined by $\sqrt{(1+x)/(1-x)}$.

Solution. We have,

$$y = \sqrt{(1+x)/(1-x)}$$

Let $u = \frac{1+x}{1-x}$, $y = u^{1/2}$

so that $y = \sqrt{\left(\frac{1+x}{1-x}\right)}$

We have $\frac{du}{dx} = \frac{(1-x)\frac{d}{dx}(1+x) - (1+x)\frac{d}{dx}(1-x)}{(1-x)^2}$

$$= \frac{(1-x)1 - (1+x)(-1)}{(1-x)^2}$$

$$= \frac{2}{(1-x)^2}$$

$$\frac{dy}{du} = \frac{1}{2}u^{-1/2}$$

$$= \frac{1}{2}\left(\frac{1+x}{1-x}\right)^{-1/2}$$

Hence, $\frac{dy}{dx} = \frac{dy}{du}.\frac{du}{dx}$

$$= \frac{1}{2}\left(\frac{1+x}{1-x}\right)^{-1/2} \times \frac{2}{(1-x)^2}$$

$$= \frac{1}{(1+x)^{1/2}(1-x)^{3/2}}$$

Example 2: If $y = 1 + \frac{x}{1!} + \frac{x^2}{2!} + \frac{x^3}{3!} + ... + \frac{x^n}{n!}$, show that $\frac{dy}{dx} - y + \frac{x^n}{n!} = 0$.

Solution. We have,

$$y = 1 + \frac{x}{1!} + \frac{x^2}{2!} + \frac{x^3}{3!} + \frac{x^4}{4!} .... + \frac{x^{n-1}}{(n-1)!} \frac{x^n}{n!}$$

$$\therefore \frac{dy}{dx} = \frac{d}{dx}(1) + \frac{d}{dx}\left(\frac{x}{1!}\right) + \frac{d}{dx}\left(\frac{x^2}{2!}\right) + \frac{d}{dx}\left(\frac{x^3}{3!}\right) + ... + \frac{d}{dx}\left(\frac{x^n}{n!}\right)$$

$$\Rightarrow \frac{dy}{dx} = 0 + \frac{1}{1!} + \frac{1}{2!}(2x) + \frac{1}{3!}(3x^2) + ... + \frac{1}{n!}(nx^{n-1})$$

$$\Rightarrow \frac{dy}{dx} = 1 + \frac{x}{1!} + \frac{x^2}{2!} + ... + \frac{x^{n-1}}{(n-1)!}$$

$$\rightarrow \frac{dy}{dx} = \left(1 + \frac{x}{1!} + \frac{x^2}{2!} + ... + \frac{x^{n-1}}{(n-1)!} + \frac{x^n}{n!}\right) - \frac{x^n}{n!}$$

$$\Rightarrow \frac{dy}{dx} = y - \frac{x^n}{n!}$$

$$\Rightarrow \frac{dy}{dx} - y + \frac{x^n}{n!} = 0$$

Example 3: Differentiate w.r.t. x.

$$\sqrt{\log\left\{\sin\left(\frac{x^2}{3}-1\right)\right\}}$$

Solution. Let, $y=\sqrt{\log\left\{\sin\left(\frac{x^2}{3}-1\right)\right\}}$

Putting $\frac{x^2}{3}-1=v$

$$\sin\left(\frac{x^2}{3}-1\right)=\sin v=u$$

and $\log\left\{\sin\left(\frac{x^2}{3}-1\right)\right\}=\log u=z$, we get

$$y=\sqrt{z},\quad z=\log u,\quad u=\sin v \text{ and } v=\frac{x^2}{3}-1$$

$$\therefore \frac{dy}{dz}=\frac{1}{2\sqrt{z}}$$

$$\frac{dz}{du}=\frac{1}{u}$$

$$\frac{du}{dv}=\cos v$$

and $\frac{dv}{dx}=\frac{2x}{3}$

Now, $\frac{dy}{dx}=\frac{dy}{dz}.\frac{dz}{du}.\frac{du}{dv}.\frac{dv}{dx}$

$$\Rightarrow \frac{dy}{dx}=\frac{1}{2\sqrt{z}}.\frac{1}{u}.\cos v.\frac{2x}{3}$$

$$= \frac{x}{3} \cdot \frac{\cos v}{u\sqrt{\log u}}$$

$$= \frac{x}{3} \cdot \frac{\cos\left(\frac{x^2}{3} - 1\right)}{\sin\left(\frac{x^2}{3} - 1\right)\sqrt{\log\left\{\sin\left(\frac{x^2}{3} - 1\right)\right\}}}$$

$$= \frac{x \cot\left(\frac{x^2}{3} - 1\right)}{3\sqrt{\log\left\{\sin\left(\frac{x^2}{3} - 1\right)\right\}}}$$

Example 4: Differentiate $\tan^{-1}\left(\frac{\sqrt{1+\sin x} + \sqrt{1-\sin x}}{\sqrt{1+\sin x} - \sqrt{1-\sin x}}\right)$

Solution. Let $y = \tan^{-1}\left(\frac{\sqrt{1+\sin x} + \sqrt{1-\sin x}}{\sqrt{1+\sin x} - \sqrt{1-\sin x}}\right)$

Since, $\sqrt{1 \pm \sin x} = \sqrt{\sin^2 \frac{x}{2} + \cos^2 \frac{x}{2} \pm 2\sin\frac{x}{2}\cos\frac{x}{2}}$

$$= \sqrt{\left(\cos\frac{x}{2} \pm \sin\frac{x}{2}\right)^2}$$

$$= \cos\frac{x}{2} \pm \sin\frac{x}{2}$$

$$\therefore y = \tan^{-1}\left\{\frac{\left(\cos\frac{x}{2} + \sin\frac{x}{2}\right) + \left(\cos\frac{x}{2} - \sin\frac{x}{2}\right)}{\left(\cos\frac{x}{2} + \sin\frac{x}{2}\right) - \left(\cos\frac{x}{2} - \sin\frac{x}{2}\right)}\right\}$$

$$= \tan^{-1}\left(\frac{2\cos\frac{x}{2}}{2\sin\frac{x}{2}}\right)$$

$$= \tan^{-1}\left(\cot\frac{x}{2}\right)$$

$$= \tan^{-1}\left\{\tan\left(\frac{\pi}{2}-\frac{x}{2}\right)\right\}$$

$$= \frac{\pi}{2}-\frac{x}{2}$$

$$\Rightarrow \frac{dy}{dx} = 0-\frac{1}{2} = -\frac{1}{2}$$

Example 5: Find $\frac{dy}{dx}$ if $y = \log|x|$

Solution: The function $y = \log|x|$ is defined for all real x except x = 0 and

$$\log|x| = \log x \quad \text{if } x > 0$$
$$\log(-x) \text{ if } x < 0$$

Hence $\frac{d}{dx}\log|x|$

$$= \left\{\frac{1}{x}, x > 0; = \frac{1}{-x}(-1) = \frac{1}{x} \quad \text{if } x < 0\right\}$$

Thus, $\frac{d}{dx}\log|x| = \frac{1}{x}, x \neq 0$

Example 6: If $\sqrt{1-x^2}+\sqrt{1-y^2} = a(x-y)$, show that

$$\frac{dy}{dx} = \sqrt{\left(\frac{1-y^2}{1-x^2}\right)}$$

Solution: Put $x = \sin\theta$ and $y = \sin\phi$ in the given equation.

$$\therefore \quad \cos\theta + \cos\phi = a(\sin\theta - \sin\phi)$$

$$\Rightarrow 2\cos\frac{\theta+\phi}{2}.\cos\frac{\theta-\phi}{2} = 2a\sin\frac{\theta-\phi}{2}.\cos\frac{\theta+\phi}{2}$$

$$\Rightarrow \cos\frac{\theta+\phi}{2}\left[\cos\frac{\theta-\phi}{2} - a\sin\frac{\theta-\phi}{2}\right] = 0$$

If $\quad \cos\frac{\theta+\phi}{2} = 0$, then $\frac{\theta+\phi}{2} = \frac{\pi}{2}$

$$\therefore \quad \theta = \pi - \phi \text{ or } \sin\theta = \sin\phi$$

or $\quad x = y$

But if we put $x = y$ in the given equation, it is not satisfied and hence we must have

$$\cos\frac{\theta-\phi}{2} - a\sin\frac{\theta-\phi}{2} = 0$$

$$\Rightarrow \cot\frac{\theta-\phi}{2} = a$$

$$\therefore \quad \frac{\theta-\phi}{2} = \cot^{-1} a$$

or, $\sin^{-1} x - \sin^{-1} y = 2\cot^{-1} a$

Differentiating w.r.t. x

$$\frac{1}{\sqrt{(1-x^2)}} - \frac{1}{\sqrt{(1-y^2)}}\frac{dy}{dx} = 0$$

$$\therefore \frac{dy}{dx} = \sqrt{\left(\frac{1-y^2}{1-x^2}\right)}$$

Exercises – 4.3

1. *Find the derivatives of the following functions.*

(i) $\sqrt{ax^2+2bx+c}$ (ii) $\dfrac{1-a\sqrt[3]{x}}{1+a\sqrt[3]{x}}$

(iii) $(3x-4)(x^2-6x+7)$ (iv) $\dfrac{x^3+1}{(x^2-1)(x^3-1)}$

(v) $\dfrac{x-\cos x}{x+\cos x}$ (vi) $\sqrt{(1+\sin^2 x)\sqrt{x}}$

(vii) $\sin^m x-\cos^n x$ (viii) $\sqrt[3]{\sin^2 x}$

(ix) $\sqrt{\dfrac{1-\tan x}{1+\tan x}}$ (x) $\sqrt{\dfrac{1-\cos x}{1+\cos x}}$

(xi) $\cos^{-1}\left(\dfrac{a+b\cos x}{b+a\cos x}\right)$ (xii) $\cos^{-1}\left(\dfrac{x-x^{-1}}{x+x^{-1}}\right)$

(xiii) $\log\left[x+\sqrt{x^2+a^2}\right]$ (xiv) $\log[\sin(\log x)]$

(xv) $\tan^{-1}\dfrac{\sin x}{1+\cos x}$ (xvi) $\tan^{-1}\left(\dfrac{a\cos x-b\sin x}{b\cos x+a\sin x}\right)$

(xvii) $\sin^{-1}\left[x\sqrt{1-x}+\sqrt{x}\sqrt{1-x^2}\right]$

(xviii) $\tan^{-1}\dfrac{3a^2x-x^3}{a(a^2-3x^2)}$

(xix) $\cos^{-1}\dfrac{x^{2n}-1}{x^{2n}+1}$

(xx) $\cot^{-1}\dfrac{\sqrt{1+\sin x}+\sqrt{1-\sin x}}{\sqrt{1+\sin x}-\sqrt{1-\sin x}}$

2. If $\sqrt{(1-x^6)}+\sqrt{(1-y^6)}=a^3(x^3-y^3)$, then prove that

$$\frac{dy}{dx}=\frac{x^2}{y^2}\sqrt{\frac{(1-y^6)}{(1-x^6)}}$$

3. If $y=\tan^{-1}\frac{4x}{1+5x^2}+\tan^{-1}\frac{2+3x}{3-2x}$, then prove that

$$\frac{dy}{dx}=\frac{5}{1+25x^2}$$

4. If $y=\sin^{-1}\frac{2x}{1+x^2}+\sec^{-1}\frac{1+x^2}{1-x^2}$, then prove that

$$\frac{dy}{dx}=\frac{4}{1+x^2}$$

5. *If* $y=\sin^{-1}2x\sqrt{1-x^2}+\sec^{-1}\frac{1}{\sqrt{(1-x^2)}}$, then prove that

$$\frac{dy}{dx}=\frac{3}{\sqrt{(1-x^2)}}$$

6. If $y=\sec^{-1}\left(\frac{\sqrt{x}+1}{\sqrt{x}-1}\right)+\sin^{-1}\left(\frac{\sqrt{x}-1}{\sqrt{x}+1}\right)$, then prove that

$$\frac{dy}{dx}=0$$

7. *If* $y=\sin^{-1}\left(\frac{5x+12\sqrt{1-x^2}}{13}\right)$, then prove that

$$\frac{dy}{dx}=\frac{1}{\sqrt{1-x^2}}$$

8. If $y = \log \tan \frac{x}{2} + \sin^{-1}(\cos x)$, then prove that

$$\frac{dy}{dx} = \operatorname{cosec} x - 1$$

9. If $y = \log x^x$, prove that $\frac{dy}{dx} = \log(ex)$.

10. If $y = \log_7(\log_7 x)$, prove that $\frac{dy}{dx} = \frac{\log_7 e}{x \log_e x}$

11. If $y = \frac{5x}{(1-x)^{2/3}} + \cos^2(2x+1)$, find $\frac{dy}{dx}$.

12. If $y = \frac{x}{2}\sqrt{a^2 + x^2} + \frac{a^2}{2}\log\left[x + \sqrt{x^2 + a^2}\right]$, then prove that

$$\frac{dy}{dx} = \sqrt{x^2 + a^2}$$

13. If $y = x \log\left(\frac{x}{a+bx}\right)$ show that

$$x^3 \frac{d^2 y}{dx^2} = \left(x \frac{dy}{dx} - y\right)^2.$$

14. If $y\sqrt{x^2+1} = \log\left\{\sqrt{x^2+1} - x\right\}$, show that

$$\left(x^2+1\right)\frac{dy}{dx} + xy + 1 = 0.$$

15. *If* $y = \log\left(\frac{\sqrt{(x+1)} - 1}{\sqrt{(x+1)} + 1}\right) + \frac{\sqrt{x}}{\sqrt{(x+1)}}$, find $\frac{dy}{dx}$.

16. If $\sin y = x \sin(a+y)$ then prove that $\frac{dy}{dx} = \frac{\sin^2(a+y)}{\sin a}$

17. If $\log(x^2 + y^2) = 2\tan^{-1}\left(\frac{y}{x}\right)$, show that $\frac{dy}{dx} = \frac{x+y}{x-y}$

18. If $\sqrt{y+x} + \sqrt{y-x} = c$ show that

$$\frac{dy}{dx} = \frac{y}{x} - \sqrt{\frac{y^2}{x^2} - 1}$$

4.16. DERIVATIVE OF HYPERBOLIC FUNCTIONS

(1) Derivative of sin*hx*

Let $y = \sin hx; x \in R$

$$= \frac{e^x - e^{-x}}{2}$$

$$\Rightarrow \frac{dy}{dx} = \frac{d}{dx}\left[\frac{e^x - e^{-x}}{2}\right]$$

$$= \frac{1}{2}\left(e^x + e^{-x}\right)$$

$$= \cosh x$$

Thus, $\boldsymbol{\frac{d}{dx}(\sin hx) = \cos hx}$ for all $x \in R$

(2) Derivative of cos*hx*

Let $y = \cos hx$

$$= \frac{e^x + e^{-x}}{2}$$

$$\Rightarrow \frac{dy}{dx} = \frac{e^x - e^{-x}}{2}$$

$$= \sin hx$$

Thus, $\boldsymbol{\frac{d}{dx}(\cos hx) = \sin hx}$ for all $x \in R$

(3) Derivative of tan*hx*

Let $y = \tan hx$

$$= \frac{\sin hx}{\cos hx}; \cos hx \neq 0$$

$$\Rightarrow \frac{dy}{dx} = \frac{\cos hx.\frac{d(\sin hx)}{dx} - \sin hx.\frac{d(\cos hx)}{dx}}{\cos h^2 x}$$

$$= \frac{\cos hx.\cos hx - \sin hx.\sin hx}{\cos h^2 x}$$

$$= \frac{\cos h^2 x - \sin h^2 x}{\cos h^2 x}$$

$$= \frac{1}{\cos h^2 x} = \sec h^2 x$$

Thus, $\frac{d}{dx}(\tan hx) = \text{sech}^2 x$

(4) Derivative of coth*x*

Let $y = \cot hx$

$$= \frac{\cos hx}{\sin hx}; \sin hx \neq 0$$

$$\Rightarrow \frac{dy}{dx} = \frac{\sin hx.\frac{d(\cos hx)}{dx} - \cos hx.\frac{d(\sin hx)}{dx}}{\sin h^2 x}$$

$$= \frac{\sin hx.\sin hx - \cos hx.\cos hx}{\sin h^2 x}$$

$$= \frac{\sin h^2 x - \cos h^2 x}{\sin h^2 x}$$

$$= \frac{-1}{\sin h^2 x}$$

$$= -\text{cosech}^2 x$$

Thus, $\frac{d}{dx}(\text{coth}x) = -\text{cosech}^2 x$

(5) Derivative of sech*x*

Let $y = \sec hx$

$$= \frac{1}{\cos hx}$$

$$\Rightarrow \frac{dy}{dx} = \frac{\cos hx.0 - 1.\sin hx}{\cos h^2 x}$$

$$= \frac{-\sin hx}{\cos h^2 x}$$

$$= -\tan hx.\sec hx \text{ for all } x \in R$$

Thus, $\frac{d}{dx}(\text{sech}x) = -\tanh x.\text{sech}x$

(6) Derivative of cosec*hx*

Let $y = \cos echx$

$$= \frac{1}{\sinh x}$$

$$\Rightarrow \frac{dy}{dx} = \frac{\sinh x.0 - 1.\cosh x}{\sinh^2 x}$$

$$= \frac{-\cosh x}{\sinh^2 x}$$

$$= -\coth x.\cos echx$$

Thus, $\frac{d}{dx}(\text{cosech}x) = -\coth x.\text{cosech}x$

4.17. DERIVATIVE OF INVERSE HYPERBOLIC FUNCTIONS

(1) Derivative of $\sin h^{-1} x$

Let $y = \sin h^{-1} x$

$x = \sin hy$

$$\therefore \quad \frac{dx}{dy} = \cos hy$$

$$\Rightarrow \frac{dy}{dx} = \frac{1}{\cos hy}$$

$$= \pm \frac{1}{\sqrt{1+\sin h^2 y}}$$

$$= \pm \frac{1}{\sqrt{1+x^2}}$$

where the sign of the radical is the same as that of cos*hy* which we know, is always positive.

Hence, $\frac{d}{dx}(\sinh^{-1} x) = \frac{1}{\sqrt{1+x^2}}$

(2) Derivative of $\cosh^{-1} x$

Let $y = \cos h^{-1} x$

$\Rightarrow x = \cos hy$

$\Rightarrow \frac{dx}{dy} = \sin hy$

where the sign of the radical is the same as that of sinhy is positive.

Now, $\cosh^{-1} x$ is always positive.

Hence, $\frac{d}{dx}(\cosh^{-1} x) = \frac{1}{\sqrt{x^2-1}}$

(3) Derivative of $\tanh^{-1} x$ v

Let $y = \tanh^{-1} x$

$\Rightarrow x = \tanh y$

$\Rightarrow \frac{dx}{dy} = \sec h^2 y$

$\Rightarrow \frac{dy}{dx} = \frac{1}{\sec h^2 y} = \frac{1}{1-\tanh^2 y}$

$$= \frac{1}{1-x^2}$$

Thus, $\frac{d}{dx}(\tanh^{-1}x) = \frac{1}{1-x^2}$

(4) Derivative of $\coth^{-1}x$

Let $y = \cot h^{-1}x$

$$\Rightarrow x = \cot hy$$

$$\Rightarrow \frac{dx}{dy} = -\operatorname{cosech}^2 y$$

$$\Rightarrow \frac{dy}{dx} = \frac{-1}{\sin h^2 y}$$

$$= \frac{-1}{\cot h^2 y - 1} = \frac{-1}{x^2 - 1}$$

Thus, $\frac{d}{dx}(\coth^{-1}x) = \frac{-1}{x^2-1}$

(5) Derivative of $\operatorname{sech}^{-1}x$

Let $y = \sec h^{-1}x$

$$\Rightarrow x = \sec hy$$

$$\Rightarrow \frac{dx}{dy} = -\sec hy \tanh y$$

$$\Rightarrow \frac{dy}{dx} = -\frac{1}{\sec hy \tanh y}$$

$$= \pm\frac{-1}{\sec hy\sqrt{(1-\sec h^2 y)}}$$

$$= \pm\frac{-1}{x\sqrt{(1-x^2)}}$$

where the sign of the radical is the same as that of tan*hy*. $\sec h^{-1}x$ is always positive, so that tanh*y* is always positive.

Hence, $\frac{d}{dx}(\operatorname{sech}^{-1}x) = \frac{-1}{x\sqrt{(1-x^2)}}$

(6) Derivative of cosech^{-1}x

Let $y = \operatorname{cosech}^{-1}x$

$$\Rightarrow x = \operatorname{cosech} y$$

$$\Rightarrow \frac{dx}{dy} = -\operatorname{cosech} y.\coth y$$

$$\Rightarrow \frac{dy}{dx} = -\frac{1}{\operatorname{cosech} y \coth y}$$

$$= \pm\frac{-1}{\operatorname{cosech} y\sqrt{(\operatorname{cosech}^2 y + 1)}}$$

$$= \pm\frac{-1}{x\sqrt{(x^2+1)}}$$

where the sign of the radical is the same as that of coth y.

Now, y and therefore coth y is positive or negative according as x is positive or negative.

$$\therefore \quad \frac{dy}{dx} = \frac{-1}{x\sqrt{(x^2+1)}} \text{ if } x > 0$$

$$= \frac{-1}{-x\sqrt{(x^2+1)}} \text{ if } x < 0$$

Thus, $\frac{d}{dx}(\operatorname{cosech}^{-1}x) = \frac{-1}{|x|\sqrt{(x^2+1)}}$

4.18. DERIVATIVE OF PARAMETRIC FUNCTION

Let $x = \phi(t)$, $y = \Psi(t)$ are two functions and t is a variable. In such a case x and y are called **parametric functions** or **parametric equations** and t is called the **parameter**.

To find $\frac{dy}{dx}$ in case of parametric functions, we first obtain the relationship between x and y by eliminating the parameter t and then we differentiate it with respect to x. But every time it is not convenient to eliminate the parameter. Therefore $\frac{dy}{dx}$ can also be obtained by the following formula.

$$\frac{dy}{dx} = \frac{dy/dt}{dx/dt}$$

4.19. DIFFERENTIATION OF IMPLICIT FUNCTION

Let $y = f(x)$

If the variables x and y are connected by a relation of the form $f(x, y) = 0$ and it is not possible or convenient to express y as a function x in the form $y = \phi(x)$, then y is said to be an **implicit function** of x.

To find $\frac{dy}{dx}$ in such a case, we differentiate both sides, of the given relation with respect to x, keeping in mind that the derivative of $\phi(y)$ w.r.t. x is

$$\frac{d\phi}{dy} \cdot \frac{dy}{dx}$$

4.20. LOGARITHMIC DIFFERENTIATION

Let $y = [f(x)]^{g(x)}$

Taking logarithm of both sides, we have

$$\log y = g(x).\log\{f(x)\}$$

Differentiating w.r.t. x, we get

$$\frac{1}{y} \cdot \frac{dy}{dx} = g(x).\frac{1}{f(x)} \cdot \frac{df(x)}{dx} + \log\{f(x)\}.\frac{dg(x)}{dx}$$

$$= y\left[\frac{g(x)}{f(x)} \cdot \frac{df(x)}{dx} + \log\{f(x)\}.\frac{dg(x)}{dx}\right]$$

$$= [f(x)]^{g(x)}\left[\frac{g(x)}{f(x)}.\frac{df(x)}{dx} + \log\{f(x)\}.\frac{dg(x)}{dx}\right]$$

This process is known as **logarithm differentiation**.

4.21. DIFFERENTIATION OF INFINITE SERIES

When the value of y is given as an infinite series and we are asked to find $\frac{dy}{dx}$. In such cases we use the fact that if a term is deleted from an infinite series, it remains unaffected The method of finding $\frac{dy}{dx}$ is explained in the following examples.

Examples

Example 1: Differentiate the following functions with respect to x.

(i) $(\sin x)^{\log x}$ (ii) $x^{\sqrt{x}}$ (iii) $(\sin x)^{\cos^{-1} x}$

Solution:

(i) Let $y = (\sin x)^{\log x}$. Then,

$$y = e^{\log x.\log\sin x}$$

On differentiating both sides w.r.t. x, we get

$$\frac{dy}{dx} = e^{\log x.\log\sin x}\frac{d}{dx}\{\log x.\log\sin x\}$$

$$= (\sin x)^{\log x}\left\{\frac{\log\sin x}{x} + \log x.\frac{1}{\sin x}.\cos x\right\}$$

$$= (\sin x)^{\log x}\left\{\frac{\log\sin x}{x} + \cot x.\log x\right\}$$

(ii) Let $y = x^{\sqrt{x}}$

Taking logarithm of both sides, we get

$$\log y = \log x^{\sqrt{x}}$$

$$= \sqrt{x}\log x$$

On differentiating both sides w.r.t. x, we have

$$\frac{1}{y}.\frac{dy}{dx} = \log x.\frac{d}{dx}\left(\sqrt{x}\right) + \sqrt{x}.\frac{d}{dx}(\log x)$$

$$= \log x.\frac{1}{2\sqrt{x}} + \sqrt{x}.\frac{1}{x}$$

$$\Rightarrow \frac{dy}{dx} = x^{\sqrt{x}}\left(\frac{\log x}{2\sqrt{x}} + \frac{1}{\sqrt{x}}\right)$$

(iii) Let $y = (\sin x)^{\cos^{-1} x}$

Taking logarithm of both sides, we have

$$\log y = \cos^{-1} x \log \sin x$$

On differentiating bothsides w.r.t x, we get

$$\frac{1}{y}.\frac{dy}{dx} = \cos^{-1} x.\frac{d}{dx}(\log \sin x) + \log \sin x \frac{d}{dx}(\cos^{-1} x)$$

$$= \cos^{-1} x.\frac{1}{\sin x}.\cos x + \log \sin x\left(\frac{-1}{\sqrt{1-x^2}}\right)$$

$$\Rightarrow \frac{dy}{dx} = (\sin x)^{\cos^{-1} x}\left(\cos^{-1} x.\cot x - \frac{\log \sin x}{\sqrt{1-x^2}}\right)$$

Exmaple 2: Differentiate the following functions w.r.t. x

(i) $(\sin x)^{\tan x} + (\cos x)^{\sec x}$ (ii) $(\log x)^x + x^{\log x}$

Solution:

(i) Let $y = (\sin x)^{\tan x} + (\cos x)^{\sec x}$

$= \mathrm{u} + \mathrm{v}$

Where $u = (\sin x)^{\tan x}$ and $v = (\cos x)^{\sec x}$

$$\therefore \frac{dy}{dx} = \frac{du}{dx} + \frac{dv}{dx}$$

Now, $u = (\sin x)^{\tan x}$

$\Rightarrow \log u = \tan x \log \sin x$

On differentiating both sides w.r.t. x, we have

$$\frac{1}{u}.\frac{du}{dx} = \tan x.\frac{d}{dx}(\log \sin x) + \log \sin x.\frac{d}{dx}(\tan x)$$

$$= \tan x \frac{1}{\sin x}(\cos x) + \log \sin x.\sec^2 x$$

$$\Rightarrow \frac{du}{dx} = (\sin x)^{\tan x}\left\{1 + \sec^2 x \log \sin x\right\}$$

$$v = (\cos x)^{\sec x}$$

$$\Rightarrow \log v = \sec x \log \cos x$$

On differentiating both sides, we get

$$\frac{1}{v}.\frac{dv}{dx} = \sec x.\frac{d}{dx}(\log \cos x) + \frac{d}{dx}(\sec x).\log \cos x$$

$$= \sec x.\frac{1}{\cos x}(-\sin x) + \sec x.\tan x.\log \cos x$$

$$= -\sec x.\tan x + \sec x.\tan x.\log \cos x$$

$$\Rightarrow \frac{dv}{dx} = (\cos x)^{\sec x}(\sec x.\tan x.\log \cos x - \sec x.\tan x)$$

$$\therefore \frac{dy}{dx} = \frac{du}{dx} + \frac{dv}{dx}$$

$$= (\sin x)^{\tan x}\left\{1 + \sec^2 x \log \sin x\right\}$$

$$+ (\cos x)^{\sec x}\{\sec x.\tan x.\log \cos x - \sec x.\tan x\}$$

(ii) Let $y = (\log x)^x + x^{\log x}$

Let $u = (\log x)^x$ and $v = x^{\log x}$

Now, $y = u + v$

$$\Rightarrow \frac{dy}{dx} = \frac{du}{dx} + \frac{dv}{dx} \qquad \ldots(1)$$

$$u = (\log x)^x$$

$$\Rightarrow \log u = x \log x$$

Differentiating both sides w.r.t. x, we have

$$\frac{1}{u}.\frac{du}{dx} = x.\frac{d}{dx}(\log x) + \log x.\frac{d}{dx}(x)$$

$$= x.\frac{1}{x} + \log x$$

$$\Rightarrow \frac{du}{dx} = (\log x)^x (1 + \log x)$$

And, $v = x^{\log x}$

$$\Rightarrow \log v = \log x.\log x = (\log x)^2$$

Differentiating both sides w.r.t. x, we get

$$\frac{1}{v}.\frac{dv}{dx} = \frac{d}{dx}\left\{(\log x)^2\right\}$$

$$= 2\log x.\frac{1}{x} = \frac{2\log x}{x}$$

$$\Rightarrow \frac{dv}{dx} = x^{\log x}\left\{\frac{2\log x}{x}\right\}$$

Putting the value of $\frac{du}{dx}$ and $\frac{dv}{dx}$ in (1), we get

$$\frac{dy}{dx} = (\log x)^x \{1 + \log x\} + x^{\log x}\left\{\frac{2\log x}{x}\right\}$$

Example 3: If $x^y = e^{x-y}$, prove that

$$\frac{dy}{dx} = \frac{\log x}{(1+\log x)^2}$$

Solution: We have,

$$x^y = e^{x-y}$$

$$\Rightarrow \log x^y = \log e^{x-y}$$

$$\Rightarrow y\log x = x - y$$

$$\Rightarrow y\log x + y = x$$

$$\Rightarrow y(\log x + 1) = x$$

$$\Rightarrow y = \frac{x}{(\log x + 1)}$$

On differentiating both sides w.r.t. *x*, we get

$$\frac{dy}{dx} = \frac{(1+\log x).1 - x.\left(0+\frac{1}{x}\right)}{(1+\log x)^2}$$

$$= \frac{\log x}{(1+\log x)^2}$$

Example 4: If $x^y + y^x = 2$, find $\frac{dy}{dx}$.

Solution: We have $x^y + y^x = 2$

$$\Rightarrow e^{y\log x} + e^{x\log y} = 2$$

On differentiating both sides w.r.t. x, we get

$$\Rightarrow \frac{d}{dx}(e^{y\log x}) + \frac{d}{dx}(e^{x\log y}) = \frac{d}{dx}(2)$$

$$\Rightarrow e^{y\log x}.\frac{d}{dx}(y\log x) + e^{x\log y}.\frac{d}{dx}(x\log y) = 0$$

$$\Rightarrow x^y\left(\frac{dy}{dx}.\log x + y.\frac{1}{x}\right) + y^x\left(1.\log y + x.\frac{1}{y}\frac{dy}{dx}\right) = 0$$

$$\Rightarrow\left(x^y.\log x + y^x.\frac{x}{y}\right)\frac{dy}{dx} + \left(x^y.\frac{x}{y} + y^x.\log y\right) = 0$$

$$\Rightarrow\left(x^y.\log x + xy^{x-1}\right)\frac{dy}{dx} + \left(yx^{y-1} + y^x.\log y\right) = 0$$

$$\Rightarrow\frac{dy}{dx} = -\left(\frac{yx^{y-1} + y^x.\log y}{x^y.\log x + xy^{x-1}}\right)$$

Example 5: If $y = \sqrt{\sin x + \sqrt{\sin x + \sqrt{\sin x + ...+\infty}}}$, prove that

$$\frac{dy}{dx} = \frac{\cos x}{2y-1}$$

Solution: We have,

$$y = \sqrt{\sin x + \sqrt{\sin x + \sqrt{\sin x + ...+\infty}}},$$

We can write

$$y = \sqrt{\sin x + y}$$

$$\Rightarrow y^2 = \sin x + y$$

Differentiating both sides w.r.t. x, we get

$$2y\frac{dy}{dx} = \cos x + \frac{dy}{dx}$$

$$\Rightarrow\frac{dy}{dx}(2y-1) = \cos x$$

$$\Rightarrow\frac{dy}{dx} = \frac{\cos x}{2y-1}$$

Example 6: If $y = a^{x^{a^{x^{\cdots x}}}}$, prove that

$$\frac{dy}{dx} = \frac{y^2 \log y}{x(1 - y\log x.\log y)}$$

Solution: The given series may be written as

$$y = a^{\left(x^y\right)}$$

$$\Rightarrow \log y = x^y \log a$$

$$\Rightarrow \log\log y = y\log x + \log\log a$$

Differentiating both sides w.r.t. x, we get

$$\frac{1}{\log y}\frac{d}{dx}(\log y) = \frac{dy}{dx}.\log x + y.\frac{d}{dx}(\log x) + 0$$

$$\Rightarrow \frac{1}{\log y}\frac{1}{y}\frac{dy}{dx} = \frac{dy}{dx}.\log x + y.\frac{1}{x}$$

$$\Rightarrow \frac{dy}{dx}\left\{\frac{1 - y\log y.\log x}{y\log y}\right\} = \frac{y}{x}$$

$$\Rightarrow \frac{dy}{dx} = \frac{y^2 \log y}{x[1 - y\log x.\log y]}$$

Example 7: Find $\frac{dy}{dx}$ in the following cases.

$$x = a\left[\cos t + \frac{1}{2}\log\tan^2\frac{t}{2}\right] \text{ and } y = a\sin t$$

Solution. (i) We have,

$$x = a\left[\cos t + \frac{1}{2}\log\tan^2\frac{t}{2}\right] \text{ and } y = a\sin t$$

$$\Rightarrow x = a\left[\cos t + \frac{1}{2}.2\log\tan\frac{t}{2}\right] \text{ and } y = a\sin t$$

$$\Rightarrow x = a\left[\cos t + \log\tan\frac{t}{2}\right] \text{ and } y = a\sin t$$

Differentiating both sides w.r.t. x, we have

$$\frac{dx}{dt} = a\left[-\sin t + \frac{1}{\tan t/2}\sec^2\frac{t}{2}.\frac{1}{2}\right] \text{ and } \frac{dy}{dt} = a\cos t$$

$$\Rightarrow \frac{dx}{dt} = a\left[-\sin t + \frac{1}{2\sin(t/2)\cos(t/2)}\right] \text{ and } \frac{dy}{dt} = a\cos t$$

$$\Rightarrow \frac{dx}{dt} = a\left[-\sin t + \frac{1}{\sin t}\right] \text{ and } \frac{dy}{dt} = a\cos t$$

$$\Rightarrow \frac{dx}{dt} = a\left[\frac{-\sin^2 t + 1}{\sin t}\right] \text{ and } \frac{dy}{dt} = a\cos t$$

$$\Rightarrow \frac{dx}{dt} = \frac{a\cos^2 t}{\sin t} \text{ and } \frac{dy}{dt} = a\cos t$$

$$\therefore \quad \frac{dy}{dx} = \frac{dy}{dt}\bigg/\frac{dx}{dt}$$

$$= \frac{a\cos t}{\dfrac{a\cos^2 t}{\sin t}}$$

$$= \tan t$$

Example 8: Differentiate

$$x\coth^{-1}(x^2+1)$$

Solution:

Let $y = x\coth^{-1}(x^2+1)$

$$\therefore \frac{dy}{dx} = x\frac{d}{dx}\left\{\coth^{-1}(x^2+1)\right\} + \coth^{-1}(x^2+1).\frac{d}{dx}(x)$$

$$= x\frac{1}{1-(x^2+1)^2}.2x + \coth^{-1}(x^2+1)$$

$$= \coth^{-1}(x^2+1) - \frac{2}{x^2+2}$$

Exercise – 4.4

1. *Find the derivative of*

(i) $\left(x^x\right)^x$ (ii) $x^{\left(x^x\right)}$

(iii) 10^{10^x} (iv) $\left(\dfrac{x}{1+x}\right)^x$

(v) $(x\log x)^{\log\log x}$ (vi) $(\tan x)^{\log x}$

(vii) $y=(\cot x)^{\sin x}+(\tan x)^{\cot x}$

(viii) $e^{x\sin x^3}+(\tan x)^x$

(ix) $y=(1+1/x)^x+x^{1+1/x}$

2. If $y=x^{(\log x)^{\log\log x}}$, prove that

$$\frac{dy}{dx}=\frac{y}{x}(\log x)^{\log\log x}.(2\log\log x+1)$$

3. If $y=e^{x+e^{x+e^{x \ldots\ldots \infty}}}$, prove that $\dfrac{dy}{dx}=\dfrac{y}{1-y}$

4. If $y=\sqrt{x}^{\sqrt{x}^{\sqrt{x}\ldots\ldots\infty}}$ then, $x\dfrac{dy}{dx}=\dfrac{y^2}{2-y\log x}$

5. If $y=(\sin x)^{(\sin x)^{(\sin x)\ldots\ldots\infty}}$, then $\dfrac{dy}{dx}=\dfrac{y^2\cot x}{1-y\log\sin x}$

6. If $x^y=e^{x-y}$, then prove that

$$\frac{dy}{dx}=\frac{\log x}{(1+\log x)^2}=\log x(\log ex)^{-2}$$

7. If $x^2+y^2=t-\dfrac{1}{t}$, $x^4+y^4=t^2+\dfrac{1}{t^2}$

Prove that $x^3y\dfrac{dy}{dx}=1$

8. If $x = \frac{3at}{1+t^3}, y = \frac{3at^2}{1+t^3}$

$$\frac{dy}{dx} = \frac{t(2-t^3)}{1+2t^3}$$

9. If $\cos\frac{x}{2}.\cos\frac{x}{2^2}........\cos\frac{x}{2^n} = \frac{\sin x}{2^n \sin(x/2^n)}$ prove that

$$\frac{1}{2}\tan\frac{x}{2} + \frac{1}{2^2}\tan\frac{x}{2^2} + ... + \frac{1}{2^n}\tan\frac{x}{2^n} = \frac{1}{2^n}\cot\frac{x}{2^n} - \cot x$$

10. If $x = \sec\theta - \cos\theta$, and $y = \sec^n\theta - \cos^n\theta$ then show that

$$(x^2+4)\left(\frac{dy}{dx}\right)^2 = n^2(y^2+4)$$

11. If $y = \sqrt{\tan x + \sqrt{\tan x + \sqrt{\tan x + ...\infty}}}$, prove that

$$(2y-1)\frac{dy}{dx} = \sec^2 x$$

12. Differentiate $\sin^{-1}\left(\frac{1-x^2}{1+x^2}\right)$ with respect to $\sin^{-1}\left(\frac{2x}{1+x^2}\right)$

13. Differentiate $\log\sec x$ with respect to $\tan x$

14. If $x = a\sec^3\theta, y = a\tan^3\theta$, find $\frac{dy}{dx}$

15. Differentiate $x^{\sin x}$ with respect to $(\sin x)^x$

16. Differentiate

(i) $\cot x \coth y$

(ii) $xa^x \sinh x$

17. If $x = 3\cos\theta - \cos 3\theta$ and $y = 3\sin\theta - \sin 3\theta$, then find $\frac{dy}{dx}$

18. Differentiate $(\tan x)^{\cot x} + (\cot x)^{\tan x}$

19. Find $\frac{dy}{dx}$, when

(i) $x = b\sin^2\theta$ and $y = a\cos^2\theta$

(ii) $x = ae^{\theta}(\sin\theta - \cos\theta)$ and $y = ae^{\theta}(\sin\theta + \cos\theta)$

(iii) $x = \frac{e^t + e^{-t}}{2}$ and $y = \frac{e^t - e^{-t}}{2}$

(iv) $x = \cos^{-1}\frac{1}{\sqrt{1+t^2}}$ and $y = \sin^{-1}\frac{1}{\sqrt{1+t^2}}$

(v) $x = a(1 - \cos\theta)$ and $y = a(\theta + \sin\theta)$ at $\theta = \frac{\pi}{2}$

20. If $x = 2\cos\theta - \cos 2\theta$ and $y = 2\sin\theta - \sin 2\theta$, prove that $\frac{dy}{dx} = \tan\left(\frac{3\theta}{2}\right)$

21. If $x = e^{\cos 2t}$ and $y = e^{\sin 2t}$, prove that $\frac{dy}{dx} = -\frac{y\log x}{x\log y}$

22. If $x = \sin^{-1}\left(\frac{2t}{1+t^2}\right)$ and $y = \tan^{-1}\left(\frac{2t}{1-t^2}\right)$, prove that $\frac{dy}{dx} = 1$

23. If $(\sin x)^y = x + y$, prove that $\frac{dy}{dx} = \frac{1-(x+y)y\cot x}{(x+y)\log\sin x - 1}$

5

Successive Differentiation

5.1. DEFINITION

If $y = f(x)$, then $\frac{dy}{dx}$ is the derivative of y with respect to x. The derivative of $\frac{dy}{dx}$ w. r. t. x is the second order derivative of y w. r. t. x and will be denoted by $\frac{d^2y}{dx^2}$. Similarly the derivative of $\frac{d^2y}{dx^2}$ w. r. t. x will be termed as the third order derivative of y w. r. t. x and will be denoted by $\frac{d^3y}{dx^3}$ and so on. The nth order derivative of y w. r. t. x will be denoted by $\frac{d^ny}{dx^n}$.

The successive derivatives of y are represented by the symbol.

$$\frac{dy}{dx}, \frac{d^2y}{dx^2}, \frac{d^3y}{dx^3}, \ldots\ldots\ldots\ldots\ldots, \frac{d^ny}{dx^n}$$

or, $y_1, y_2, y_3, \ldots\ldots\ldots\ldots\ldots\ldots\ldots, y_n$

or, $y', y'', y''', \ldots\ldots\ldots\ldots\ldots\ldots\ldots, y^n$

or, $Dy, D^2y, D^3y, \ldots\ldots\ldots\ldots, D^ny$

or, $f'(x), f''(x), f'''(x), \ldots\ldots\ldots\ldots, f^n(x)$

The value of these derivative at a are denoted by

$$y_n(a),\ y^n(a),\ D^n y(a),\ f^n(a) \text{ or } \left(\frac{d^n y}{dx^n}\right)_{x=a}$$

Examples

Example 1: Find $\frac{d^2y}{dx^2}$, if $y = x^3 + \tan x$

Solution: Given that, $y = x^3 + \tan x$

$$\Rightarrow \frac{dy}{dx} = 3x^2 + \sec^2 x$$

$$\Rightarrow \frac{d^2y}{dx^2} = \frac{d}{dx}\left(3x^2 + \sec^2 x\right)$$

$$= 6x + 2\sec.\sec x \tan x$$

$$= 6x + 2\sec^2 x \tan x$$

Example 2: If $y = 3e^{2x} + 2e^{3x}$, then prove that $\frac{d^2y}{dx} - 5\frac{dy}{dx} + 6y = 0$

Solution: We have, $y = 3e^{2x} + 2e^{3x}$

Therefore, $\frac{dy}{dx} = 6e^{2x} + 6e^{3x}$

$$= 6\left(e^{2x} + e^{3x}\right)$$

and $\frac{d^2y}{dx^2} = 12e^{2x} + 18e^{3x}$

$$= 6\left(2e^{2x} + 3e^{3x}\right)$$

Now,

$$\frac{d^2y}{dx^2} - 5\frac{dy}{dx} + 6y = 6\left(2e^{2x} + 3e^{3x}\right) - 5 \times 6\left(e^{2x} + e^{3x}\right) + 6\left(3e^{2x} + 2e\ \right)$$

$$= 0$$

Example 3: If $y = \sin^{-1} x$, then show that $\left(1 - x^2\right)\frac{d^2y}{dx^2} - x\frac{dy}{dx} = 0$

Solution: Given that $y = \sin^{-1} x$, we have

$$\frac{dy}{dx} = \frac{1}{\sqrt{1 - x^2}}$$

$$\Rightarrow \left(\sqrt{1 - x^2}\right)\frac{dy}{dx} = 1$$

Squaring bothsides, we get

$$\left(1 - x^2\right)\left(\frac{dy}{dx}\right)^2 = 1$$

Differentiating bothsides, we have

$$\left(1 - x^2\right)\frac{dy}{dx} \times \frac{d^2y}{dx^2} + \left(\frac{dy}{dx}\right)^2 (0 - 2x) = 0$$

$$\Rightarrow \left(1 - x^2\right)\frac{dy}{dx} - x\frac{dy}{dx} = 0$$

Example 4: If $y = \tan x + \sec x$, prove that $\frac{d^2y}{dx^2} = \frac{\cos x}{(1 - \sin x)^2}$

Solution: We have, $y = \tan x + \sec x$

$$\therefore \quad \frac{dy}{dx} = \sec^2 x + \sec x \tan x$$

$$= \frac{1}{\cos^2 x} + \frac{1}{\cos x} \times \frac{\sin x}{\cos x}$$

$$= \frac{1+\sin x}{\cos^2 x} = \frac{1+\sin x}{1-\sin^2 x}$$

$$= \frac{1+\sin x}{(1+\sin x)(1-\sin x)} = \frac{1}{1-\sin x}$$

$$\Rightarrow \frac{dy}{dx} = \frac{1}{1-\sin x}$$

$$\Rightarrow \frac{d^2 y}{dx^2} = \frac{d}{dx}\left\{\frac{1}{1-\sin x}\right\}$$

$$= \frac{d}{dx}\left\{(1-\sin x)^{-1}\right\}$$

$$= (-1)(1-\sin x)^{-2} \frac{d}{dx}(1-\sin x)$$

$$= \frac{-1}{(1-\sin x)^2} \times -\cos x$$

$$= \frac{\cos x}{(1-\sin x)^2}$$

Example 5: If $y = x\log\left(\frac{x}{a+bx}\right)$, prove that $x^3 \frac{d^2 y}{dx^2} = \left(x\frac{dy}{dx} - y\right)^2$

Solution: We have, $y = x\log\left(\frac{x}{a+bx}\right)$

$$\Rightarrow y = x\{\log x - \log(a+bx)\}$$

$$\Rightarrow \frac{y}{x} = \log x - \log(a+bx)$$

Differentiating both side w.r.t. x., we get

$$\frac{x\frac{dy}{dx} - y}{x^2} = \frac{1}{x} - \frac{1}{a+bx} - \frac{d}{dx}(a+bx)$$

$$\Rightarrow x\frac{dy}{dx}-y=x^2\left\{\frac{1}{x}-\frac{b}{a+bx}\right\}$$

$$\Rightarrow x\frac{dy}{dx}-y=\frac{ax}{a+bx} \qquad \ldots\text{(i)}$$

Again, differenting both side of (i) w.r.t. x., we find

$$x\frac{d^2y}{dx^2}+\frac{dy}{dx}-\frac{dy}{dx}=\frac{(a+bx).a-ax(o+b)}{(a+bx)^2}$$

$$\Rightarrow x\frac{d^2y}{dx^2}=\frac{a^2}{(a+bx)^2}$$

$$\Rightarrow x^3\frac{d^2y}{dx^2}=\frac{a^2x^2}{(a+bx)^2} \qquad \text{[Multiplying both sides by } x^2\text{]}$$

$$\Rightarrow x^3\frac{d^2y}{dx^2}=\left(\frac{ax}{a+bx}\right)^2 \qquad \ldots\text{(ii)}$$

From (i) & (ii), we have

$$x^3\frac{d^2y}{dx^2}=\left(x\frac{dy}{dx}-y\right)^2$$

Example 6: If $y=\left[x+\sqrt{x^2+1}\right]^m$,

show that $\left(x^2+1\right)y_2+xy_1-m^2y=0$

Solution: We have, $y=\left[x+\sqrt{x^2+1}\right]^m$

$$\therefore \frac{dy}{dx}=m\left[x+\sqrt{x^2+1}\right]^{m-1}.\frac{d}{dx}\left(x+\sqrt{x^2+1}\right)$$

$$\Rightarrow \frac{dy}{dx} = m\left[x + \sqrt{x^2+1}\right]^{m-1} \cdot \left[1 + \frac{2x}{2\sqrt{x^2+1}}\right]$$

$$\Rightarrow \frac{dy}{dx} = m\frac{\left(\sqrt{x^2+1}+x\right)^m}{\sqrt{x^2+1}}$$

$$\Rightarrow y_1 = \frac{my}{\sqrt{x^2+1}}$$

$$\Rightarrow y_1\sqrt{x^2+1} = my$$

$$\Rightarrow y_1^2\left(x^2+1\right) = m^2y^2$$

Differentiating w.r.t. x, we get

$$2y_1y_2\left(1+x^2\right) + y_1^2(2x) = 2m^2yy_1$$

$$\Rightarrow y_2\left(1+x^2\right) + xy_1 - m^2y = 0$$

Example 7: If $x = a\cos\theta + b\sin\theta$ and $y = a\sin\theta - b\cos\theta$, prove that

$$y^2\frac{d^2y}{dx^2} - x\frac{dy}{dx} + y = 0$$

Solution: We have, $x^2 + y^2 = (a\cos\theta + b\sin\theta)^2 + (a\sin\theta - b\cos\theta)^2$

$$\Rightarrow x^2 + y^2 = a^2\left(\cos^2\theta + \sin^2\theta\right) + b^2\left(\sin^2\theta + \cos^2\theta\right)$$

$$\Rightarrow x^2 + y^2 = a^2 + b^2$$

Differentiating both side, w.r.t. x, we get

$$2x + 2y\frac{dy}{dx} = 0$$

$$\Rightarrow \frac{dy}{dx} = \frac{-x}{y} \qquad \ldots\text{(i)}$$

Again, differentiating w.r.t. x, we find

$$\frac{d^2y}{dx^2} = -\left\{\frac{y.1 - x\frac{dy}{dx}}{y^2}\right\}$$

$$= -\frac{y - x\left(\frac{-x}{y}\right)}{y^2}$$

$$= -\frac{\left(x^2 + y^2\right)}{y^3} \qquad \ldots\text{(ii)}$$

Now, $y^2\frac{d^2y}{dx^2} - x\frac{dy}{dx} + y = -y^2\left(\frac{x^2+y^2}{y^3}\right) - x\left(-\frac{x}{y}\right) + y$

$= 0$ [Using (i)&(ii).]

Example 8: If $ax^2 + 2hxy + by^2 = 1$, prove that

$$\frac{d^2y}{dx^2} = \frac{h^2 - ab}{(hx+by)^3}$$

Solution: We have, $ax^2 + 2hxy + by^2 = 1$

Differentiating both side w.r.t. x, we get

$$2ax + 2hy + 2hx.\frac{dy}{dx} + 2by.\frac{dy}{dx} = 0$$

$$\Rightarrow 2(ax+hy) + 2(hx+by)\frac{dy}{dx} = 0$$

$$\Rightarrow \frac{dy}{dx} = -\frac{ax+hy}{hx+by}$$

Again, differentiating both sides, we find

$$\frac{d^2y}{dx^2} = -\frac{(hx+by)\left(a+h\frac{dy}{dx}\right)-(ax+hy)\left(h+b\frac{dy}{dx}\right)}{(hx+by)^2}$$

$$= \frac{\left(h^2-ab\right)\left(y-x\frac{dy}{dx}\right)}{(hx+by)^2}$$

$$= \frac{h^2-ab}{(hx+by)^2}\left\{y+\frac{x(ax+hy)}{hx+by}\right\}$$

$$= \frac{h^2-ab}{(hx+by)^3}.ax^2+2hxy+by^2$$

$$= \frac{h^2-ab}{(hx+by)^3}, \text{ since } ax^2+2hxy+by^2=1.$$

Example 9: If $y=\sin(\sin x)$, prove that $\frac{d^2y}{dx^2}+\tan x\frac{dy}{dx}+y\cos^2 x=0$

Solution: We have, $y=\sin(\sin x)$

$$\therefore \frac{dy}{dx}=\cos(\sin x)\cos x$$

$$\Rightarrow \frac{d^2y}{dx^2}=-\sin(\sin x).\cos x.\cos x+\cos(\sin x).(-\sin x)$$

$$= -\sin(\sin x)\cos^2 x-\cos(\sin x)\sin x$$

$$= -y\cos^2 x-\cos(\sin x)\cdot\frac{\sin x}{\cos x}\cdot\cos x$$

$$= -y\cos^2 x-\tan x.\frac{dy}{dx}$$

$$\Rightarrow \frac{d^2y}{dx^2}+y\cos^2 x+\tan x\frac{dy}{dx}=0$$

Example 10: Change the independent variable to z in the equation

$$\frac{d^2y}{dx^2}+\cot x\frac{dy}{dx}+4y\operatorname{cosec}^2 x=0$$

by means of the transformation $z=\log\tan {x/2}$

Solution: We have,

$$z=\log\tan {x/2}$$

$$\Rightarrow \tan {x/2}=e^z$$

$$\Rightarrow \frac{x}{2}=\tan^{-1}e^z$$

$$\Rightarrow x=2\tan^{-1}e^z$$

$$\therefore \frac{dy}{dx}=\frac{dy}{dz}\times\frac{dz}{dx}$$

$$\Rightarrow \frac{dy}{dx}=\frac{dy}{dz}.\frac{1}{\tan {x/2}}.\frac{1}{2}\sec^2 {x/2}$$

$$\Rightarrow \frac{dy}{dx}=\frac{\sec^2 {x/2}}{2\tan {x/2}}.\frac{dy}{dz}=\operatorname{cosec}x\frac{dy}{dz}$$

Again, $$\frac{d^2y}{dx^2}=-\operatorname{cosec}x.\cot x.\frac{dy}{dz}+\frac{d^2y}{dz^2}.\frac{dz}{dx}.\operatorname{cosec}x$$

$$=-\operatorname{cosec}x.\cot x.\frac{dy}{dz}+\operatorname{cosec}^2 x.\frac{d^2y}{dz^2}$$

Substituting the value of ${dy/dx}$ and ${d^2y/dx^2}$ in the given differential equation, we have

$$-\operatorname{cosec}x.\cot x\frac{dy}{dz}+\operatorname{cosec}^2 x\frac{d^2y}{dz^2}+\cot x\operatorname{cosec}x\frac{dy}{dz}+4y.\operatorname{cosec}^2 x=0$$

$$\Rightarrow \frac{d^2y}{dz^2}+4y=0$$

Exercise – 5.1

1. If $p^2 = a^2\cos^2\theta + b^2\sin^2\theta$, prove that

$$p + \frac{d^2p}{d\theta^2} = \frac{a^2b^2}{p^3}$$

2. If $y = a\cos(\log x) + b\sin(\log x)$, show that

$$x^2\frac{d^2y}{dx^2} + x\frac{dy}{dx} + y = 0$$

3. If $y = e^{ax}\sin(bx+c)$, show that

$$\frac{d^2y}{dx^2} = r^2e^{ax}\sin(bx+c+2\Phi),$$

Where $r = \sqrt{a^2+b^2}, \Phi = \tan^{-1}\left(b/a\right)$

4. If $y = x^x$, prove that

$$\frac{d^2y}{dx^2} - \frac{1}{y}\left(\frac{dy}{dx}\right)^2 - \frac{y}{x} = 0$$

5. $y = \frac{\log x}{x}$,, show that $\frac{d^2y}{dx^2} = \frac{2\log x - 3}{x^3}$

6. If $y = x^3\log x$, prove that $\frac{d^4y}{dx^4} = \frac{6}{x}$

7. If $x = a(\theta+\sin\theta), y = a(1+\cos\theta)$, prove that $\frac{d^2y}{dx^2} = -\frac{a}{y^2}$

8. If $y = ae^{2x} + be^{-x}$, show that

$$\frac{d^2y}{dx^2} - \frac{dy}{dx} - 2y = 0$$

9. If $y = Ae^{-kt}\cos(pt + c)$, prove that

$$\frac{d^2y}{dt^2} + 2k\frac{dy}{dt} - n^2y = 0, \text{ where } n^2 = p^2 + k^2$$

10. Find A and B so that $y = A\sin 3x + B\cos 3x$ satisfies the equation

$$\frac{d^2y}{dx^2} + 4\frac{dy}{dx} + 3y = 10\cos 3x$$

11. If $x = \sin t$, $y = \sin pt$, prove that

$$\left(1 - x^2\right)\frac{d^2y}{dx^2} - x\frac{dy}{dx} + p^2y = 0$$

12. If $\log y = \tan^{-1}x$, show that $\left(1 + x^2\right)y_2 + (2x - 1)y_1 = 0$

13. If $y = e^{a\cos^{-1}x}$, show that

$$\left(1 - x^2\right)\frac{d^2y}{dx^2} - x\frac{dy}{dx} - a^2y = 0$$

14. If $p^2 = a^2\cos^2\theta + b^2\sin^2\theta$, show that

$$\frac{d^2p}{d\theta^2} + p = \frac{a^2b^2}{p^3}$$

15. If $x = a\sin 2\theta(1 + \cos 2\theta)$, $y = a\cos 2\theta(1 - \cos 2\theta)$, prove that

$$\left[1 + \left({dy}/{dx}\right)^2\right]^{3/2} = 4a\cos 3\theta$$

16. If $y = x\log\left[(ax)^{-1} + a^{-1}\right]$, prove that

$$x(x + 1)\frac{d^2y}{dx^2} + x\frac{dy}{dx} = y - 1$$

17. If $y = \left(\tan^{-1}x\right)^2$, prove that

$$\left(x^2+1\right)^2\frac{d^2y}{dx^2}+2x\left(x^2+1\right)\frac{dy}{dx}-2=0$$

18. If $ax=\sin(x+y)$, prove that

$$y\left(1+\left(\frac{dy}{dx}\right)\right)^3=0$$

19. If $y=Ae^{px}+Be^{qx}$, show that

$$\frac{d^2y}{dx^2}-(p+q)\frac{dy}{dx}+pqy=0$$

20. Change the independent variable to θ in the equation

$$\frac{d^2y}{dx^2}+\frac{2x}{1+x^2}\frac{dy}{dx}+\frac{y}{\left(1+x^2\right)^2}=0$$

by means of transformation $x=\tan\theta$

5.2. nth DERIVATIVES OF SOME STANDARD FUNCTIONS

5.2.1. Determination of the nth Derivative of $(ax+b)^m$

Let $y = (ax+b)^m$

Then, $y_1 = ma(ax+b)^{m-1}$

$$y_2 = m(m-1)a^2(ax+b)^{m-2}$$

$$y_3 = m(m-1)(m-2)a^3(ax+b)^{m-3}$$

In general, $\boldsymbol{y_n = m(m-1)(m-2) \ldots\ldots (m-n+1)a^n(ax+b)^{m-n}}$

Note: In case, m is a positive integer, y_n can be written as

$$y_n = \frac{m!}{(m-n)!}a^n(ax+b)^{m-n}$$

Cor.1. If $y = x^m$, then

$$y_n = m(m-1)(m-2) \ldots\ldots (m-n+1)x^{m-n}$$

Cor.2. If $m = -1$, then

$$y_n = (-1)(-2)(-3)\ldots\ldots(-n)a^n(ax+b)^{-1-n}$$

$$\Rightarrow \quad \boldsymbol{\frac{d^n}{dx^n}\left(\frac{1}{ax+b}\right) = \frac{(-1)^n(n!)a^n}{(ax+b)^{n+1}}}$$

5.2.2. Determination of the nth derivative of $\ln(ax+b)$

Let $y = \ln(ax+b)$

Then, $\quad y_1 = \dfrac{a}{ax+b} = (ax+b)^{-1}$

$$y_2 = (-1)a^2(ax+b)^{-2}$$

$$y_3 = (-1)^2 2a^3(ax+b)^{-3}$$

... ...

... ...

In general, $y_n = (-1)^{n-1} \dfrac{a^n(n-1)!}{(ax+b)^n}$

Cor. If $y = \ln x$, then

$$y_n = (-1)^{n-1} \frac{(n-1)!}{x^n}$$

5.2.3. Determination of the nth derivative of a^{mx}.

Let $y = a^{mx}$

Then, $y_1 = ma^{mx} \ln a$

$$y_2 = m^2 a^{mx} (\ln a)^2$$

$$y_3 = m^3 a^{mx} (\ln a)^3$$

... ...

... ...

$$y_n = m^n a^{mn} (\ln a)^n$$

Cor.1. If $y_n = m^n a^{mn} (\ln a)^n$ then

$$y_n = a^x (\ln a)^n$$

Cor.2. If $y = e^x$, then

$$y_n = e^x$$

Cor.3. $y = e^{nx}$, then

$$y_n = m^n e^{Mx}$$

5.2.4. Determination of the nth Derivative of sin(ax+b)

Let $y = \sin(ax+b)$,

Then, $y_1 = a\cos(ax+b) = a\sin\left(ax+b+\frac{1}{2}\pi\right)$

$$y_2 = a^2\cos\left(ax+b+\frac{1}{2}\pi\right) = a^2\sin\left(ax+b+\frac{2}{2}\pi\right)$$

$$y_3 = a^3\cos\left(ax+b+\frac{2}{2}\pi\right) = a^3\sin\left(ax+b+\frac{3}{2}\pi\right)$$

... ...

... ...

$$\boldsymbol{y_n = a^n\sin\left(ax+b+\frac{np}{2}\right)}$$

Cor. If $y = \sin x$, then

$$\boldsymbol{y_n = \sin\left(x+\frac{np}{2}\right)}$$

5.2.5. Determination of the nth derivative of cos(ax+b)

Let $y = \cos(ax+b)$

Then $y_1 = -a\sin(ax+b) = a\cos\left(ax+b+\frac{1}{2}\pi\right)$

$$y_2 = -a^2\sin\left(ax+b+\frac{1}{2}\pi\right) = a^2\cos\left(ax+b+\frac{2}{2}\pi\right).$$

$$y_3 = -a^3\sin\left(ax+b+\frac{2}{2}\pi\right) = a^3\cos\left(ax+b+\frac{3}{2}\pi\right)$$

... ...

... ...

$$y_n = a^n \cos\left(ax + b + \frac{np}{2}\right)$$

Cor. If $y = \cos x$, then

$$y_n = \cos\left(x + \frac{np}{2}\right)$$

5.2.6. *Determination of the nth derivative of $e^{ax}\sin(bx + c)$*

Let $y = e^{ax}\sin(bx + c)$

Then, $y_1 = ae^{ax}\sin(bx + c) + e^{ax} b\cos(bx + c)$

$$= e^{ax}\left[a\sin(bx + c) + b\cos(bx + c)\right]$$

Put $a = r\cos\theta$ and $b = r\sin\theta$

Then, $r = \sqrt{a^2 + b^2}$ and $\theta = \tan^{-1}(b/a)$

Hence, we have

$$y_1 = re^{ax}\sin(bx + c + \theta)$$

$$y_2 = r^2 e^{ax}\sin(bx + c + 2\theta)$$

$$y_3 = r^3 e^{ax}\sin(bx + c + 3\theta)$$

… … … …

… … … …

In general, $\boldsymbol{y_n = r^n e^{ax}\sin(bx + c + nq)}$

$$= \left(a^2 + b^2\right)^{n/2} e^{ax}\sin\left\{bx + c + n\tan^{-1}(b/a)\right\}$$

Similarly, if $y = e^{ax}\cos(bx + c)$, then

$$\boldsymbol{y_n = r^n e^{ax}\cos(bx + c + nq),}$$

where $r = \sqrt{a^2 + b^2}$ and $\theta = \tan^{-1}(b/a)$

5.3. DETERMINATION OF nTH DERIVATIVE OF RATIONAL FUNCTIONS (DE MOIVRE'S THEOREM)

In order to determine the nth derivative of a rational function, we have to decompose it into partial fractions. We can also apply *Demoivre's theorem* whenever necessary; which states

$$(\cos\theta \pm i\sin\theta)^n = \cos n\theta \pm i\sin n\theta$$

where n is any integer, positive or negative and $i = \sqrt{-1}$.

5.4. THE DERIVATIVES OF THE PRODUCTS OF THE POWERS OF SINES AND COSINES

In order to determine the nth derivative of a product of the powers of sin*es* and cosines we express it as the sum of the sines and cosines of multiples of the independent veriable.

For example, Let $y = \sin^4 x$

We have,

$$\sin^2 x = \frac{1-\cos 2x}{2}$$

$$\Rightarrow \sin^4 x = \left(\frac{1-\cos 2x}{2}\right)^2$$

$$= \frac{1}{4} - \frac{\cos 2x}{2} + \frac{\cos^2 2x}{4}$$

$$= \frac{1}{4} - \frac{1}{2}\cos 2x + \frac{1}{8}(1+\cos 4x)$$

$$= \frac{3}{8} - \frac{1}{2}\cos 2x + \frac{1}{8}\cos 4x$$

$$\Rightarrow \frac{d^n(\sin^4 x)}{dx^n} = -\frac{1}{2}2^n\cos\left(2x+\frac{n\pi}{2}\right) + \frac{1}{8}4^n\cos\left(4x+\frac{n\pi}{2}\right)$$

Examples

Example 1: If $y = \frac{1}{(3x+8)^2}$, find y_n.

Solution: We have,

$$y = \frac{1}{(3x+8)^2} = \frac{1}{(3x+8)^{1+1}}$$

$$y_n = \frac{(-1)^n (n+1)^n .3^n}{(3x+8)^{n+2}}$$

Example 2: If $y = e^{2x} \sin^3 \theta$, find y_n.

Solution: $y = e^{2x} \sin^3 \theta$

$$= e^{2x}\left[\frac{1}{4}(3\sin\theta - \sin 3\theta)\right]$$

$$= \frac{3}{4}e^{2x}\sin\theta - \frac{1}{4}e^{2x}\sin 3\theta$$

Let $p = \frac{3}{4}e^{2x}\sin\theta$ and $q = \frac{1}{4}e^{2x}\sin 3\theta$

Now, $p_n = \frac{3}{4}\left(\sqrt{5}\right)^n \sin\left(\theta + n\tan^{-1}\frac{1}{2}\right)$

and, $q_n = \frac{1}{4}\left(\sqrt{13}\right)^n \sin\left(3\theta + n\tan^{-1}\frac{3}{2}\right)$

Since, $y = p - q$

Therefore, $y_n = p_n - q_n$

$$= \frac{3}{4}\left(\sqrt{5}\right)^n \sin\left(\theta + n\tan^{-1}\frac{1}{2}\right) - \frac{1}{4}\left(\sqrt{13}\right)^n \sin\left(3\theta + n\tan^{-1}\frac{1}{2}\right)$$

Example 3: Find y_n, if $y = \frac{x}{(x-1)(x-2)}$

Solution: By the method of partial fractions, we have

$$\frac{x}{(x-1)(x-2)} = \frac{2}{x-2} - \frac{1}{x-1}$$

$$y_n = \frac{2(-1)^n . n!}{(x-2)^{n+1}} + \frac{(-1)^{n+1} . n!}{(x-1)^{n+1}}$$

Example 4: Find y_n, if $y = x\log\left(\frac{x-1}{x+1}\right)$

Solution: We have

$$y = x\log\left(\frac{x-1}{x+1}\right)$$

$$\Rightarrow \quad \frac{dy}{dx} = \log\frac{x-1}{x+1} + x\left(\frac{x+1}{x-1}\right) \times \frac{2}{(x+1)^2}$$

$$= \log\frac{x-1}{x+1} + \frac{2x}{(x-1)(x+1)}$$

$$= \log\frac{x-1}{x+1} + \frac{1}{x-1} + \frac{1}{x+1}$$

$$\Rightarrow \quad \frac{d^2y}{dx^2} = \frac{2}{x^2-1} - \frac{1}{(x-1)^2} - \frac{1}{(x+1)^2}$$

$$= \frac{1}{x-1} - \frac{1}{x+1} - \frac{1}{(x-1)^2} - \frac{1}{(x+1)^2}$$

Differentiating $(n-2)$ times *w.r.t.* x, we get

$$\frac{d^n y}{dx^n} = (-1)^{n-2}\left[\frac{(n-2)!}{(x-1)^{n-1}} - \frac{(n-2)!}{(x+1)^{n-1}} - \frac{(n-1)!}{(x-1)^n} - \frac{(n-1)!}{(x+1)^n}\right]$$

$$= (-1)^{n-2}\left[\frac{(n-2)!}{(x-1)^n}\{x-1-(n-1)\} - \frac{(n-2)!}{(x+1)^n}\{(x+1)+(n-1)\}\right]$$

$$= (-1)^{n-2}(n-2)!\left[\frac{x-n}{(x-1)^n} - \frac{x+n}{(x+1)^n}\right]$$

$$= (-1)^n(n-2)!\left[\frac{x-n}{(x-1)^n} - \frac{x+n}{(x+1)^n}\right]$$

$$\left[\because (-1)^{n-2} = (-1)^n\cdot(-1)^{-2} = (-1)^n\cdot 1 = (-1)^n\right]$$

Example 5: Find y_n, if $y = \dfrac{x}{x^2+1}$

Solution: Now, $y = \dfrac{x}{x^2+1} = \dfrac{x}{(x+i)(x-i)}$

$$= \frac{1}{2}\left[\frac{1}{x+i} + \frac{1}{x-i}\right]$$

$$\therefore y_n = \frac{1}{2}\left[\frac{(-1)^n.(n!)}{(x+i)^{n+1}} + \frac{(-1)^n(n!)}{(x-i)^{n+1}}\right]$$

$$= \frac{1}{2}(-1)^n.n!\left[\frac{1}{(x+i)^{n+1}} + \frac{1}{(x-i)^{n+1}}\right]$$

Let $x = r\cos\theta$ and $1 = r\sin\theta$

Then, $r = \sqrt{1+x^2}$ and $\theta = \tan^{-1}(1/x)$

Also, $y_n = \dfrac{1}{2}\dfrac{(-1)^n.n!}{r^{n+1}}\left[\dfrac{1}{(\cos\theta + i\sin\theta)^{n+1}} + \dfrac{1}{(\cos\theta - i\sin\theta)^{n+1}}\right]$

$$=\frac{1}{2}\frac{(-1)^n.n!}{r^{n+1}}\left[(\cos\theta+i\sin\theta)^{-(n+1)}+(\cos\theta-i\sin\theta)^{-(n+1)}\right]$$

$$=\frac{1}{2}\frac{(-1)^n.n!}{r^{n+1}}\left[\cos(n+1)\theta-i\sin(n+1)\theta\right.$$

$$\left.+\cos(n+1)\theta+i\sin(n+1)\theta\right]$$

$$=\frac{(-1)^n.n!}{r^{n+1}}.\cos(n+1)\theta$$

where $r=\sqrt{1+x}$ and $\theta=\tan^{-1}(1/x)$

Example 6: Show that the *nth* derivative of $y=\left(1+x+x^2+x^3\right)^{-1}$ is $\frac{1}{2}(-1)^n n!\sin^{n+1}\theta\left[\sin(n+1)\theta-\cos(n+1)\theta+(\sin\theta+\cos\theta)^{-n-1}\right]$, where $\theta=\cot^{-1}x$

Solution: We have,

$$y=\frac{1}{(1+x)\left(1+x^2\right)}$$

$$\Rightarrow\quad y=\frac{1}{(1+x)(x+i)(x-i)}$$

By the method of partial fractions

$$y=\frac{1}{2}\left[\frac{1}{1+x}+\frac{i-1}{2}.\frac{1}{x+i}-\frac{i+1}{2}.\frac{1}{x-i}\right]$$

$$y_n=\frac{1}{2}(-1)^n n!\left[\frac{1}{(1+x)^{n+1}}+\frac{i-1}{2}.\frac{1}{(x+i)^{n+1}}-\frac{(i+1)}{2}.\frac{1}{(x-i)^{n+1}}\right]$$

Put, $x=r\cos\theta,\quad 1=r\sin\theta$

so that $r = \sqrt{x^2 + 1}$ and $\theta = \cot^{-1} x$

Now, $$\frac{1}{(1+x)^{n+1}} = \frac{1}{r^{n+1}}(\cos\theta + i\sin\theta)^{-n-1}$$

$$\frac{1}{x+i} = \frac{1}{r^{n+1}}(\cos\theta + i\sin\theta)^{-(n+1)}$$

$$= \frac{1}{r^{n+1}}\left[\cos(n+1)\theta - i\sin(n+1)\theta\right]$$

and $$\frac{1}{(x-i)^{n+1}} = \frac{1}{r^{n+1}}(\cos\theta - i\sin\theta)^{-(n+1)}$$

$$= \frac{1}{r^{n+1}}\left[\cos(n+1)\theta + i\sin(n+1)\theta\right]$$

$$\therefore \frac{d^n y}{dx^n} = \frac{1}{2}\frac{(-1)^n n!}{r^{n+1}}\Big[(\cos\theta + i\sin\theta)^{-n-1} + \tfrac{1}{2}\big\{(i-1)\big(\cos(n+1)\theta \cdot$$

$$- i\sin(n+1)\theta\big) - (i+1)\big(\cos(n+1)\theta + i\sin(n+1)\theta\big)\big\}\Big]$$

$$= \tfrac{1}{2}\frac{(-1)^n n!}{r^{n+1}}\left[(\cos\theta + \sin\theta)^{-n-1} + \sin(n+1)\theta - \cos(n+1)\theta\right]$$

Now, $$r^{n+1} = \left(x^2 + 1\right)^{(n+1)/2}$$

$$= \left(\cot^2\theta + 1\right)^{(n+1)/2}$$

$$= \operatorname{cosec}^{n+1}\theta$$

$$\Rightarrow \frac{1}{r^{n+1}} = \sin^{n+1}\theta$$

$$\therefore \frac{d^n y}{dx^n} = \frac{1}{2}(-1)^n n! \sin^{n+1}\theta\left[(\cos\theta + \sin\theta)^{-n-1} + \sin(n+1)\theta - \cos(n+1)\theta\right]$$

Exercise – 5.2

1. *Find the nth derivative:*

(i) $\cos^4 x$ (ii) $e^x \cos 2x.\cos 4x$

(iii) $e^{3x} \sin x.\sin 2x.\sin 3x$ (iv) $\sin^2 x \cos^3 x$

(v) $e^{ax} \cos^2 x.\sin x$ (vi) $e^x \sin^4 x$

(vii) $\cos x.\cos 2x.\cos 3x$ (viii) $\sin^3 x$

(ix) $\sin 5x.\sin 3x$ (x) $\sqrt{9x+8}$

(xi) $\ln(x+2)$ (xii) $\dfrac{1}{a-bx}$

2. *Find the nth derivative :*

(i) $\dfrac{x^4}{(x-1)(x-2)}$ (ii) $\dfrac{x+1}{x^2+4}$

(iii) $\dfrac{1}{(x^2+1)(x^2+2)}$ (iv) $\dfrac{1}{x^3+1}$

(v) $\dfrac{4x}{(x-1)^2(x+1)}$ (vi) $x\tan^{-1} x$

(vii) $\tan^{-1}\left(\dfrac{x\sin\alpha}{1-x\cos\alpha}\right)$ (viii) $\dfrac{x^2}{(x-a)(x-b)(x-c)}$

(ix) $\dfrac{x}{1+3x+2x^2}$ (x) $\dfrac{1}{x^4-a^4}$

(xi) $\tan^{-1}\dfrac{1-x}{1+x}$ (xii) $\tan^{-1}\dfrac{\sqrt{1+x^2}-1}{x}$

3. If $y = x(x+1)\log(x+1)^3$, prove that

$$\frac{d^n y}{dx^n} = \frac{3(-1)^{n-1}(n-3)!(2x+n)}{(x+1)^{n-1}}$$

provided that $n \geq 3$.

4. Show that the nth derivative of $y = \tan^{-1} x$ is

$$(-1)^{n-1}(n-1)!\sin\left(\tfrac{\pi}{2}-y\right)\sin^{n}\left(\tfrac{\pi}{2}-y\right)$$

5. Prove that the value of the nth derivative of $y = x^3/(x^2-1)$ for $x = 0$ is zero, if n is even and $-(n!)$ if n is odd and greater than 1.

5.5. LEIBNITZ'S THEOREM

If $y = uv$, where u and v are functions of x having derivatives of nth order, then

$$y_n = u_n v + {}^n c_1 u_{n-1} v_1 + {}^n c_2 u_{n-2} v_2 + \cdots + {}^n c_r u_{n-r} v_r + \cdots + {}^n c_n u v_n$$

Proof: This theorem will be proved by mathematical induction

Step I: We have, $y = uv$

$$y_1 = u_1 v + v_1 u$$

and $$y_2 = u_2 v + u_1 v_1 + v_1 u_1 + v_2 u$$

$$= u_2 v + 2c_1 u_1 v_1 + 2c_2 u v_2$$

Step II: Assume the theorm is true for a particular value m of n, so that, we have

$$y_m = u_m v + {}^m c_1 u_{m-1} v_1 + {}^m c_2 u_{m-2} v_2 + \cdots\cdots + {}^m c_{r-1} u_{m-r+1} v_{r-1}$$
$$+ {}^m c_r u_{m-r} v_r + \ldots. + {}^m c_m u v_m$$

Differentiating bothsides w.r.t x, we get

$$y_{m+1} = u_{m+1} v + u_m v_1 + {}^m c_1 u_m v_1 + {}^m c_1 u_{m-1+1} v_2 + {}^m c_2 u_{m-1} v_2 + \cdots$$
$$+ {}^m c_{r-1} u_{m-r+2} v_{r-1} + {}^m c_{r-1} u_{m-r-1} v_r + \ldots. + mc_m u v_{m+1}$$

$$= u_{m+1} v + \left(1 + {}^m c_1\right) u_m v_1 + \left({}^m c_1 + {}^m c_2\right) u_{m-1} v_2 + \cdots$$
$$+ \left({}^m c_{r-1} + {}^m c_r\right) v_{m-r+1} v_r + \cdots + {}^m c_m u v_{m+1}$$

We know that

$${}^m c_{r-1} + {}^m c_r = {}^{m+1} c_r$$

$$1 + {}^m c_1 = 1 + m = {}^{m+1} c_1$$

and $${}^m c_m = 1 = {}^{m+1} c_{m+1}$$

$$\therefore \ (uv)_{m+1} = u_{m+1} v + {}^{m+1} c_1 u_m v_1 + {}^{m+1} c_2 u_{m-1} v_2 + \cdots$$
$$+ {}^{m+1} c_r u_{m-r+1} v_r + \ldots. + {}^{m+1} c_{m+1} u v_{m+1}$$

Thus, the theorem is true for $n = m$ and $n = m + 1$.

Examples

Example 1: If $y = x^3 e^{ax}$, find y^n.

Solution: Let $u = e^{ax}$ and $v = x^3$

Then,

$$y_n = \left(e^{ax}\right)_n x^3 +^n c_1 \left(e^{ax}\right)_{n-1} \left(x^3\right)_1 +^n c_2 \left(e^{ax}\right)_{n-2} \left(x^3\right)_2 +^n c_3 \left(e^{ax}\right)_{n-3} \left(x^3\right)_3$$

Notice that $V_4, V_5, \ldots$ are all zero.

Thus,

$$y_n = a^n e^{ax} x^3 + na^{n-1} e^{ax} \left(3x^2\right) + \frac{n(n-1)}{2} a^{n-2} e^{ax} (6x)$$

$$+ \frac{n(n-1)(n-2)}{6} a^{n-3} e^{ax} (6)$$

$$= a^{n-3} e^{ax} \left[a^3 x^3 + 3na^2 x^2 + 3n(n-1)ax + n(n-1)(n-2)\right]$$

Example 2: If $y^{1/m} + y^{-1/m} = 2x$, prove that

$$\left(x^2 - 1\right) y_{n+2} + (2n+1) x y_{n+1} + \left(n^2 - m^2\right) y_n = 0$$

Solution: We have,

$$x^{1/m} + y^{-1/m} = 2x$$

$$\Rightarrow y^{2/m} - 2xy^{1/m} + 1 = 0$$

$$\therefore y^{1/m} = \frac{2x \pm \sqrt{4x^2 - 4}}{2}$$

$$= x \pm \sqrt{x^2 - 1}$$

$$\Rightarrow y = \left(x \pm \sqrt{x^2 - 1}\right)^m$$

$$\therefore y_1 = m\left(x \pm \sqrt{x^2 - 1}\right)^{m-1} \left(1 \pm \frac{x}{\sqrt{x^2 - 1}}\right)$$

Squaring both sides, we get

$$\left(x^2-1\right)y_1^2=m^2y^2$$

Again, differentiating both sides w.r.t. x, we get

$$2\left(x^2-1\right)y_1y_2+2xy_1^2=2m^2yy_1$$

$$\Rightarrow\left(x^2-1\right)y_{n+2}+xy_1-m^2y=0$$

Applying Leibnitz's theorem, we get

$$\left(x^2-1\right)y_{n+2}+{}^nc_1y_{n+1}.2x+{}^nc_2y_n\cdot 2+2y_{n+1}+{}^nc_1y_n-m^2y_n=0$$

$$\Rightarrow\left(x^2-1\right)y_{n+2}+(2n+1)xy_{n+1}-n^2y_n=0$$

Example 3: If $y=\cos^{-1}x$, then show that

$$\left(1-x^2\right)y_n+2-(2n+1)xy_{n+1}-n^2y_n=0$$

Solution: Now, $y=\cos^{-1}x$

$$\therefore y_1=\frac{-1}{\sqrt{1-x^2}}$$

Squaring both sides, we get

$$y_1^2=\frac{1}{1-x^2}$$

$$\Rightarrow\left(1-x^2\right)y_1^2=1$$

Differentiating both sides, we get

$$-2xy_1^2+\left(1-x^2\right)2y_1y_2=0$$

$$\Rightarrow\left(1-x^2\right)y_2-xy_1=0$$

Applying Leibnitz's we have

$$\left(1-x^2\right)y_{n+2}+n(-2x)y_{n+1}+\frac{n(n-1)}{2}(-2)y_n-xy_{n+1}-ny_n=0$$

$$\Rightarrow \left(1-x^2\right)y_{n+2} - (2n+1)(-2)xy_{n+1} - n^2 y_n = 0$$

Example 4: If $y = \dfrac{\log_e x}{x}$, prove that

$$y_n = \frac{(-1)^n n!}{x^{n+1}}\left[\log_e x - 1 - \frac{1}{2} - \frac{1}{3} - \ldots\ldots - \frac{1}{n}\right]$$

Solution: We have,

$$y = \frac{\log x}{x} = x^{-1}\log_e x$$

Let, $u = x^{-1} v = \log_e x$

Then $u_1 = (-1)x^{-2}$

$u_2 = (-1)(-2)x^{-3}$

... ...

... ...

$u_n = (-1)(-2)x^{-n-1}$

$= (-1)^n (n!) x^{-n-1}$

$v = \log_e x$

$v_1 = x^{-1}$

$v_2 = (-1)x^{-2}$

... ...

... ...

$v_n = (-1)(-2)\ldots(-n+1)x^{-n}$

$= (-1)^{n-1}(n-1)!\,x^{-n}$

Applying Leibnitz's theorem, we have

$$\frac{d^n}{dx^n}\left(x^{-1}\log_e x\right) = (-1)^n n!\,x^{-n-1}.\log_e x + {}^n c_1 (-1)^{n-1}(n-1)!\,x^{-n}\cdot x^{-1}$$

$$+....+^{n}c_{r}(-1)^{n-r}(n-r)!x^{-n+r-1}$$

$$+....+^{n}c_{n}x^{-1}\cdot(-1)^{n-1}(n-1)!x^{-n}$$

$$=(-1)^{n}(n!)x^{-n-1}\left\{\log_{e}x-1+.....\frac{(r-1)!}{r!}.......\frac{(n-1)!}{n!}\right\}$$

$$=\frac{(-1)^{n}(n!)}{x^{n+1}}\left\{\log_{e}x-1-\frac{1}{2}-\frac{1}{3}......-\frac{1}{n}\right\}$$

Example 5: If $y=\cos\left(m\sin^{-1}x\right)$, show that

$\left(1-x^{2}\right)y_{n+2}-(2n+1)xy_{n+1}+\left(m^{2}-n^{2}\right)y_{n}=0$ and hence find $y_{n}(0)$.

Solution: We have,

$$y=\cos\left(m\sin^{-1}x\right) \quad ...(i)$$

$$y_{1}=-\sin\left(m\sin^{-1}x\right)\times\frac{m}{\sqrt{1-x^{2}}} \quad ...(ii)$$

$$\Rightarrow\left(1-x^{2}\right)y_{1}^{2}=m^{2}\sin^{2}\left(m\sin^{-1}x\right)$$

$$\Rightarrow\left(1-x^{2}\right)y_{1}^{2}=m^{2}\left(1-y^{2}\right)$$

Differentiating both sides, we get

$$\left(1-x^{2}\right)2y_{1}y_{2}-2xy_{1}^{2}=-2m^{2}yy_{1}$$

$$\Rightarrow\left(1-x^{2}\right)y_{2}-xy_{1}+m^{2}y=0 \quad ...(iii)$$

Applying Leibnitz's theorem, we have

$$\left(1-x^{2}\right)y_{n}+2+^{n}c_{1}.(-2_{x}).y_{n+1}+^{n}c_{2}(-2).y_{n}-xy_{n+1}-^{n}c_{1}y_{n}+m^{2}y_{n}=0$$

$$\Rightarrow\left(1-x^{2}\right)y_{n+2}-(2n+1)2y_{n+1}+\left(m^{2}-n^{2}\right)y_{n}=0$$

Putting $x=0$, we get

$$y_{n+2}(0) = \left(n^2 - m^2\right) y_n(0) \qquad \ldots(iv)$$

From (i), (ii) and (iii), we have

$$y(0) = 1, \qquad y(1) = 0, \qquad y_2(0) = m^2$$

Putting $n = 1, 2, 3, 4$, etc. in (iv), we get

$$y_3(0) = \left(1^2 - m^2\right) y_1(0) = 0$$

$$y_4(0) = \left(2^2 - m^2\right) y_2(0) = m^2\left(2^2 - m^2\right)$$

$$y_5(0) = \left(3^2 - m^2\right) y_3(0) = 0$$

$$y_6(0) = \left(4^2 - m^2\right) y_4(0) = m\left(2^2 - m^2\right)\left(4^2 - m^2\right)$$

In general, $y_n(0) = 0$, If n is odd,

$= m^2\left(2^2 - m^2\right)\left(4^2 - m^2\right)\ldots\ldots\left((n-2)^2 - m^2\right)$, if n is even.

Exercises – 5.3

1. *Find the nth derivative:*

   (i) $\left(x^2 e^x \cos x\right)_n$ (ii) $x^4 e^{ax}$

   (iii) $x^3 \cos x$ (iv) $x^3 \sin ax$

   (v) $e^x \ln x$ (vi) $x \tan^{-1} x$

   (vii) $x^n e^{ax} \sin bx$ (viii) $e^{ax}\left[a^2x^2 - 2nax + n(n+1)\right]$

2. If $\cos^{-1}\left(\frac{y}{b}\right) = \log\left(\frac{x}{n}\right)$, prove that

   $$x^2 y_{n+2} + (2n+1)xy_{n+1} - 2n^2 y_n = 0$$

3. If $y = \sin^{-1} x$, then show that

   $$\left(1-x^2\right)y_{n+2} - (2n+1)xy_{n+1} - n^2 y_n = 0$$

4. If $x + y = 1$, show that

   $$\frac{d^n}{dx^n}\left(x^n y^n\right) = n!\left\{y^n - \left({}^n c_1\right)^2 y^{n-1} + n\left({}^n c_2\right)^2 y^{n-2} x^2 + \ldots + (-1)^n x^n\right\}$$

5. If $y = e \tan^{-1} x$, then show that

   $$\left(1+x^2\right)y_{n+2} + \left\{2(n+1)x - 1\right\}y_{n+1} + n(n+1)y_n = 0$$

6. If $y = \frac{\sin^{-1} x}{\sqrt{1-x^2}}$, show that

   $$\left(1-x^2\right)y_{n+2} - (2n+3)xy_{n+1} - (n+1)^2 y_n = 0$$

7. If $y = x^2 \sin x$, prove that

   $$\frac{d^n y}{dx^n} = \left(x^2 - n^2 + n\right)\sin\left(x + \frac{n\pi}{2}\right) - 2nx\cos\left(x + \frac{n\pi}{2}\right)$$

8. If $y = \left(\sin^{-1} x\right)^2$, show that

$$\left(1 - x^2\right) y_{n+2} - (2n+1) x y_{n+1} - n^2 y_n = 0$$

9. If $y = a\cos(\log_e x) + b\sin(\log_e x)$, show that

$$x^2 y_{n+2} + (2n+1) x y_{n+1} + y_n = 0$$

10. If $y = \left(x^2 - 1\right)^n$, prove that

$$\left(x^2 - 1\right) y_{n+2} + 2x y_{n+1} - n(n+1) y_n = 0$$

11. If $x(1-x)y_2 - (4-12x)y_1 - 36y = 0$, then

$$x(1-x)y_{n+2} - \left[4 - n - (12-2n)x\right]y_{n+1} - (4-n)(9-n)y_n = 0$$

12. If $y = \left(1 + x^2\right)^{m/2} \sin\left(m\tan^{-1} x\right)$, show that

$$y_{2n}(0) = y_{2n+1}(0) = (-1)^n m(m-1)(m-2)......(m-2n)$$

13. Find the value of the nth derivative of

$$y = \frac{x^2 - x}{\left(x^2 - 4\right)^2}$$

for $x = 0$.

14. If $y = \ln\left[x + \sqrt{1 + x^2}\right]$, find the value of y_n at $x = 0$.

15. If $y = \left(\sinh^{-1} x\right)^2$, find the value of y_n at $x = 0$.

6

Tangents and Normals

6.1. EQUATION OF A TANGENT

If ψ is the angle which the tangent at any point (x, y) on the curve $y = f(x)$ makes with the x-axis, then

$$\tan\psi = \frac{dy}{dx} = f'(x)$$

Thus, the equation of the tangent at the point (x, y) on the curve $y = f(x)$ is

$$Y - y = f'(x)(X - x)$$

where (x, y) is an arbitrary point on the tangent.

6.1.1. Tangent to the curve x = f(t), y = g(t), at the Point 't'

The slope of the tangent of the curve $x = f(t)$, $y = g(t)$, at which $f'(t) \neq 0$, at any point 't' is

$$\frac{dy}{dx} = \frac{dy}{dt} \times \frac{dt}{dx} = \frac{g'(t)}{f'(t)}$$

Thus, the equation of the tangent to the curve $x = f(t)$, $y = g(t)$ at the point 't' is

$$[Y - g(t)]f'(t) = [X - f(t)]g'(t)$$

6.1.2. Tangent to the Curve f(x, y) = 0 at any Point (x, y)

If the equation of the curve is given in the implicit form $f(x, y) = 0$, then at any point (x, y), where $\frac{df}{dy} \neq 0$, we have $f'(x) = -\frac{\partial f}{\partial x} \Big/ \frac{\partial f}{\partial y}$

Thus, the equation of the tangent at any point (x, y) on the curve $f(x, y) = 0$, where $\frac{df}{dy} \neq 0$, will be given by

$$Y - y = -\left(\frac{\partial f}{\partial x} \Big/ \frac{\partial f}{\partial y}\right)(X - x)$$

or,
$$(X - x)\frac{\partial f}{\partial x} + (Y - y)\frac{\partial f}{\partial y} = 0$$

6.2. EQUATION OF THE NORMAL AT A POINT

The normal to a curve at a point is the straight line which passes through the point and is perpendicular to the tangent to the curve at the point.

Let m be the slope of the normal since the normal and the tangent are perpendicular to each other, then at any point (x, y)

$$m \cdot \frac{dy}{dx} = -1$$

$$\Rightarrow \quad m = -1 \Big/ \frac{dy}{dx}$$

Hence, the equation of the normal at any point (x, y) is

$$Y - y = \left(\frac{-1}{dy/dx}\right)(X - x)$$

or,
$$\frac{dy}{dx}(Y - y) + (X - x) = 0$$

where (x, y) is an arbitary point on the slope.

6.2.1. Slope to the Curve x = f(t), y = g(t), at the Point 't'

The equation of the normal at any point 't' on the curve, $x = f(t)$, $y = g(t)$ will be given by

$$[X - f(t)]f'(t) + [Y - g(t)]g'(t) = 0$$

6.2.2. Slope to the Curve f(x, y) = 0 at any Point (x, y)

If the equation of the curve is given in the implicit form $f(x, y) = 0$, then the equation of the normal at any point (x, y) will be given by

$$(X - x)\frac{\partial f}{\partial y} - (Y - y)\frac{\partial f}{\partial x} = 0$$

Examples

Example 1: Find the equation of the tangent and normal to the curve at any point (x, y) if

$$\frac{x^m}{a^m} + \frac{y^m}{b^m} = 1$$

Solution: The given equation is

$$\frac{x^m}{a^m} + \frac{y^m}{b^m} = 1 \qquad \text{...(i)}$$

Differentiating w.r.t. x, we get

$$m\frac{x^{m-1}}{a^m} + m\frac{y^{m-1}}{b^m}.\frac{dy}{dx} = 0$$

$$\therefore \quad \frac{dy}{dx} = -\frac{1}{a}\left(\frac{x}{a}\right)^{m-1}.b\left(\frac{b}{y}\right)^{m-1}$$

Equation of the tangent is

$$Y - y = \frac{dy}{dx}(X - x)$$

$$\Rightarrow \quad Y - y = -\frac{1}{a}\left(\frac{x}{a}\right)^{m-1}.b\left(\frac{b}{y}\right)^{m-1}(X - x)$$

$$\Rightarrow \quad \frac{X}{a}\left(\frac{x}{a}\right)^{m-1} - \left(\frac{x}{a}\right)^m = -\frac{Y}{b}\left(\frac{y}{b}\right)^{m-1} + \left(\frac{y}{b}\right)^m$$

$$\Rightarrow \quad \frac{X}{a}\left(\frac{x}{a}\right)^{m-1} + \frac{Y}{b}\left(\frac{y}{b}\right)^{m-1} = \left(\frac{x}{a}\right)^{m} + \left(\frac{y}{b}\right)^{m}$$

$$\Rightarrow \quad \frac{X}{a}\left(\frac{x}{a}\right)^{m-1} + \frac{Y}{b}\left(\frac{y}{b}\right)^{m-1} = 1 \qquad \text{[From (i)]}$$

Example 2: Find the equation of the tangent and normal to the curve at any point (x, y), if

$$x = a\sin^3\theta, \quad y = a\cos^3\theta$$

Solution: We have,

$$x = a\sin^3\theta, \quad y = a\cos^3\theta$$

Thus,

$$\frac{dx}{d\theta} = 3a\sin^2\theta\cos\theta$$

$$\frac{dy}{d\theta} = -3a\cos^2\theta\sin\theta$$

$$\therefore \quad \frac{dy}{dx} = \frac{dy}{d\theta} \div \frac{dx}{d\theta} = -\frac{\cos\theta}{\sin\theta}$$

Thus, the equation of the tangent is

$$y - a\cos^3\theta = -\frac{\cos\theta}{\sin\theta}\left(x - a\sin^3\theta\right)$$

$$\Rightarrow \quad x\cos\theta + y\sin\theta = a\sin\theta.\cos\theta\left(\cos^2\theta + \sin^2\theta\right)$$

$$\Rightarrow \quad x\cos\theta + y\sin\theta = \left(a/2\right)\sin 2\theta$$

Slope of the tangent is $-\cot\theta$ and hence slope of the normal will be

$$\tan\theta = \frac{\sin\theta}{\cos\theta}$$

Therefore, the equation of the normal will be

$$y - a\cos^3\theta = \frac{\sin\theta}{\cos\theta}\left(x - a\sin^3\theta\right)$$

$$\Rightarrow \qquad y\cos\theta - x\sin\theta = a\left(\cos^4\theta - \sin^4\theta\right)$$

$$\Rightarrow \qquad y\cos\theta - x\sin\theta = a\cos 2\theta$$

Example 3: Find the equation of the normal to the curve

$$y = (1+x)^y + \sin^{-1}\left(\sin^2 x\right) \text{ at } x = 0.$$

Solution: We have,

$$y = (1+x)^y + \sin^{-1}\left(\sin^2 x\right)$$

Let $u = (1+x)^y$ and $v = \sin^{-1}\left(\sin^2 x\right)$

When $x = 0, y = 1$

$\therefore$ Point P is (0, 1)

Also, $y = u + v$

$$\therefore \quad \frac{dy}{dx} = \frac{du}{dx} + \frac{dv}{dx}$$

Clearly, $\dfrac{dv}{dx} = 0$ at P

$$u = (1+x)^y$$

$$\Rightarrow \qquad \log u = y\log(1+x)$$

$$\therefore \qquad \frac{1}{u}\frac{du}{dx} = \frac{dy}{dx}\log(1+x) + \frac{y}{1+x}$$

$$\Rightarrow \qquad \frac{du}{dx} = u\left[\frac{dy}{dx}\log(1+x) + \frac{y}{1+x}\right]$$

Also $u = 1$ at $P(0, 1)$

$$\log(1+x) = \log 1 = 0$$

and

$$\frac{y}{1+x} = 1$$

$$\therefore \qquad \frac{du}{dx} = 1$$

Hence, $\frac{dy}{dx} = \frac{du}{dx} + \frac{dv}{dx}$

$= 1 + 0 = 1$

$=$ slope of tangent

$\therefore$ Slope of the normal will be -1.

Hence, the equation of the normal at $P(0, 1)$ is

$$y - 1 = -1(x - 0)$$

$$\Rightarrow \quad x + y = 1$$

Example 4: Tangents are drawn from origin to the curve $y = \sin x$. Prove that their points of contact lie on $x^2y^2 = x^2 - y^2$.

Solution: Let the tangent be drawn at the point (x, y)

The equation of the tangent at (x, y) is

$$Y - y = \frac{dy}{dx}(X - x)$$

But $y = \sin x$

$$\Rightarrow \quad \frac{dy}{dx} = \cos x$$

$$\therefore \quad Y - y = \cos x(X - x)$$

Since it passes through $(0, 0)$

$$\therefore \quad -y = -x\cos x$$

$$\Rightarrow \quad \frac{y}{x} = \cos x \text{ and } y = \sin x$$

$$\therefore \quad \frac{y^2}{x^2} + y^2 = \cos^2 x + \sin^2 x = 1$$

$$\Rightarrow \quad y^2 + x^2y^2 = x^2$$

$$\Rightarrow \quad x^2 - y^2 = x^2y^2$$

Hence the points of contact lie on

$$x^2 - y^2 = x^2y^2$$

Example 5: Show that the line $\frac{x}{a}+\frac{y}{b}=1$ touches the curve, $y=be^{-x/a}$ at the point where the curve crosses y-axis.

Solution: We have

$$y=be^{-x/a}$$

It cuts y-axis at $(0, b)$

$$\therefore \quad \frac{dy}{dx}=-\frac{b}{a}e^{-x/a}$$

$$\Rightarrow \quad \left(\frac{dy}{dx}\right)_{(0,\,b)}=-\frac{b}{a}.1=-\frac{b}{a}$$

The equation of the tangent at $(0, b)$ is

$$y-b=-(b/a).(x-0)$$

$$\Rightarrow \quad bx+ay=ab$$

$$\Rightarrow \quad \frac{x}{a}+\frac{y}{b}=1$$

Example 6: The curve, $y=ax^3+bx^2+cx+5$ touches the x-axis at $P(-2, 0)$ and cuts the y-axis at a point Q where its gradient is 3. Find a, b, c.

Solution: The curve cuts y-axis at $(0, 5)$ at which

$$\frac{dy}{dx}=3$$

$$\Rightarrow \quad 3ax^2+2bx+c=3 \text{ at } (0, 5)$$

$$\therefore \quad c=3 \qquad \ldots(1)$$

Again, $\frac{dy}{dx}=0$ at $P(-2, 0)$ as it touches x-axis.

$$\therefore \quad 12a-4b+c=0$$

$$\Rightarrow \quad 12a-4b+3=0 \qquad \ldots(2)$$

Also, P lies on the curve, $-8a+4b-2c+5=0$

$$\Rightarrow \quad -8a+4b-1=0 \qquad \ldots(3)$$

Solving (2) and (3), we get

$$a = -\tfrac{1}{2}, \quad b = -\tfrac{3}{4}$$

Example 7: Find the equation(s) of tangent(s) to the curve $y = x^3 + 2x + 6$ which is perpendicular to the line $x + 14y + 4 = 0$,

Solution: Let the co-ordinates of the point of contact be (x_1, y_1).

Then $y_1 = x_1^3 + 2x_1 + 6$ ...(i)

$\left[\because (x_1, y_1) \text{ lies on } y = x^3 + 2x + 6\right]$

The equation of the curve is

$$y = x^3 + 2x + 6 \quad \text{...(ii)}$$

Differentiating bothsides w.r.t. *x*, we get

$$\frac{dy}{dx} = 3x^2 + 2$$

$$\Rightarrow \quad \left(\frac{dy}{dx}\right)_{(x_1, y_1)} = 3x_1^2 + 2$$

Since the tangent at (x_1, y_1) is perpendicular to the line $x + 14y + 4 = 0$, therefore,

Slope of the tangent at (x_1, y_1) × Slope of the line = −1

$$\Rightarrow \quad \frac{dy}{dx} \times \frac{-1}{14} = -1$$

$$\Rightarrow \quad \left(3x_1^2 + 2\right)\left(\frac{-1}{14}\right) = -1$$

$$\Rightarrow \quad 3x_1^2 + 2 = 14 \quad \Rightarrow \quad x_1 = \pm 2$$

When $x_1 = 2, \quad y_1 = 2^3 + 2 \times 2 + 6 = 18$ [Using (i)]

When $x_2 = -2, \; y_1 = (-2)^3 + 2(-2) + 6 = -6$ [Using (ii)]

So, the co-ordinates of the points of contact are (2, 18) and (–2, –6).

The equation of the tangent at (2, 18) is

$$y - 18 = 14(x - 2)$$

$$\Rightarrow \quad 14x - y - 10 = 0$$

The equation of the tangent at (–2, –6) is

$$y - (-6) = 14(x - (-2))$$

$$\Rightarrow \quad 14x - y + 22 = 0$$

Example 8: For the curve $y = 4x^3 - 2x^5$ find all points at which the tangent passes through the origin.

Solution: Let (x_1, y_1) be the required point on $y = 4x^3 - 2x^5$

Then, $\quad y_1 = 4x_1^3 - 2x_1^5$ ...(i)

The equation of the given curve is

$$y = 4x^3 - 2x^5$$

Differentiating w.r.t. x, we get

$$\frac{dy}{dx} = 12x^2 - 10x^4$$

$$\Rightarrow \quad \left(\frac{dy}{dx}\right)_{(x_1, y_1)} = 12x_1^2 - 10x_1^4$$

So, the equation of the tangent at (x_1, y_1) is

$$y - y_1 = \left(\frac{dy}{dx}\right)_{(x_1, y_1)} (x - x_1)$$

$$\Rightarrow \quad y - y_1 = \left(12x_1^2 - 10x_1^4\right)(x - x_1)$$

This passes through the origin, therefore

$$0 - y_1 = \left(12x_1^2 - 10x_1^4\right)(0 - x_1)$$

$$\Rightarrow \quad y_1 = 12x_1^3 - 10x_1^5 \qquad \text{...(ii)}$$

Subtracting (ii) from (i), we get

$$0 = -8x_1^3 + 8x_1^5$$

$$\Rightarrow \qquad 8x_1^3\left(x_1^2 - 1\right) = 0$$

$$\Rightarrow \qquad x_1 - 0 \text{ or } x_1 = \pm 1$$

When $x_1 = 0$, $y_1 = 0$.

When $x_1 = 1$, $y_1 = 12 - 10 = 2$.

When $x_1 = -1$, $y_1 = -12 + 10 = -2$.

Hence, the required points are (0, 0), (1, 2) and (–1, –2).

Exercise – 6.1

1. *Find the equation of the tangents and normal to the curves at given points:*

   (i) $x^2 + y^2 = r^2$ at (x', y')

   (ii) $y = 2x^3 - x^2 + 3$ at $(1, 4)$

   (iii) $x = a(\theta + \sin\theta),\ y = a(1 - \cos\theta)$ at θ

   (iv) $x^2/a^2 - y^2/b^2 = 1$ at $\left(a\sqrt{2}, b\right)$

   (v) $y^2 = 4ax$ at (x', y')

   (vi) $xy = c^2$ at $\left(cp, c/p\right)$

   (vii) $x^2(x - y) + a^2(x + y) = 0$ at $(0, 0)$

   (viii) $x = \dfrac{2at^2}{1+t^2},\ y = \dfrac{2at^3}{1+t^2}$ for $t = \dfrac{1}{2}$

   (ix) $c^2\left(x^2 + y^2\right) = x^2y^2$ at $(c/\cos\theta,\ c/\sin\theta)$

   (x) $\left(x^2 + y^2\right)x - a\left(a^2 - y^2\right) = 0$ for $x = -3a/5$

2. Prove that the straight line $x/a + y/b = 2$ touches the curve

   $$\left(x/a\right)^n + \left(y/b\right)^n = 2$$

   at the point (a, b), whatever be the value of n.

3. Prove that the portion of the tangent to the curve

   $$\frac{x + \sqrt{a^2 - y^2}}{a} = \log\frac{a + \sqrt{a^2 - y^2}}{y}$$

   intercepted between the point of contact and the x-axis is constant.

4. Find the condition for the line $x\cos\theta + y\sin\theta = p$ to touch the curve $x^m/a^m + y^m/b^m = 1$.

5. Prove that the equation of the normal to the astroid $x^{2/3} + y^{2/3} = a^{2/3}$ may be written in the form $x \sin\phi - y\cos\phi + a\cos 2\phi = 0$.

6. Prove that the equation of the tangent at any point $(4m^2, 8m^3)$ of the semi-cubical parabola $x^3 - y^2 = 0$ is $y = 3mx - 4m^3$ and show that it meets the curve again at $(m^2, -m^3)$, where it is normal if $9m^2 = 2$.

7. Show that the tangent at any point (*x*, *y*) on the curve $y^m = ax^{m-1} + x^m$, makes intercepts $\dfrac{ax}{(m-1)a + mx}$ and $\dfrac{ay}{m(a+x)}$ on the co-ordinate axes.

8. Prove that the sum of the intercepts on the co-ordinate axes of any tangent to $\sqrt{x} + \sqrt{y} = \sqrt{a}$ is constant.

9. Show that the distance from origin of the normal at any point of the curve

$$x = ae^{\theta}\left(\sin \tfrac{\theta}{2} + 2\cos \tfrac{\theta}{2}\right),\ y = ae^{\theta}\left(\cos \tfrac{\theta}{2} - 2\sin \tfrac{\theta}{2}\right)$$

is twice the distance of the tangent at the point from the origin.

10. Show that the line $x\cos\theta + y\sin\theta = p$ will touch the curve

$x^m y^n - a^{m+n} = 0$ if $p^{m+n} m^n n^m = (m+n)^{m+n} a^{m+n} \cos^m\theta \sin^n\theta$.

11. Show that $x\cos\alpha + y\sin\alpha = p$ the line touches the curve

$x^m y^n + a^{m+n}$, provided $p^{m+n} m^n n^m = (m+n)^{m+n} a^{m+n} \cos^m \alpha \sin^n \alpha$.

12. Prove that all points of the curve $y^2 = 4a\{x + a\sin(x/a)\}$ at which the tangent is parallel to the axis of *x* lie on a parabola.

13. If the line $x\cos\alpha + y\sin\alpha = p$ touches the curve

$\left(\frac{x}{a}\right)^{n/(n-1)} + \left(\frac{y}{b}\right)^{n/(n-1)} = 1$, prove that $(a\cos\alpha)^n + (b\sin\alpha)^n = p^n$

14. The tangent at any point on the curve $x^3 + y^3 = 2a^3$ cuts off lengths *p* and *q* on the co -ordinate axes; show that $p^{-3/2} + q^{-3/2} = 2^{-1/2} a^{-3/2}$

15. Determine the equation(s) of tangent(s) line to the curve $y = 4x^3 - 3x + 5$ which are perpendicular to the line $9y + x + 3 = 0$.

16. Find the equation of a normal to the curve $y = x \log_e x$ which is parallel to $2x - 2y + 3 = 0$.

17. Show that the normal at the point $(3t, 4/t)$ of the curve $xy = 12$ cuts the curve again at the point whose parameter t_1 is given by $t_1 = \dfrac{-16}{9t^2}$.

18. Find the points on the curve $x^2 + y^2 - 2x - 3 = 0$ the tangents at which are parallel to x-axis.

19. Show that the normal at any point of the curve $x = a\cos\theta + a\theta\sin\theta$, $y = a\sin\theta + \theta\cos\theta$ is at a constant distance from the origin.

6.3. ANGLE OF INTERSECTION OF TWO CURVES

Definition: *The angle of intersection of two curves at a point of intersection is the angle between the tangents to the curves at that point.*

Consider two curves,

$$y = f(x) \text{ and } y = g(x)$$

Let m_1, m_2 be the slopes of the tangents at P to $y = f(x)$ and $y = g(x)$ respectively.

If α, β be the angle that the tangents at P to $y = f(x)$ and $y = g(x)$ respectively make with the x-axis, then

$$m_1 = \tan\alpha = f'(x_1)$$

$$m_2 = \tan\beta = g'(x_1)$$

The angle θ, between the tangents is given by

$$\tan\theta = \frac{m_1 - m_2}{1 + m_1 m_2}$$

$$= \frac{f'(x) - g'(x)}{1 + f'(x_1)g'(x_1)}$$

Note: 1. *The two curves cut each other orthogonally at* (x_1, y_1) *if*

$$f'(x_1)g'(x_1) = -1$$

2. *The two curves touch each other at* (x_1, y_1) *if*

$$f'(x_1) = g'(x_1)$$

6.4. LENGTH OF THE TANGENT, SUB-TANGENT, NORMAL AND SUB-NORMAL AT ANY POINT OF A CURVE

Let $P(x, y)$ be any point on the curve $y = f(x)$. Let the tangent and the normal to the curve at P meet the x-axis at T and G respectively.

The length PT is called the **length of the tangent** at P and the length PG is called the **length of the normal** at P. TM, the projection of the tangent on the x-axis, is called **length of the sub-tangent**, and MG, the projection of the normal on the x-axis is called the **length of the sub-normal**.

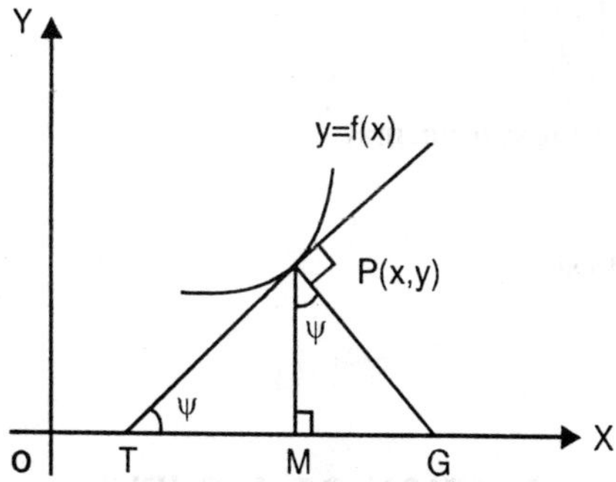

Figure 6.1

$$\tan\psi = \frac{dy}{dx}, \text{ where } \psi = \angle MPG$$

From the figure 6.1,

(i) Length of the tangent

$$TP = MP\,\text{cosec}\,\psi$$

$$= y\sqrt{\left(1+\cot^2\psi\right)}$$

$$= y\sqrt{\left[1+\left(\frac{dx}{dy}\right)^2\right]}$$

(ii) Length of the sub-tangent

$$= TM$$

$$= MP\cot\psi$$

$$= y\frac{dx}{dt}$$

(iii) Length of the normal

$$= GP$$

$$= MP\sec\psi$$

$$= y\sqrt{\left(1+\tan^2\psi\right)}$$

$$= y\sqrt{1+\left(\frac{dy}{dx}\right)^2}$$

(iv) Length of the sub-normal

$$= MG$$

$$= MP \tan\psi$$

$$= y\frac{dy}{dx}$$

6.5. PEDAL EQUATION OF A CURVE

Definition: *A relation between the distance, r, of any point on the curve from the origin and the length of the perpendicular P from the origin to the tangent at the point is called Pedal equation of the curve.*

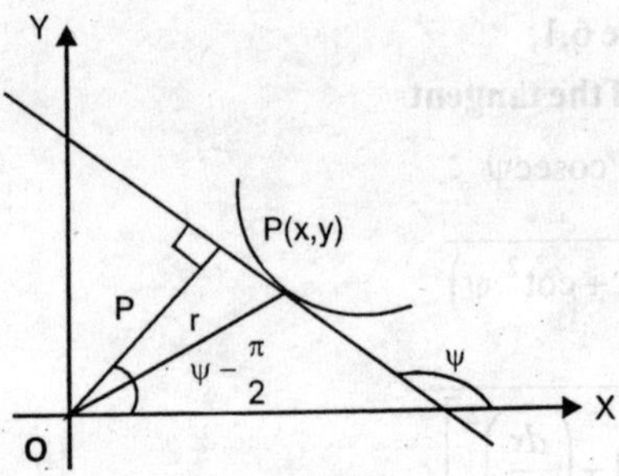

Figure 6.2

Let the equation of the curve be

$y = f(x)$ ...(i)

Equation of the tangent at $P(x, y)$ is

$$Y - y = f'(x)(X - x)$$

$$\Rightarrow \quad Xf'(x) - Y + y - xf'(x) = 0 \qquad \text{...(ii)}$$

If P_0, be the length of the perpendicular from (0, 0) to this tangent, we have

$$P = \frac{y - xf'(x)}{\sqrt{\left(1 + f'^2(x)\right)}}$$

Also, $r^2 = x^2 + y^2$ ...(iii)

Eliminating x and y between (i), (ii) and (iii), we obtain the required Pedal equation of the given curve.

Examples

Example 1: Find the angle of intersection of the curves $y^2 = 2x$ and $x^2 = 16y$.

Solution: The two curves meet at (0, 0) and (8, 4) tangents are x-axis and y-axis at (0, 0).

So, the angle of intersection will be $\frac{\pi}{2}$.

Example 2: Find the condition that the following conics may cut orthogonally: $ax^2 + by^2 = 1$ and $a'x^2 + b'y^2 = 1$.

Solution: We have,

$$ax^2 + by^2 = 1$$

$$\Rightarrow \quad \frac{dy}{dx} = \frac{-ax}{by} = m_1 \text{ (say)}$$

Also $a'x^2 + b'y^2 = 1$

$$\Rightarrow \quad \frac{dy}{dx} = \frac{-a'x}{b'y} = m_2 \quad \text{(say)}$$

If the curves cut orthogonally, then $m_1 m_2 = -1$

$$\Rightarrow \quad \left(-\frac{ax}{by}\right)\left(-\frac{a'x}{b'y}\right) = -1$$

$$\Rightarrow \quad aa'x^2 + bb'y^2 = 0 \quad \text{...(i)}$$

We have now to put the values of x^2 and y^2 corresponding to the points of intersection of

$$ax^2 + by^2 - 1 = 0$$

$$a'x^2 + b'y^2 - 1 = 0$$

$$\therefore \frac{x^2}{b'-b} = \frac{y^2}{a-a'} = \frac{1}{ab'-a'b}$$

Putting the values of x^2 and y^2 in (i)

$$aa'\frac{(b'-b)}{ab'-a'b} + bb'.\frac{(a-a')}{ab'-a'b} = 0$$

$$\Rightarrow \frac{b'-b}{bb'} + \frac{a-a'}{aa'} = 0$$

$$\Rightarrow \left(\frac{1}{b} - \frac{1}{b'}\right) + \left(\frac{1}{a'} - \frac{1}{a}\right) = 0$$

$$\Rightarrow \frac{1}{a} - \frac{1}{b} = \frac{1}{a'} - \frac{1}{b'} \text{ is the required condition.}$$

Example 3: Show that the angle between the tangents at any point *P* and the line joining *P* to the origin 0 is the same at all points of the curve.

$$\log\left(x^2 + y^2\right) = k \tan^{-1}\left({}^{y}/_{x}\right)$$

Solution: We have,

$$\log\left(x^2 + y^2\right) = k \tan^{-1}\left({}^{y}/_{x}\right)$$

$$\therefore \frac{1}{x^2+y^2}\left(2x + 2y\frac{dy}{dx}\right) = k.\frac{1}{1+{}^{y^2}/_{x^2}}.\frac{x\,{}^{dy}/_{dx} - y.1}{x^2}$$

$$\Rightarrow 2\left(x + y\frac{dy}{dx}\right) = k\left(x\frac{dy}{dx} - y\right) \qquad \ldots\text{(i)}$$

If the tangent makes an angle ψ with x-axis, then

$$\tan\psi = \frac{dy}{dx}$$

Let *OP* where *P* is (x, y) makes an angle θ with x-axis, then,

$$\tan\theta = {}^{y}/_{x}$$

If α be the angle between the tangent and *OP*, then

$$\alpha = \psi - \theta$$

$$\therefore \quad \tan\alpha = \tan(\psi - \theta) = \frac{\tan\psi - \tan\theta}{1 + \tan\psi . \tan\theta}$$

$$\Rightarrow \quad \tan\alpha = \frac{\frac{dy}{dx} - \frac{y}{x}}{1 + \left(\frac{y}{x}\right)\left(\frac{dy}{dx}\right)}$$

$$= \frac{x\left(\frac{dy}{dx}\right) - y}{x + y\left(\frac{dy}{dx}\right)}$$

$$= \frac{2}{k} \quad \text{by (i)}$$

$$\therefore \quad \alpha = \tan^{-1}\left(\tfrac{2}{k}\right) \text{ i.e. constant.}$$

Example 4: Find the condition that the line $x\cos\alpha + y\sin\alpha = p$ may touch the curve $\frac{x^m}{a^m} + \frac{y^m}{b^m} = 1$.

Solution: The tangent to the given curve at any point (x, y) is

$$\frac{X}{a}\left(\frac{x}{a}\right)^{m-1} + \frac{Y}{b}\left(\frac{y}{b}\right)^{m-1} = 1 \qquad \ldots(1)$$

Compare with $X\cos\alpha + Y\sin\alpha = p$ ...(2)

$$\frac{\left(\frac{x}{a}\right)^{m-1}}{a\cos\alpha} = \frac{\left(\frac{y}{b}\right)^{m-1}}{b\sin\alpha} = \frac{1}{p}$$

$$\therefore \quad \frac{x}{a} = \left(\frac{a\cos\alpha}{p}\right)^{1/(m-1)}$$

$$\frac{y}{b} = \left(\frac{b\sin\alpha}{p}\right)^{1/(m-1)} \qquad (3)$$

But the point (x, y) lies on

$$\left(\frac{x}{a}\right)^m + \left(\frac{y}{b}\right)^m = 1$$

Putting for $\frac{x}{a}$ and $\frac{y}{b}$ from (3) in (4), we get

$$\left(\frac{a\cos\alpha}{p}\right)^{m/(m-1)}+\left(\frac{b\sin\alpha}{p}\right)^{m/(m-1)}=1$$

$$\Rightarrow \quad (a\cos\alpha)^{m/(m-1)}+(b\sin\alpha)^{m/(m-1)}=p^{m/(m-1)}$$

Example 5: Find the lengths of the tangent, normal, subtangent and sub-normal at the point (2, 4) of the parabola $y^2=8x$.

Solution: For the parabola,

$$y^2=8x$$

$$\Rightarrow \quad 2y\frac{dy}{dx}=8$$

$$\Rightarrow \quad \frac{dy}{dx}=\frac{8}{2y}$$

$$\Rightarrow \quad \left(\frac{dy}{dx}\right)_{y=4}=1$$

Therefore, the length of the tangent at (2, 4)

$$=y\sqrt{\left\{1+\left(\frac{dy}{dx}\right)^2\right\}}$$

$$=4\sqrt{1+1}$$

$$=4\sqrt{2}$$

The length of the normal at (2, 4)

$$=y\sqrt{1+\left(\frac{dy}{dx}\right)^2}$$

$$=4\sqrt{(1+1)}$$

$$=4\sqrt{2}$$

The length of the sub-normal at (2, 4)

$$= y\frac{dy}{dx} = 4$$

Example 6: Find the pedal equation of the parabola

$$y^2 = 4a(x+a)$$

Solution: The equation of the tangent at any point (x, y) of the given parabola is

$$Y - y = \frac{2a}{y}(X - x)$$

$$\Rightarrow \quad 2aX - yY + y^2 - 2ax = 0 \qquad \text{...(i)}$$

The length, p, of the perpendicular from the origin to this tangent is given by

$$p = \frac{y^2 - ax}{\sqrt{(4a^2 + y^2)}}$$

$$= \frac{4a(x+a) - 2ax}{\sqrt{[4a^2 + 4a(x+a)]}}$$

$$= \frac{2a(x+2a)}{\sqrt{[4a(x+2a)]}}$$

$$= \sqrt{a(x+2a)} \qquad \text{...(ii)}$$

Also, $x^2 + y^2 = r^2$

Thus, $r^2 = x^2 + 4a(x+a) = (x+2a)^2$

From (i) and (ii), we get

$$p^2 = ar$$

which is the required pedal equation of the given parabola.

Exercise – 6.2

1. *Find the angle of intersection of the following curves.*

   (i) $x^2 + y^2 = a^2\sqrt{2}$ and $x^2 - y^2 = a^2$

   (ii) $x^2 y = 1$ and $y = x^2$

   (iii) $x^2 = 4ay$ and $2y^2 = ax$

   (iv) $y^2 = ax$ and $x^3 + y^3 = 3axy$

   (v) $y = \sin x$ and $y = \cos x$.

2. *Find the condition that the following conics may cut orthogonally.*

   (i) $\dfrac{x^2}{a^2} + \dfrac{y^2}{b^2} = 1$ and $\dfrac{x^2}{a'} + \dfrac{y^2}{b'} = 1$

   (ii) $\dfrac{x^2}{a^2 + k_1} + \dfrac{y^2}{b^2 + k_1} = 1$ and $\dfrac{x^2}{a^2 + k_2} + \dfrac{y^2}{b^2 + k_2} = 1$

   (iii) $x^3 - 3xy^2 = -2$ and $3x^2 y - y^3 = 2$

3. Prove that the curves

   $x^3 + 2xy^2 - 10a^2 x + 12a^2 y + 3a^3 = 0$

   and

   $y^3 + 2xy^2 - 5a^2 x - a^3 = 0$

   intersect at an angle $\tan^{-1}(88/73)$ at the point (39, –29).

4. If x_1, y_1 be the parts of the axes intercepted by the tangent at any point (x, y) on the curve $\left(x/a\right)^{2/3} + \left(y/b\right)^{2/3} = 1$, show that $x_1^2/a^2 + y_1^2/b^2 = 1$.

5. Find the lengths of the tangent, normal, sub-tangent and sub-normal at the point 'θ'.

   (i) $x = a\cos^3\theta$ and $y = a\sin^3\theta$

   (ii) $x = a(\theta + \sin\theta)$ and $y = a(1 - \cos\theta)$

6. In the curve $x^{m+n} = a^{m-n} y^{2n}$, prove that the mth power of the

subtangent varies as the nth power of the sub-normal.

7. Show that in the curve $y = a\log\left(x^2 - a^2\right)$ the sum of the tangent and the sub-tangent varies as the product of the co-ordinates of the point.

8. For the catenary, $y = c\cos h(x/c)$, prove that the length of the normal is y^2/c.

9. Show that for the curve $x = a + b\log\left[b + \sqrt{\left(b^2 - y^2\right)}\right] - \left[\sqrt{\left(b^2 - y^2\right)}\right]$ sum of the sub-normal and sub-tangent is constant.

10. Show that for the curve $y = hn^{-\frac{1}{2}}e^{-h^2n^2}$ the product of the abscissa and the sub-tangent is constant.

11. Show that the pedal equation of the curve

$c^2\left(x^2 + y^2\right) = x^2y^2$ is $1/p^2 + 3/r^2 = 1/c^2$.

12. Show that the pedal equation of ellipse $x^2/a^2 + y^2/b^2 = 1$

is $1/p^2 = 1/a^2 + 1/b^2 - r^2/\left(a^2b^2\right)$

13. Show that the pedal equation of the curve

$x = ae^{\theta}(\sin\theta - \cos\theta),\ y = ae^{\theta}(\sin\theta + \cos\theta)$

is $r = \sqrt{2}p$

14. Show that the pedal equation of the curve

$x = a\left(3\cos\theta - \cos^3\theta\right),\ y = a\left(3\sin\theta - \sin^3\theta\right)$

is $3p^2\left(7a^2 - r^2\right) = \left(10a^2 - r^2\right)^2$.

6.6. ANGLE BETWEEN RADIUS VECTOR AND TANGENT

Let $p(r, \theta)$ be a given point on the curve, $r = f(\theta)$.

We transform to cartesian co-ordinates and obtain

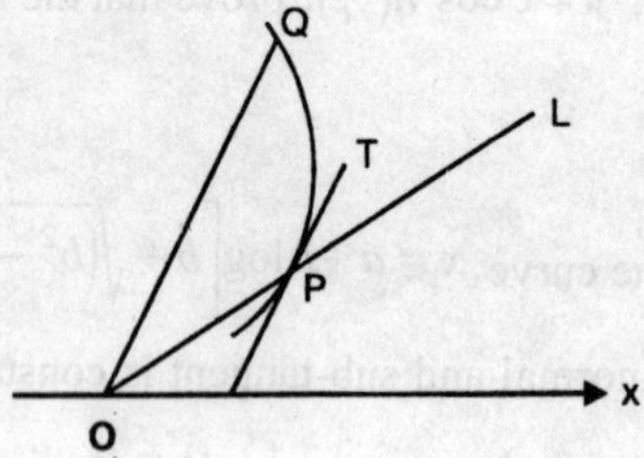

$$x = r\cos\theta = f(\theta)\cos\theta, \quad y = r\sin\theta = f(\theta)\sin\theta$$

$$\tan\psi = \frac{dy}{dx} = \frac{dy}{d\theta} \div \frac{dx}{d\theta}$$

$$= \frac{f'(\theta)\sin\theta + f(\theta)\cos\theta}{f'(\theta)\cos\theta - f(\theta)\sin\theta}$$

Dividing the numerator and denominator by $f'(\theta)\cos\theta$, we obtain

$$\tan\psi = \frac{\tan\theta + f(\theta)/f'(\theta)}{1 - [f(\theta)/f'(\theta)]\tan\theta} \qquad \ldots\text{(i)}$$

Denoting the angle between the radius vector and the tangent by ϕ_1, we have

$$\psi = \theta + \phi$$

Comparing (i) and (ii), we see

$$\tan\phi = \frac{f(\theta)}{f'(\theta)} = r\frac{d\theta}{dr}$$

Corollary: *Angle of Intersection of two curves*

If ϕ_1, ϕ_2 be the angle between the common radius vector and the tangents to the two given curves at a point of intersection, then the angle of intersection of the curve is

$$[\phi_1 - \phi_2]$$

6.7. LENGTH OF THE PERPENDICULAR FROM POLE TO THE TANGENT

Let $r = f(\theta)$ be a given curve and let $P(r, \theta)$ be any point on the curve.

Let $OM = p$

From the triangle OMP, we get

$$p = r\sin\phi$$

To obtain the result in terms of r and θ,

$$\frac{1}{p^2} = \frac{1}{r^2\sin^2\phi} = \frac{1}{r^2}\text{cosec}^2\phi = \frac{1}{r^2}\left[1+\frac{1}{r^2}\left(\frac{dr}{d\theta}\right)^2\right]$$

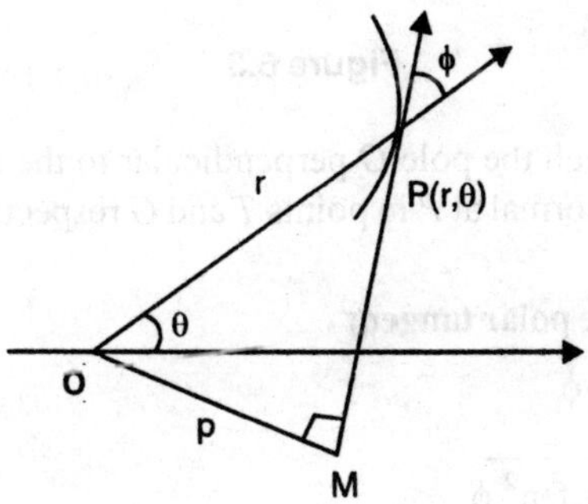

Hence, $\frac{1}{p^2} = \frac{1}{r^2} + \frac{1}{r^4}\left(\frac{dr}{d\theta}\right)^2$ ...(i)

Putting $r = \frac{1}{u}$, we get

$$\frac{dr}{d\theta} = -\frac{1}{u^2}\frac{du}{d\theta}$$

Substituting these values in (i), we get

$$\frac{1}{p^2} = u^2 + \left(\frac{du}{d\theta}\right)^2$$

6.8. LENGTH OF THE POLAR TANGENT, POLAR NORMAL, POLAR SUB-TANGENT AND POLAR SUB-NORMAL

Let $P(r, \theta)$ be any point on the curve $r = f(\theta)$

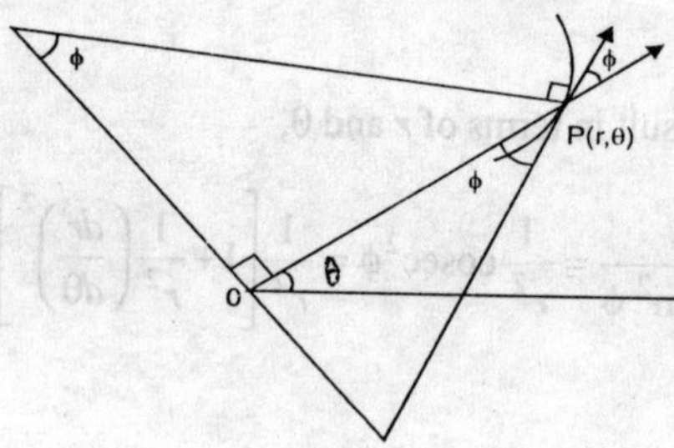

Figure 6.3

Draw a line through the pole O perpendicular to the radius OP and let it meet the tangent and normal at P in points T and G respectively. From fig. 6.3, we have

(i) Length of the polar tangent

$$= PT = r\sec\phi$$

$$= r\sqrt{1+\tan^2\phi}$$

$$= r\sqrt{1+\left(r\frac{d\theta}{dr}\right)^2}$$

(ii) Length of polar normal

$$= PG = r\operatorname{cosec}\phi$$

$$= r\sqrt{1+\cot^2\phi}$$

$$= r\sqrt{1+\left(\frac{1}{r}\frac{dr}{d\theta}\right)^2}$$

$$= \sqrt{\left\{r^2+\left(\frac{dr}{d\theta}\right)^2\right\}}$$

(iii) Length of the polar sub-tangent

$$= OT = OP \tan\phi$$

$$= r.r\frac{d\theta}{dr}$$

$$= r^2\frac{d\theta}{dr}$$

$$= -\frac{d\theta}{du}, \text{ where } u = \frac{1}{r}$$

(iv) Length of polar sub-normal

$$= OG = OP \cot\phi$$

$$= r.\frac{1}{r}\frac{dr}{d\theta}$$

$$= \frac{dr}{d\theta}$$

$$= -\frac{1}{u^2}\frac{du}{d\theta}, \text{ where } u = \frac{1}{r}.$$

6.9. PEDAL EQUATION OF A CURVE WHOSE POLAR EQUATION IS GIVEN

Let, $r = f(\theta)$ be the given curve. ...(i)

We have,

$$\frac{1}{p^2} = \frac{1}{r^2} + \frac{1}{r^4}\left(\frac{dr}{d\theta}\right)^2 \qquad \text{...(ii)}$$

Eliminating θ between (i) and (ii), we obtain the required pedal equation of the curve.

The pedal equation is sometimes more conveniently obtained by eliminating θ and ϕ between (i)

$$\tan\phi = r\frac{d\theta}{dr}$$

and $\quad p = r\sin\phi$.

Examples

Example 1: Find the angle of intersection of the cardioides

$$r = a(1+\cos\theta), \quad r = b(1-\cos\theta)$$

Solution: Let $P(r_1, \theta_1)$ be a point of intersection.

Let ϕ_1, ϕ_2 be the angles which OP makes with the two tangents to the two curves at P.

For the curve, $r = a(1+\cos\theta)$, we have

$$\frac{dr}{d\theta} = -a\sin\theta$$

$$\therefore \quad r\frac{d\theta}{dr} = -\frac{a(1+\cos\theta)}{a\sin\theta}$$

$$= \frac{-2\cos^2 \theta/2}{2\sin\theta/2 . \cos\theta/2} = -\cot\theta/2$$

$$\Rightarrow \quad \tan\phi = \tan\left(\frac{\pi}{2}+\frac{\theta}{2}\right)$$

$$\Rightarrow \phi = \frac{\pi}{2}+\frac{\theta}{2}$$

$$\text{Hence,} \quad \phi_1 = \frac{\pi}{2}+\frac{\theta_1}{2}$$

For the curve, $r = b(1-\cos\theta)$, we have

$$\frac{dr}{d\theta} = b\sin\theta$$

$$\Rightarrow \quad \tan\phi = r\frac{d\theta}{dr}$$

$$= \frac{b(1-\cos\theta)}{b\sin\theta}$$

$$= \tan\frac{\theta}{2}$$

$$\Rightarrow \quad \phi = \frac{\theta}{2}$$

Hence, $\phi_2 = \dfrac{\theta_1}{2}$

Therefore, $\phi_1 - \phi_2 = \pi/2$ and hence the curves cut eachother at right angles.

Example 2: Show that the parabolas $\dfrac{2a}{r} = 1 - \cos\theta$ and $\dfrac{2b}{r} = 1 + \cos\theta$ intersect orthogonally.

Solution: Let ϕ_1, ϕ_2 be the angles between the common radius vector and the tangents to the given parabolas at a point of intersection. We have,

$$\frac{2a}{r} = 1 - \cos\theta$$

Differentiating bothsides, we get

$$\frac{-2a}{r^2} = \sin\theta \frac{d\theta}{dr}$$

$$\therefore \quad r\frac{d\theta}{dr} = -\frac{2a}{r}.\frac{1}{\sin\theta}$$

i.e., $\tan\phi_1 = -\tan\dfrac{\theta}{2}$

Similarly, for the parabola,

$$\frac{2b}{r} = 1 + \cos\theta$$

Differentiating, we get

$$\frac{-2b}{r^2} = -\sin\theta \frac{d\theta}{dr}$$

$$\therefore \quad r\frac{d\theta}{dr} = \frac{2b}{r}.\frac{1}{\sin\theta}$$

$$= \frac{1+\cos\theta}{\sin\theta}$$

$$= \cot\frac{\theta}{2}$$

i.e., $\tan\phi_2 = \cot\dfrac{\theta}{2}$

Thus, $\tan\phi_1 . \tan\phi_2 = -\tan\dfrac{\theta}{2} . \cot\dfrac{\theta}{2}$

$$= -1$$

Hence the two parabolas intersect orthogonally.

Example 3: For the cardioid $r = a(1-\cos\theta)$, find the length of the (i) polar subtangent, (ii) polar sub-normal (iii) polar tangent and (iv) polar normal at any point.

Solution: We have,

$$r = a(1-\cos\theta)$$

Differentiating, we get

$$\frac{dr}{d\theta} = a\sin\theta$$

(i) length of polar sub-tangent $= r^2 \dfrac{d\theta}{dr}$

$$= \frac{a^2(1-\cos\theta)^2}{a\sin\theta}$$

$$= 4a\left(\sin^2\frac{\theta}{2}\right) . \frac{1}{2\cos\frac{\theta}{2} . \sin\frac{\theta}{2}}$$

$$= 2a\tan\frac{\theta}{2} . \sin^2\frac{\theta}{2}$$

(ii) length of polar sub-normal $= \dfrac{dr}{d\theta}$

$$= a\sin\theta$$

(iii) length of polar tangent $= r\sqrt{\left\{1+\left(r\dfrac{d\theta}{dr}\right)^2\right\}}$

$$= a(1-\cos\theta)\sqrt{\left\{1+\left(\frac{a(1-\cos\theta)}{a\sin\theta}\right)^2\right\}}$$

$$= 2a\sin^2\frac{\theta}{2}\sqrt{\left\{1+\tan^2\frac{\theta}{2}\right\}}$$

$$= 2a\sin^2\frac{\theta}{2}\sec\frac{\theta}{2}$$

$$= 2a\tan\frac{\theta}{2}.\sin\frac{\theta}{2}$$

(iv) length of polar normal $= \sqrt{\left\{r^2+\left(\dfrac{dr}{d\theta}\right)^2\right\}}$

$$= \sqrt{\left\{a^2(1-\cos\theta)^2+a^2\sin^2\theta\right\}}$$

$$= a\sqrt{\{2-2\cos\theta\}}$$

$$= 2a\sin\frac{\theta}{2}$$

Example 4: Show that for the equiangular spiral $r = ae^{\theta\cot\alpha}$, the length of the perpendicular p from the origin on the tangent at any point is $a\sin\alpha e^{\theta\cot\alpha}$.

Solution: We have,

$$\frac{dr}{d\theta} = a\cot\alpha e^{\theta\cot\alpha}$$

$$\therefore \quad \tan\phi = r\frac{d\theta}{dr} = \frac{ae^{\theta\cot\alpha}}{a\cot\alpha e^{\theta\cot\alpha}} = \tan\alpha$$

$$\therefore \quad \phi = \alpha$$

$$\therefore \quad p = r\sin\phi = r\sin\alpha = a\sin\alpha e^{\theta\cot\alpha}$$

Example 5: Find the pedal equation of the curve

$$r^m = a^m \sin m\theta.$$

Solution: Taking logarithm of bothsides, we have

$$m\ln r = m\ln a + \ln\sin m\theta$$

Differentiating bothsides, we have

$$\frac{m}{r}\cdot\frac{dr}{d\theta} = m\cot m\theta$$

i.e., $\tan\phi = r\dfrac{d\theta}{dr} = \tan m\theta$

$$\therefore \quad \phi = m\theta$$

Since, $p = r\sin\phi$

$$\therefore \quad p = r\sin m\theta$$

$$= r.\frac{r^m}{a^m} = \frac{r^{m+1}}{a^m}$$

Thus, $p = \dfrac{r^{m+1}}{a^m}$ is the required pedal equation.

Exercise – 6.3

1. *Find the angle between the radius vector and the tangent for each of the following curves:*

   (i) $\frac{l}{r} = 1 + e\cos\theta$ (ii) $r^m = a^m \cos m\theta$

   (iii) $r = a(1 - \cos\theta)$ (iv) $r = a\,\mathrm{cosec}^2 \frac{\theta}{2}$ at $\theta = \frac{\pi}{2}$

   (v) $r^m = a^m(\cos m\theta + \sin m\theta)$ (vi) $2a/\pi = 1 + \cos\theta$

2. *Find the angle between the curves*

   (i) $r = ae^\theta$, $re^\theta = b$

   (ii) $r = a\theta/(1+\theta)$, $r = a/(1+\theta^2)$

   (iii) $r = a(1 + \sin\theta)$, $r = a(1 - \sin\theta)$

   (iv) $r = a\,\mathrm{cosec}^2 \frac{\theta}{2}$, $r = b\sec^2 \frac{\theta}{2}$

   (v) $r = a\sin 2\theta$, $r = a\cos 2\theta$

   (vi) $r = a\log\theta$, $r = a/\log\theta$

   (vii) $r = b$, $r^2 = a^2\cos 2\theta + b^2$

   (viii) $r^2 \sin 2\theta = 4$, $r^2 = 16\sin 2\theta$

   (ix) $r^2 = a^2\cos^2 2\theta$, $r = a(1 + \cos\theta)$

   (x) $r = a\theta$, $r\theta = a$.

3. Show that the angle ϕ between the radius vector on the tangent at any point on the Lemniscate $r^2 = a^2 \cos 2\theta$ is $\frac{\pi}{2} \times 2\theta$.

4. Show that the radius vector is inclined at a constant angle to the tangent at any point on the equiangular spiral.

5. Show that the curves $r^n = a^n \cos n\theta$ and $r^n = a^n \sin n\theta$ intersect orthogonally.

6. Show that in the case of the curve $r = a(\sec\theta \pm \tan\theta)$, if a radius vector OPP' be drawn cutting the curve in P and P' and if the tangents at P, P' meet in T, then $TP = TP'$.
7. Show that the tangents drawn at the end of chords of the cardiod $r = a(1+\cos\theta)$ which pass through the pole are perpendicular to each other.
8. Show that the two curves $r^2 = a^2\cos 2\theta$ and $r = a(1+\cos\theta)$ intersect at an angle $3\sin^{-1}(3/4)^{1/4}$.
9. If two tangents to the cardioid $r = a(1+\cos\theta)$ are parallel, show that the line joining their points of contact subtends an angle $2\pi/3$ at the pole.
10. Show that for the curve $\theta = \cos^{-1}\frac{r}{k} - \sqrt{\left[\frac{(k^2 - r^2)}{r^2}\right]}$ the length of polar tangent is constant.
11. Show that for the spiral $r = aq$, the polar sub-normal is constant.
12. Show that for the curve: $r = ae^{b\theta^2}$, ratio of polar sub-normal to the polar sub-tangent varies as θ^2.
13. *Find the pedal equations of*

(i) $r^m = a^m \cos m\theta$ (ii) $r = ae^{\theta}\cot\alpha$

(iii) $r = a(1-\cos\theta)$ (iv) $r = a\sin m\theta$

(v) $r^2 = a^2\cos^2\theta$ (vi) $r\theta = a$

(vii) $r\left(1-\sin\frac{\theta}{2}\right)^2 = a$ (viii) $r^m = a^m \sin m\theta + b^m \cos m\theta$

14. Show that the pedal equation of the spiral $r = a \sec h\, n\theta$ is of the form $\frac{1}{p^2} = \frac{A}{r^2} + B$.

7

Mean Value Theorems

7.1. ROLLE'S THEOREM

If a function f is

(i) contiuous in a closed interval $[a,b]$

(ii) derivable in the open interval $]a,b[$ and

(iii) $f(a)=f(b)$,

then there exists at least one value 'c' $\in\ ]a, b[$, such that $f'(c)=0$.

Proof. Since f is continuous on $[a,b]$, therefore, f must be bounded on $[a, b,]$.

Let M be a greatest value of f and m be a least value of f. Two defferent cases arise :

(i) $M=m$

Then f is constant over $[a,b]$ and consequently $f(x)=0$, fore all x in $[a, b]$.

(ii) $M \neq m$

Since $f(a)=f(b)$, therefore, at least one of the numbers M and m is different from $f(a)$ and therefore, also from $f(b)$.

When M and m are unequal, as in case (ii), at least one of them must be different from them. The number c being different from a and b, lies within the interval $[a, b]$ and as such belongs to the open interval $]a, b[$. The function f which is derivable in the open interval $]a, b[$ is, in particular, derivable at c, so that

$$\lim_{h \to 0} \frac{f(c+h)-f(c)}{h}$$

exists and is the same when $h \to 0$ through positive or negative values.

Also $f(c)$0 is the greatest value of the function, we have

$$f(c+h) \le f(c)$$

What ever positive or negative values h has. thus

$$\frac{f(c+h)-f(c)}{h} \le o \text{ fore } h > 0 \qquad \text{... (iii)}$$

$$\frac{f(c+h)-f(c)}{h} \ge o \text{ for } h < 0 \qquad \text{... (iv)}$$

Let $h \to 0$ through positive values from (iii), we get

$$f'(c) \le 0 \qquad \text{... (v)}$$

Let $h \to 0$ through negative values. From (iv) , we get

$$f'(c) \ge 0$$

The same conclusion would be similarly reached if, it is the least value which differs from $f(a)$ and $f(b)$.

Note 1: Interpreted gemetrically, Rolle's theorem says that the curve representing the greaph of the function f must have a tangent parallel to x-axix, at least at one point between a and b.

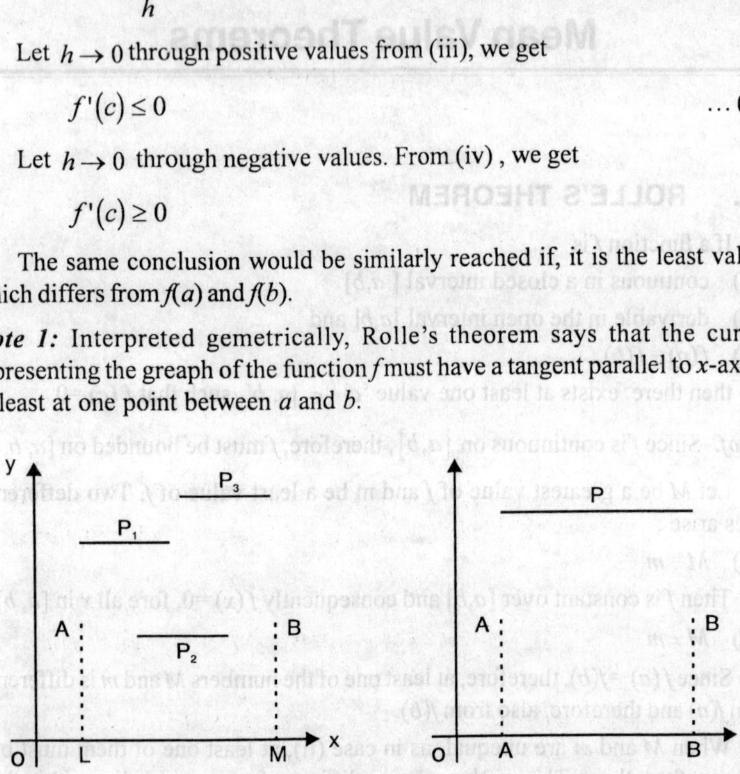

Figure 7.1

Note 2: The conclusion of Rolle's theorem may not hold good for a function which does not satisfy any of it conditions. Consider the function given by $y = f(x) = |x|$ in the interval [-1,1].

The function is continuous in [-1,1] and $f(1)=f(-1)$ As for the derivative, f being not derivable at 0, the function is not derivable in]-1,1[. Thus not all the three conditions of the Rolle's theorm are satisfied. Also the conclusion is not valid in that $f'(x)$ vanishes for no value of x.

7.2. LAGRANGE'S MEAN VALUE THEOREM

If a function f defined on $[a,b]$, such that

i) f is continuous on $[a,b]$

ii) f is derivable on $]a,b[$

then there exists a real number $c \in]a, b[$ such that $f(b) - f(a) = (b-a)f'(c)$

Proof: Let F be a function defined on $[a, b]$ by setting

$$F(x) = f(x) + Ax, \text{ for all } x \text{ in } [a, b] \quad \ldots(i)$$

where A is a constant to be suitably chosen.

Now,

(1) Since f is continuous on $[a, b]$ and $x \to Ax$ is continuous on $[a, b]$, therefore, F is continuous on $[a, b]$.

(2) Also, since f is derivable on $]a, b[$ and $x \to Ax$ is derivable on $]a, b[$, therefore, F is derivable on $]a, b[$.

(3) Let us choose A so that $F(a) = F(b)$. This givens us

$$-A = \frac{f(b)-f(a)}{b-a} \quad \ldots(ii)$$

From (1), (2) and (3) above, we find that F satisfies all the conditions of Rolle's theorem on $[a, b]$. and consequently, there exists a real number c in $]a, b[$ such that $F'(c)=0$

From (ii) and (iii), we have

$$\frac{f(b)-f(a)}{b-a} = f'(c)$$

$$f(b)-f(a) = (b-a)f'(c)$$

Another form of statement of Lagrange's Mean Value Theorem

If a function f is continuous in a closed interval $[a, a+h]$ and derivable in the open interval $]a, a+h[$, then there exists at least one number 'θ' ∈]0,1[, such that

$$f(a+h) = f(a) + hf'(a+\theta h)$$

$$f(a+h) = f(a) + hf'(a+\theta h)$$

We write $b - a = h$ so that h denotes the length of the interval $[a, b]$ which may now be written as $[a, a+h]$. The number, 'c' which lies between a and $a+h$ is greater than a and less than $a+h$ so that we may write $c = a+h\theta$, where θ is some number between 0 and 1.

Thus the equation (i) becomes

$$\frac{f(a+h)-f(a)}{h} = f'(a+\theta h)$$

$$= f(a+h) = f(a) + hf'(a+\theta h) \quad (o < \theta < 1)$$

Note 1: Now $[f(b) - f(a)]$ is the change in the function f as x changes from a to b so that $\dfrac{f(b)-f(a)}{b-a}$ is the average rate of change of the function over the interval $[a, b]$.

Note 2: If we draw some curves satisfying the conditions of the theorem, we will realise that the theorem, as stated in the geometrical form, is almost self-evident.

Geometrical Interpretation of Lagranges theorem

Geometrically interpreted, Lagrange's Mean value theorem says that the tangent to the graph of f at some suitable point between a and b is parallel to the chord joining the points on the graph with abscissae a and b.

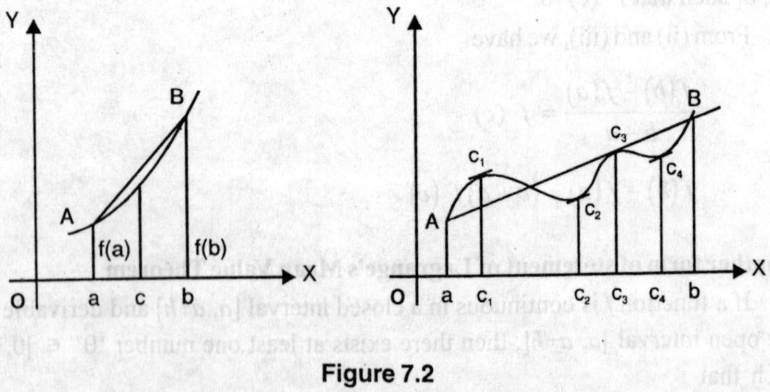

Figure 7.2

7.3. GRAPHS OF HYPERBOLIC FUNCTIONS

7.3.1. Graph of y = sin hx

We have,

$$\frac{d(\sinh x)}{dx} = \cosh x \quad x \in R$$

Now $x \in]-\infty, \infty[$

$$e^x > 0 \text{ and } \bar{e}^x > 0$$

so that

$$\cos hx = \frac{1}{2}\left(e^x + e^{-x}\right) > 0$$

Thus, sin hx is strictly increasing in $]-\infty, 0[$

Also $\sin h.\ 0 = 0$ and

$\sin hx < 0$ for all $x \in]-\infty, 0[$

> 0 for all $x \in]-0, \infty[$

The graph of y = sin hx is given in fig. 7.3.

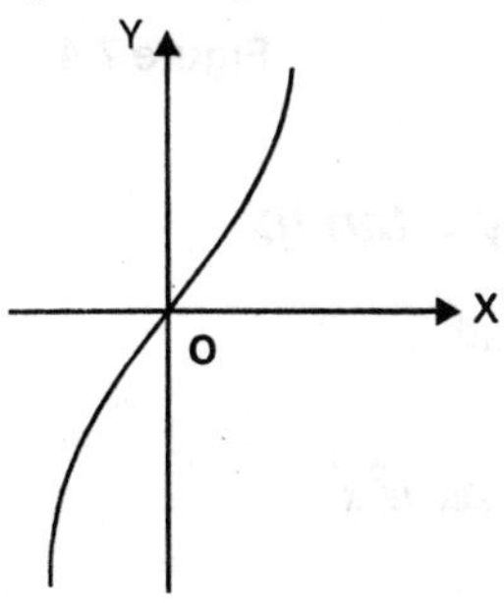

Figure 7.3

7.3.2. Graph of y = cos hx

We have for all $x \in R$

$$\frac{d(\cos hx)}{dx} = \sin hx$$

Now, sin hx < 0 for all x $\in]-\infty,0[$

and sin hx > 0 for all x $\in]0,\infty[$

Thus cos hx is strictly increasing in $[-\infty,0]$

and strctly decreasing in $[\infty,0]$.

Also, $\lim\limits_{x \to \infty} \cos hx = \infty = \lim\limits_{x \to +\infty} \cos hx$

and $\cos h\theta = 1$.

The graph of y = cos hx is given in fig. - 7.4

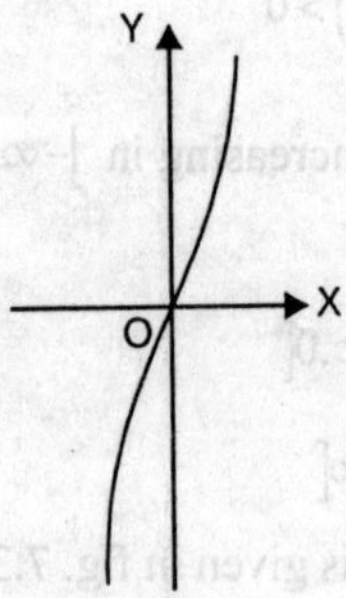

Figure 7.4

7.3.3. Graph of y = tan hx

We have for all $x \in R$

$$\frac{d(\tan hx)}{dx} = \sec h^2 x$$

Thus tan hx is strictly increasing in $]-\infty, \infty[$

Also tan $h0 = 0$ and

$$\lim_{x \to +\infty} \tan hx = \lim_{x \to +\infty} \frac{e^x - e^{-x}}{e^x + e^{-x}}$$

$$= \lim_{x \to +\infty} \frac{1-e^{-2x}}{1+e^{-2x}}$$

$$= 1$$

$$\lim_{x \to -\infty} \tan hx = \lim_{x \to -\infty} \frac{e^{x} - e^{-x}}{e^{x} + e^{-x}}$$

$$= \lim_{x \to -\infty} \frac{e^{2x} - 1}{e^{2x} + 1}$$

$$= -1$$

The graph of $y = \tan hx$ is given in fig. 7.5

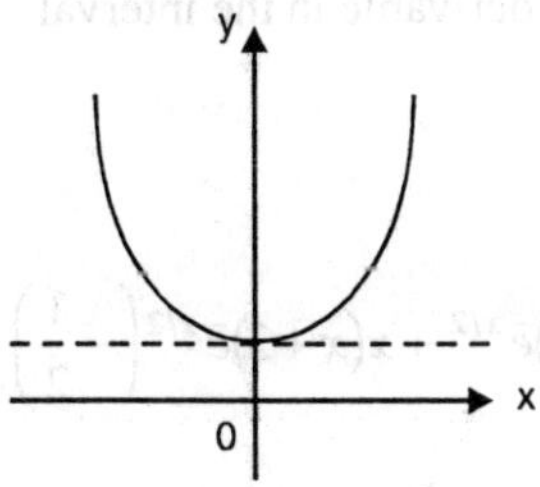

Figure 7.5

7.3.4. Graph of y = cot hx, y= sec hx, y = cosec hx

Briefly, we describe the graph without giving the details.

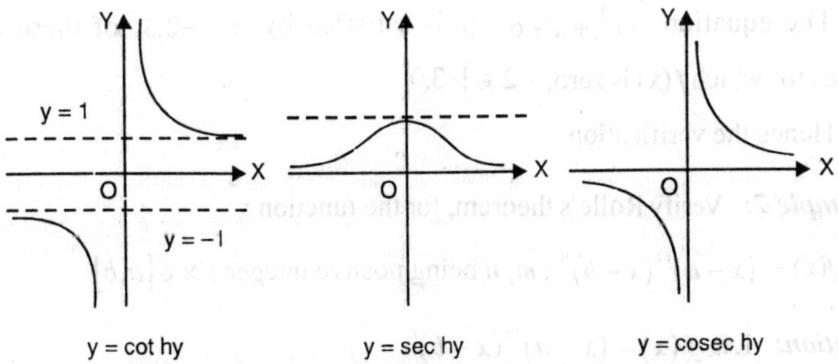

Figure 7.6

Each of these curves possesses asymptotes.

Also the asymptotes of

y = cot hx are x =0, y=1 and y=-1

y =sec hx is y=0

y = cosec hx are y = 0 and x = 0.

Examples

Example 1: Verify Rolle's theorem for $f(x) = x(x+3)e^{x/2}$ in [–3, 0]

Solution: Let $f(x) = x(x+3)\bar{e}^{x/2}$, $x \in [-3,0]$

Here $f(-3) = 0 = f(0)$

Also, the function is derivable in the interval

$[-3,0]$

We have

$$f'(x) = (2x+3)\bar{e}^{x/2} + x(x+3)\bar{e}^{x/2}\left(-\frac{1}{2}\right)$$

$$= \frac{(-x^2+x+6)}{2} e^{\frac{-x}{2}}$$

Now, $f'(x) = 0$

$\Leftrightarrow -x^2 + x + 6 = 0$

The equation $-x^2 + x + 6 = 0$ is satisfied by $x = -2,3.$ of these two values for which f(x) is zero, $-2 \in]-3,0[$.

Hence the verification.

Example 2: Verify Rolle's theorem, for the function :

$f(x) = (x-a)^m (x-b)^n$; *m*, *n* being positve integer ; $x \in [a,b]$

Solution: Let $f(x) = (x-a)^m (x-b)^n$

As f is a polynomial, so it satisfies the following conditions:

(i) f is continuous in $[a, b]$

(ii) f is derivable in $]a, b[$

(iii) $f(a) = 0 = f(b)$

Let f exists at c, $c \in]a,b[$ such that

$f'(c) = 0$

$$= m(c-a)^{m-1}(c-b)^n + n(c-a)^m(c-b)^{n-1} = 0$$

$$= (c-a)^{m-1} + (c-b)^{n-1}(mc - mb + nc - na) = 0$$

$$\Rightarrow \quad c = \frac{mb + na}{m + n}$$

Example 3: Find 'c' of Lagrange's mean value theroem if

$$f(x) = x(x-1)(x-2); \quad a = 0, b = \frac{1}{2}$$

Solution: We have,

$$f(x) = x(x-1)(x-2)$$

$$\therefore \quad f(0) = 0, f\left(\frac{1}{2}\right) = \left(\frac{1}{2} - 1\right)\left(\frac{1}{2} - 2\right) = \frac{3}{8} \qquad \ldots(1)$$

Also, $f'(x) = 3x^2 - 6x + 2$ ...(2)

Putting $a = 0, b = \frac{1}{2}$ in

$$f(b) - f(a) = (b-a)f'(c)$$

We have from (1) and (2)

$$\frac{3}{8} = \frac{1}{2}\left(3c^2 - 6c + 2\right)$$

$$\Rightarrow \quad 12c^2 - 24c + 5 = 0$$

$$\therefore \quad c = \frac{6 \pm \sqrt{21}}{6}$$

Since $\frac{6+\sqrt{21}}{6}$ does not lie in $\left]0, \frac{1}{2}\right[$, therefore this value of c has to be discarded. Hence the required value of c is $\frac{6-\sqrt{21}}{6}$.

Example 4: Prove that for any quadratic fucnction px^2+qx+r, the value of 0 in Lagrange's theorem is always $\frac{1}{2}$ whatever p,q,r,a,h may be.

Solution:

Let $f(x) = px^2+qx+r,\ x \in [a, a+h]$. As, f is a polynomial, it satisfies the conditions of Lagrange's Mean value Theorem and hence there exists

$\theta\,(0<\theta<1)$

satisfying

$$f(a+h)-f(a)=h\,f'(a+\theta h)$$

$$\Rightarrow\quad p(a+h)^2+q(a+h)+r-pa^2-qa-r=h[2p(a+\theta)h+q]$$

$$\Rightarrow\quad 2pah+ph^2+qh=2pah+2p\theta h^2+qh$$

$$\Rightarrow\quad q=\frac{1}{2}$$

Example 5: If a function f is such that its derivative f' is continuous in $[a, b]$ and derivable in $]a, b[$, then prove that there exists a number $c\,(a<c<b)$ such that

$$f(b)=f(a)+(b-a)f'(a)+\frac{1}{2}(b-a)^2 f''(c)$$

Solution: Consider a function ϕ such that

$\phi(x)=f(b)-f(x)-(b-x)f'(x)-(b-x)^2 A,$ where A is a constant to be determined such that $\phi\,(a)=\phi\,(b)$.

ϕ is continuous in $[a, b]$ and derivable in $]a, b[$. Thus ϕ satisfies all the conditions of Rolle's Theorem where A is given by

$$f(b) - f(a) - (b-a)f'(a) - (b-a)^2 A = 0 \qquad \text{...(i)}$$

$\therefore$ there exists $c \in]a, b[$ such that $\phi'(c) = 0$

We have, $\phi'(x) = -f'(x) + f'(x) - (b-x)f''(x) + 2(b-x)A$

$$= -(b-x)f''(x) + 2(b-x)A$$

$$\phi'(c) = 0$$

$$= -(b-c)f''(c) + 2(b-c)A = 0$$

$$= (b-c)\{2A - f''(c)\} = 0$$

$$= A = \frac{1}{2}f''(c) \quad (b \neq c)$$

From (i), we have

$$f(b) = f(a) + (b-a)f'(a) + \frac{1}{2}(b-a)^2 f''(c)$$

Example 6: Show that $x/(1+x) < \log(1+x) < x \ \forall \ x > 0.$

Solution: We have,

$$f(x) = \log(1+x) - \frac{x}{1+x}, x > 0$$

$$\Rightarrow f'(x) = \frac{1}{1+x} - \frac{1}{(1+x)^2}$$

$$= \frac{x}{(1+x)^2}$$

$f'(x) > 0 \ \forall \ x > 0$ and is 0 for $x = 0$

$\rightarrow$ f is strictly increasing in $[0,\infty[$,

Also $f(0)=0$

It follows that $f(x)>f(0)=0 \;\forall\; x>0$

$$= \log(1+x) > \frac{x}{1+x} \qquad \ldots(i)$$

Again

$$\Rightarrow \quad f'(x) = 1 - \frac{1}{1+x} = \frac{x}{1+x}$$

$$\Rightarrow \quad g'(x) > 0 \;\forall\; x>0 \text{ and is 0 for x=0}$$

$$\Rightarrow \quad \text{g (x) is strictly increasing in the interval } [0,\infty[$$

Also $\;g(0)=0$

Thus $\;g(x)<g(0)=0$

$$\Rightarrow \quad x > \log(1+x) \forall x > 0 \qquad \ldots(ii)$$

From (i) and (ii) we get the required result.

Example 7: Show that

$$\frac{v-u}{1+v^2} < \tan^{-1} v - \tan^{-1} u < \frac{v-u}{1+u^2}, \quad 0<u<v \text{ and deduce that}$$

$$\frac{\pi}{4} + \frac{3}{25} < \tan^{-1}\frac{4}{3} < \frac{\pi}{4} + \frac{1}{6}$$

Solution:

Let $f(x) = \tan^{-1} x, \;\; u<x<v$

$$= f'(x) = \frac{1}{1+x^2}$$

By Langarange's Mean Value Theorem, there exists

$c \in]u, v[$ such that

$$\frac{f(v)-f(u)}{v-u} = f'(c)$$

$$\Rightarrow \quad \frac{\tan^{-1} v - \tan^{-1} u}{v - u} = \frac{1}{1+c^2}$$

$$u < c \Rightarrow u^2 < c^2$$

$$\Rightarrow \quad a + u^2 < a + c^2$$

$$\Rightarrow \quad \frac{1}{1+u^2} > \frac{1}{1+c^2}$$

$$\therefore \quad \frac{\tan^{-1} v - \tan^{-1} u}{v - u} < \frac{1}{1+u^2}$$

$$\Rightarrow \quad \tan^{-1} v - \tan^{-1} u < \frac{v-u}{1+u^2} \qquad \ldots\text{(i)}$$

Again $c < u$

$$\Rightarrow \quad c^2 < v^2$$

$$\Rightarrow \quad 1 + c^2 < 1 + v^2$$

$$\Rightarrow \quad \frac{1}{1+c^2} > \frac{1}{1+v^2}$$

$$\Rightarrow \quad \frac{\tan^{-1} v - \tan^{-1} u}{v - u} > \frac{1}{1+v^2} \qquad \ldots\text{(ii)}$$

From (i) and (ii), we have

$$\frac{v-u}{1+v^2} < \tan^{-1} v - \tan^{-1} u < \frac{v-u}{1+u^2}$$

Let $v = \frac{4}{3}, \quad u = 1$

$$\therefore \quad \frac{3}{25} < \tan^{-1}\frac{4}{3} < \frac{1}{6}$$

$$= \frac{3}{25} + \frac{\pi}{4} < \tan^{-1}\frac{4}{3} < \frac{1}{6} + \frac{\pi}{4}$$

Example 8: Show that $\frac{\tan x}{x} > \frac{x}{\sin x}$ for $0 < x < \pi/2$

Solution: We have,

$$\frac{\tan x}{x} > \frac{x}{\sin x}$$

$$\Rightarrow \quad \frac{\tan x \sin x - x^2}{x \sin x} > 0 \quad \text{for } 0 < x < \pi/2$$

Since $x \sin x > 0$ for $0 < x < \pi/2$, it is enough to show that

$$\tan x \sin x - x^2 > 0, \; 0 < x < \pi/2$$

Let $f(x) = \tan x \sin x - x^2, \; 0 < x < \pi/2$

$$f'(x) = \sin x \cdot \sec^2 x + \tan x \cdot \cos x - 2x$$

$$= \sin x \sec^2 x + \sin x - 2x$$

$$f''(x) = \cos x \sec^2 x + \sin x 2 \sec^2 x \tan x + \cos x - 2$$

$$= \sec x + \cos x - 2 + 2 \sin x \tan x \sec^2 x$$

$$= \left(\sqrt{\sec x} - \sqrt{\cos x}\right)^2 + 2 \sin x \tan x \sec^2 x$$

$$> 0 \text{ for } 0 < x < \pi/2$$

$\therefore$ f' is strictly increasing in $[0, \pi/2]$

Also $f'(0) > 0$

$\Rightarrow$ $f'(x) > 0$ for $0 < x < \pi/2$

$\Rightarrow$ f is strictly increasing in $[0, \pi/2]$

Also $f(0) = 0$

$$\Rightarrow f(x) > 0 \text{ for } 0 < x < \pi/2$$

$$\Rightarrow \tan x \sin x - x^2 > 0 \text{ for } 0 < x < \pi/2$$

$$\therefore \quad \frac{\tan x \sin x - x^2}{x \sin x} > 0, 0 < x \pi/2$$

$$\Rightarrow \frac{\tan x}{x} > \frac{x}{\sin x}, 0 < x < \pi/2$$

Exercise – 7.1

1. *Verify Rolle's theorem for the function f defined on [a, b] in each of the following cases:*

   i) $f(x) = x^2, \quad a = -1, \; b = 1$

   ii) $f(x) = \log\left[\left(x^2 + ab\right)/(a+b)x\right], \; x \in [a,b]$

   iii) $f(x) = 2 + (x-1)^{2/3}, \; x \in [0,2]$

   iv) $f(x) = \sin x / e^x; \; x \in [0, \pi]$

   v) $f(x) = e^x(\sin x - \cos x); \; x \in [\pi/4, 5\pi/4]$

   vi) $f(x) = x^3 - 3x + 2, \; a = -2, \; b = 1$

   vii) $f(x) 1/x^2 + 1, \; a = -3, \; b = 3$

   viii) $f(x) = \sin x, \quad a = 0, \quad b = \pi$

2. Prove that if $a_0, \; a_1 \ldots\ldots.. a_n$ be real numbers such that

   $$\frac{a_0}{n+1} + \frac{a_1}{n} + \ldots\ldots\ldots\ldots \frac{a_{n-1}}{2} + a_n = 0,$$

   then there exists at least one real x between 0 and 1 such that

   $$a_0 x^n + a_1 x^{n-1} \ldots\ldots. + a_n = 0$$

3. Show that there is no real number k for which the equation $x^3 - 3x + k = 0$ has two distinct roots in [0, 1].

4. Show that between any two roots of $e^x \cos x = 1$, there exists al least one root of $e^x \sin x - 1 = 0$.

5. If f, ϕ, ψ are continuous in $[a,b]$ and derivable in $]a,b[$ then show that there is a value c lying between a and b such that.

   $$\begin{vmatrix} f(a) & f(b) & f(x) \\ \phi(a) & \phi(b) & \phi(x) \\ \psi(a) & \psi(b) & \psi(x) \end{vmatrix} = 0$$

6. **Verify the Mean value theorem for**

(i) $f(x) = \log x$ in $[1, e]$

(ii) $f(x) = x^3$; $a = -2$, $b = 1$

(iii) $f(x) = lx^2 + mx + n$ in $[a, b]$

(iv) $f(x) = x^n$; $a = -1$, $b = 1$

(v) $f(x) = x$; $a = -2$, $b = 1$

(vi) $f(x) = \cos x$; $a = 0$, $b = \pi/2$

7. Find 'c' so that $f'(c) = [f(b) - f(a)]/(b-a)$ in the folloing cases:

(i) $f(x) = x^2 - 3x - 1$; $a = -11/7$ $b = 13/7$

(ii) $f(x) = \sqrt{x^2 - 4}$; $a = 2$, $b = 3$

8. If f', g' are continuous and differentiable, then show that for $a < c < b$.

$$\frac{f(b) - f(a) - (b-a)f'(a)}{g(b) - g(a) - (b-a)g'(a)} = \frac{f''(c)}{g''(c)}$$

9. Explain the failure of the theorem in the interval [1, 1] when.

$f(x) = 1/x$, $(x \neq 0)$;

$f(0) = 0$.

10. Show that $x/\tan x$ decreases in the interval $]0, \pi/2[$.

11. Determine the intervals in which the function

$f(x) = (x^4 + 6x^2 + 17x^2 + 32x + 32)e^{-x}$ is increasing or decreasing.

12. Show that $f(x) = x - \sin x$ is an increasing function through out every interval. Determine for what values of $a, ax - \sin x$ is an increasing function.

13. Show that $x^{-1} \log(1 + x)$ decreasing as x increases from 0 to ∞.

14. Show that e^{-x} lies between $1-x$ and $1-x+\dfrac{x^2}{2}$ for all $x \in R$.

15. Separate the intervals in which the function

$f(x)=\left(x^2+x+1\right)/\left(x^2-x+1\right)$ increasing or decreasing.

16. The derivative of a function f is positive for every value of x in an interval $[c-h,c]$ and negative for every value of x in $[c,c+h]$; show that $f(c)$ is the greatest value of the function in the interval $[c-h,c+h]$.

17. Show that

$$x \in \left]1,\infty\right[\;\Rightarrow x-1>\log x>(x-1)x^{-1}$$

18. Show that the function $f(x)=\cosh x$ is strictly dcereasing in $\left]-\infty,0\right[$ and strictly increasing in $\left[0,\infty\right[$.

7.4. CAUCHY'S MEAN VALUE THEOREM

If two functions $f(x)$ and $g(x)$ defined on $[a,b]$ such that

(i) $f(x)$ and $g(x)$ are continuous on $[a,b]$,

(ii) $f(x)$ and $g(x)$ are derivable on $]a,b[$ and

(iii) $g'(x)$ does not vanish at any point of $]a\ ,b\ [$

Then there exists a real number $c \in]a, b[$

such that

$$\frac{f(b)-f(a)}{g(b)-g(a)}=\frac{f'(c)}{g'(c)}$$

Proof: Firstly, we observe that

$$g(b)-g(a)\neq 0$$

For, if $g(a)=g(b)$, then the function $g(x)$ would satisfy all the conditions of Rolle's theorem, and consequently for some x in$[a, b]$ we would have $g'(x)=0$.

Consider the function $\phi(x)$ defined on $[a,b]$ by setting

$$F(x)=f(x)+Ag(x)=f(b)+Ag(b)$$

where A is constant to be determined such that

$$\phi(a)=\phi(b)$$

Thus,

$$\Rightarrow\ A=-\frac{f(b)-f(a)}{g(b)-g(a)},\ \text{for}\ [g(b)-g(a)]\neq 0$$

Now, $f(x)$ and $g(x)$ are derivable in $[a,b]$. Also A is a constant. Therefore, $\phi(x)$ is derivable in $[a,b]$ and its derivative is

$$f'(x)+Ag'(x)$$

Thus, $\phi(x)$ satisfies the conditions of Rolle's theorem. Therefore, there exists a real number c in $]a,b[$ such that $g'(c)=0$

$$\Rightarrow \quad f'(c)+Ag'(c)=0$$

$$\Rightarrow \quad f'(c)=-Ag'(c)$$

$$\Rightarrow \quad \frac{f'(c)}{g'(c)}=-A$$

$$=\frac{f(b)-f(a)}{g(b)-g(a)} \qquad [\because g'(c)=0]$$

Hence the theorem.

Another form: If two functions $f(x)$ and $g(x)$ are derivable in a closed interval $[a, a+h]$ and for any x in $[a, a+h]$, then there exists at least one number α belong to the open interval $]0,1[$ such that

$$\frac{f(a+h)-f(a)}{g(a+h)-g(a)}=\frac{f'(a+\alpha h)}{g'(a+\alpha h)} \qquad (0<\theta<1)$$

7.5. GENERALISED MEAN VALUE THEOREM: TAYLOR'S THEOREM

If a function f is such that

(i) the $(n-1)$ th derivative f^{n-1} is continuous in $[a,a+h]$

(ii) the nth derivative f^n exists in $]a,a+h[$ and

(iii) p is a given positve integer.

then there exists at least one number, θ between 0 and 1 such that

$$f(a+h)=f(a)+hf'(a)+\frac{h^2}{2}f''(a)+.....+\frac{h^{n-1}}{(n-1)}f^{n-1}(a)$$

$$+\frac{h^n(1-\theta)^{n-p}}{(n-1)p}f^n(a+\theta h) \qquad ...(i)$$

Proof:

The statement of the condition (i) implies that $f, f', f'' \cdots\cdots\cdots f^{n-2}$ in $[a, a+h]$.

Let ϕ be a function defined by

$$\phi(x) = f(x) + (a+b-x)f'(x) + \frac{(a+h-x)^2}{2}f''(c) + \ldots.$$

$$+\frac{(a+h-x)^{n-1}}{(n-1)}f^{n-1}(x) + A(a+h-x)^p.$$

where A is a constant to be determined such that

$$\phi(a) = \phi(a+h)$$

Thus A is given by

$$f(a+h) = f(a) + hf'(a) + \frac{h^2}{2}f''(a)\ldots\ldots\frac{h^{n-1}}{(n-1)}f^{n-1}(a) + Ah^p \quad \ldots\text{(ii)}$$

The function is continuous in $[a, a+h]$, derivable in $]a, a+b[$ and $\phi(a) = \phi(a+h)$. Therefore, by Rolle's theorem, there exists at least one number θ, between 0 and 1 such that

$$\phi'(a+\theta h) = 0$$

But $\phi'(x) = \dfrac{(a+h-x)^{n-1}}{(n-1)!}f^n(x) - PA(a+h-x)^{p-1}$

$$\therefore \quad 0 = \phi'(a+\theta h) = \frac{h^{n-1}(1-\theta)^{n-1}}{(n-1)!}f^n(a+\theta h) - PA(1-\theta)^{p-1}h^{p-1}$$

$$\Rightarrow \quad A = \frac{h^{n-p}(1-\theta)^{n-p}}{p.(n-1)}f^n(a+\theta h), for \ (1-\theta) \neq 0 \text{ and } h \neq 0$$

Substituting the value of A in (ii), we get the required result (i)

(i) Remainder after n terms due to Schlomilch and Roche

$$R_n = \frac{h^n (1-\theta)^{n-p}}{p(n-1)!} f^n(a+\theta h)$$

Where R_n is known as *Taylor's remainder* after *n* terms.

This remainder is due to Schlomiltch and Roche.

(ii) Remainder after *n* terms due to Cauchy.

Substituting p = 1, we get

$$R_n = \frac{h^{n-1}(1-\theta)^{n-1}}{(n-1)} f^n(a+\theta h)$$

This form of remainder is due to Cauchy.

(iii) Remainder after *n* terms due to Lagrange.

Putting *P* = *n*, we obtain

$$R_n = \frac{h^n}{n!} f^n(a+\theta h)$$

Corollary: Let x be a point of the interval $[a, a+h]$. Let *f* satisfy the conditions of Taylor's theorem in $[a, a+h]$ so that it also satisfies the conditions for $[a, x]$.

Changing $(a+h)$ to *x* in (i) we get

$$f(x) = f(a) + (x-a)f'(a) + \frac{(x-a)^2}{2} f''(a) + \frac{(n-a)^3}{3} f'''(a) + ...$$

$$+ \frac{(n-a)^{n-1}}{(n-1)!} f^{n-1}(a) + \frac{(x-a)^n (1-\theta)^{n-p}}{P.(n-1)!} f^n[a+\theta(x-a)], 0 < 0 < 1.$$

This result holds gor all $x \in [a, a+h]$

7.6. MACLAURIN'S THEOREM

Putting $a = 0$, we observe that if $x \in [0, h]$, then

$$f(x) = f(0) + xf'(0) + \frac{x^2}{2} f''(0) + \frac{x^3}{3} f'''(0) + ... +$$

$$\frac{x^{n-1}}{(n-1)!}f^{n-1}(0)+\frac{x^n(1-\theta)^{n-p}}{p(n-1)!}f^n(\theta x)$$

which holds when

(i) f^{n-1} is continuous in $[0, h]$ and

(ii) f^n exists in $]0,h[$

Maclaurin's Power Series for a given function.

Let a function f possess continuous derivatives of all orders in the interval $[0,x]$ so that we have

$$f(x)=f(0)+xf'(0)+......+\frac{x^{n-1}}{(n-1)!}f^{n-1}(0)+R_n$$

Where R_n is Lagrange's or Taylor's form of remainder.

Thus, we have

$$f(x)=f(0)+xf'(0)+.....+\frac{x^n}{n!}f^n(0)+.... \qquad ...(i)$$

valid for those of the values of n for

which $\lim R_n = 0 \quad as\ n\to\infty$

The series (i) is known as Maclaurin's infinite series for the expansion of $f(x)$ as power series.

7.7. POWER SERIES EXPANSIONS OF SOME STANDARD FUNCTIONS

1. Expansion of e^x

Let $f(x)=e^x$, for all $x\in R$.

Then $f^n(x)=e^x$, for all $x\in R$.

Thus for each positive integer n, f^n is defined in the interval $[-h, h]$ whatever positive real number h may be.

Denoting by R_n the Lagrange's form of remainder after n terms, we have

$$R_n = \frac{x^n}{n!} f^n(\theta x)$$

$$= \frac{x^n}{n!} e^{\theta x}, \quad 0 < \theta < 1$$

Now $x > 0 \Rightarrow \theta x < x \Rightarrow e^{\theta n} < e^n$

$x < 0 \Rightarrow e^{\theta n} < 1$

Let us write $a_n = \frac{x^n}{n!}$, for all $n \in N$

Then $\frac{a_n + 1}{a_n} = \frac{x}{n+1}$

so that $\frac{\lim}{n \to \infty} \frac{a_n + 1}{a_n} = 0$

From above, it follows that $\frac{\lim}{n \to \infty} a_n$ exists and equals zero.

Now, $\frac{\lim}{n \to \infty} R_n(x) = e^{\theta x} \left(\frac{\lim}{n \to \infty} \frac{x^n}{n!} \right) = 0$

We see that $R_n \to 0$ as $n \to \infty$ for all $x \in R$.

Thus, we obtain the expansion

$$e^x = 1 + x + \frac{x^2}{2!} + \cdots\cdots + \frac{x^n}{n!} + \cdots\cdots$$

for all $x \in R$

2. Expansion of $\sin x$

Let $f(x) = \sin x$, for all $x \in R$

Then $f^n(x) = \sin\left(x + \frac{n\pi}{2}\right)$, for all $x \in R$

Denoting by R_n the Lagrange's form of remainder, we have

$$R_n = \frac{x^n}{n!} f^n(\theta x)$$

$$= \frac{x^n}{n!} \sin\left(\theta x + \frac{n\pi}{2}\right)$$

so that $|R_n| = \left|\frac{x^n}{n!}\right| \left|\sin\left(\theta x + \frac{n\pi}{2}\right)\right| \leq \frac{x^n}{n!}$

$\Rightarrow R_n \to 0$ as $n \to \infty$ for all $x \in R$

It follows that for all $x \in R$

$$\sin x = x - \frac{x^3}{3!} + \frac{x^5}{5!} \ldots\ldots\ldots$$

3. Expansion of $\cos x$

Let $f(x) = \cos x$, for all $x \in R$

Then $f^n(x) = \cos\left(x + \frac{n\pi}{2}\right)$, for all $x \in R$.

Denoting by the Lagrange's form of remainder, we have

$$R_n = \frac{x^n}{n!} f^n(\theta x), \text{ where } 0 < \theta < 1$$

$$= \frac{x^n}{n!} \cos\left(\theta x + \frac{n\pi}{2}\right)$$

Now for all $x \in R$,

$$|R_n| \leq \left|\frac{x^n}{n!}\right|$$

and $\lim_{n \to \infty} \frac{x^n}{n!} = 0$

Therefore $\lim_{n \to \infty} R_n = 0$

$\Rightarrow R_n \to 0$ as $n \to \infty$ for all $x \in R$

It follows that for all $x \in R$,

$$\cos x = 1 - \frac{x^2}{2!} + \frac{x^4}{4!} - \frac{x^6}{6!} + \cdots\cdots$$

for all $x \in R$,

4. Expansion of $(1+x)^m$

Let $f(x) = (1+x)^m$

Now, being any real number, possess continuous derivatives of everyorder when

$(1+x) > 0$ i,e., when $x > -1$

Also, $f^n(x) = m(m-1)(m-2)\ldots\ldots(m-n+1)(1+x)^{m-n}$

Writing Cauchy's remainder after n terms, we have

$$R_n = \frac{x^n}{(n-1)!}(1-\theta)^{n-1} f^n(\theta x) \text{ where } 0 < \theta < 1$$

$$= \frac{x^n}{(n-1)!}(1-\theta)^{n-1} m(m-1)\ldots\ldots(m-n+1)(1+\theta x)^{m-n}$$

$$= x^n \frac{m(m-1)\ldots\ldots\ldots(m-n+1)}{(n-1)!}\left(\frac{1-\theta}{1+\theta x}\right)^{n-1}(1+\theta x)^{m-1}$$

Let $|x| < 1$

$\Leftrightarrow -1 < x < 1$

Now $-1 < x$

$\Rightarrow -\theta < \theta x$

$\Rightarrow 1-\theta < 1+\theta x$

$$\Rightarrow \quad \frac{1-\theta}{1+\theta x} < 1$$

Thus, we have

$$0 < \left(\frac{1-\theta}{1+\theta x}\right)^{n-1} < 1$$

Let $(m-1)$ be positive, We have

$$0 < 1+\theta x < 1+1$$

$$\Rightarrow \quad 0 < (1+\theta x)^{m-1} < 2^m$$

Let $(m-1)$ be negative. We have

$$\theta x \geq -|x|$$

$$\Rightarrow \quad 1+\theta x \geq 1-|x|$$

$$\Rightarrow \quad (1+\theta x)^{m-1} \leq (1-|x|)^{m-1}$$

We know that

$$\lim_{n \to \infty} \frac{m(m-1)\ldots\ldots.(m-n+1)}{(n-1)!} x^n = 0$$

We observe that $R_n \to 0$ as $n \to \infty$ if $|x| < 1$.

By substitution, we get

$$(1+x)^m = 1 + mx + \frac{m(m-1)}{2} x^2 + \frac{m(m-1)(m-2)}{3} x^3 + \ldots\ldots$$

when, $-1 < x < 1$

5. Expansion of $\log(1+x)$

Let $f(x) = \log(1+x)$, whenever $-1 < x \leq 1$.

Then $f^n(x) = \dfrac{(-1)^{n-1}(n-1)!}{(1+x)^n}$, whenever $x > -1$.

Now cosider the following cases :

Case–1

Let $0 \le x \le 1$

Writing Lagrange's remainder after n terms, we have

$$R_n = \frac{x^n}{n!} f^n(\theta x)$$

$$= \frac{x^n}{n} f^n(\theta x)$$

$$= \frac{x^n}{n!}(-1)^{n-1} \frac{(n-1)!}{(1+\theta x)^n}$$

$$= \frac{(-1)^{n-1}}{n}\left(\frac{x}{1+\theta x}\right)^n$$

Since $0 \le x \le 1$, $0 < \theta < 1$, therefore.

$$0 < \frac{x}{1+\theta x} < 1$$

$$\therefore \quad |R_n| < \frac{1}{x} \text{ and } \frac{1}{n} \to 0 \text{ as } n \to \infty.$$

Therefore, $\lim\limits_{n \to \infty} R_n = 0$

Case–2

Let $-1 < x < 0$

Since in this case $\left|\frac{x}{1+\theta x}\right|$ need not be less than unity, therefore, we may not be able to show easily that $R_n \to 0$ as $x \to \infty$ by considering Lagrange's remainder.

Writing Cauchy's remainder, we have

$$R_n = \frac{x^n}{(n-1)!}(1-\theta)^{n-1} f^n(\theta x)$$

$$= (-1)^{n-1} x^n \left(\frac{1-\theta}{1+\theta x}\right)^{n-1} \frac{1}{1+\theta x}$$

Since $|x| < 1$,

therefore, $\left|\frac{1-\theta}{1+\theta x}\right| < 1,$

so that $\left|\left(\frac{1-\theta}{1+\theta x}\right)^{n-1}\right| < 1$

and $\left|\frac{1}{1+\theta x}\right| < \frac{1}{1-|x|}$

consequently,

$$|R_n| < \frac{|x|^n}{1-|x|}$$

By substitution, we have

$$\log(1+x) = x - \frac{x^2}{2} + \frac{x^3}{3} \ldots\ldots.|x| < 1.$$

7.8. FORMAL EXPANSIONS OF FUNCTIONS

Formal expansion of a function as a power series may, however, be obtained by assuming that it can be so expanded, i.e., R_n tends to 0 as n tends to infinity.

Thus, we have

$$f(x) = f(0) + xf'(0) + \frac{x^2}{2} f''(0) + \cdots\cdots \qquad \ldots(i)$$

To obtain the expression of a function, on the assumption that it is possible, we have only to calculate the values of its derivatives for $x = 0$ and substitute them in (i). This method can be illustrated by means of examples.

7.9. MONOTONE FUNCTIONS

Defination. 7.9.1 A function f defined on an interval I is said to be increasing in I if $f(x_1) \le f(x_2)$, whenever x_1, x_2 are in I and $x_1 < x_2$.

Defination. 7.9.2 A function f defined in an interval I is said to be strictly increasing in I if $f(x_1) \le f(x_2)$ whenever x_1, x_2 are in I and $x_1 < x_2$.

Definition. 7.9.3 A function f defined on an interval I is said to be decreasing in I if $f(x_1) \ge f(x_2)$, whenever x_1, x_2 are in I and $x_1 < x_2$.

Definition. 7.9.4 A function f defined on an interival I is said to be strictly decreasing in I if $f(x_1) > f(x_2)$ whenever x_1, x_2 are in I and $x_1 < x_2$.

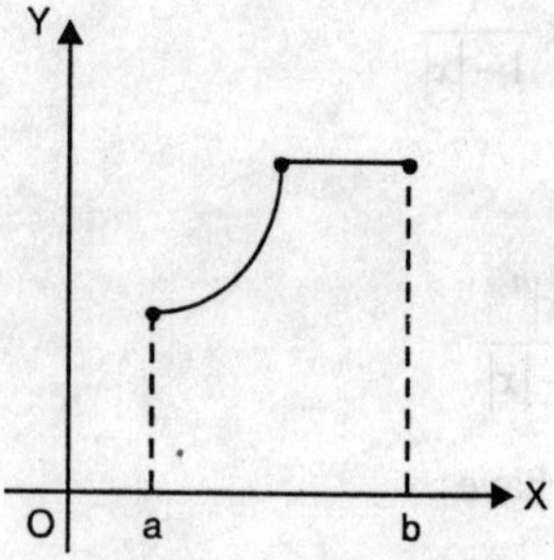

Graph of an increasing function

Figure 7.7

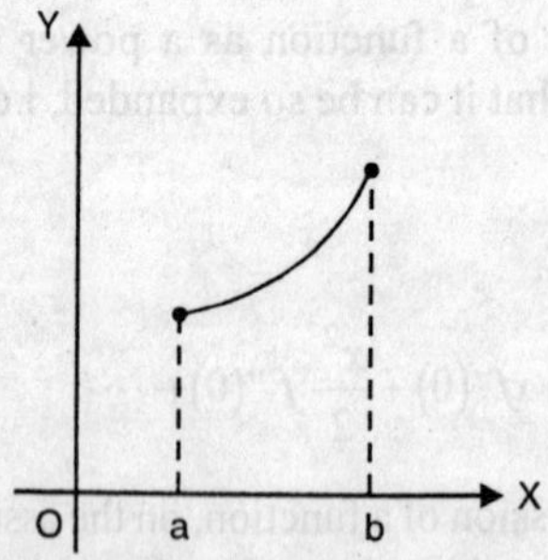

Graph of a strictly increasing function

Figure 7.8

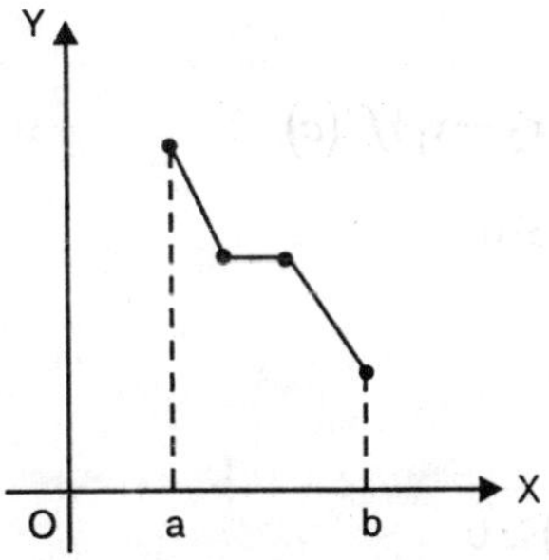

Graph of an decreasing function

Figure 7.9

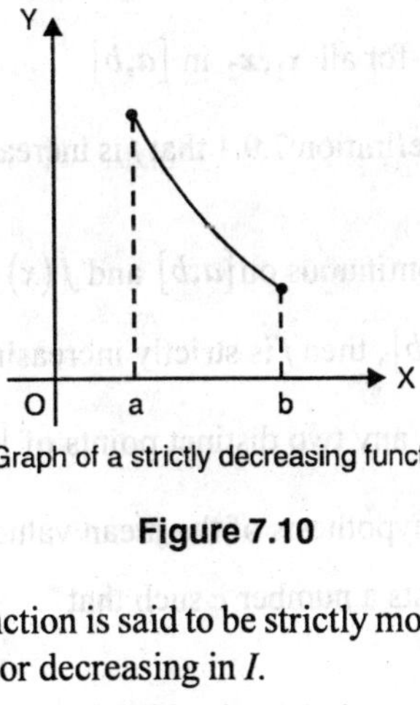

Graph of a strictly decreasing function

Figure 7.10

Definition. 7.9.5 A function is said to be strictly monotone in an interval *I* if it is either increasing in *I* or decreasing in *I*.

Definition. 7.9.6 A function is said to be strictly monotone in an interval *I* if it is increasing in *I* or strictly decreasing in *I*.

Theorem 7.9.1: If *f* is continuous on $[a,b]$ and $f'(x) \geq 0$ in $]a,b[$ then *f* is incrasing in $[a,b]$.

Proof : Let x_1 and x_2 be any two distinct points of $[a,b]$ such that $x_1 < x_2$. Then *f* satisfies the hypothesis of the mean value theorem in $[x_1, x_2]$. Therefore, there exists a number *c* such that.

$$x_1 < c < x_2$$

and $f(x_2) - f(x_1) = (x_2 - x_1) f'(c)$

Since $x_2 - x_1 > 0$

and $f'(c) \geq 0$

therefore,

$$f(x_2) - f(x_1) \geq 0$$

i.e., $f(x_1) \leq f(x_2)$

Since $x_1 < x_2$

$\Rightarrow \quad f(x_1) \leq f(x_2)$ for all x_1, x_2 in $[a,b]$

therefore, it follows by definition 7.9.1 that *f* is increasing in $[a,b]$.

Theorem 7.9.2: If *f* is continuous on $[a,b]$ and $f(x) > 0$ in $]a,b[$ then *f* is strictly increasing in $[a,b]$, then *f* is strictly increasing in $[a,b]$.

Proof: Let x_1 and x_2 be any two distinct points of $[a,b]$ such that $x_1 < x_2$.

Then *f* satisfies the hypothesis of the mean value theorem in $[x_1, x_2]$.

Therefore, there exists a number *c* such that

$$x_1 < c < x_2$$

and $\quad f(x_2) - f(x_1) = (x_2 - x_1) f'(c)$

Since $x_2 - x_1 > 0$ and $f'(c) > 0$

therefore, $f(x_2) - f(x_1) > 0$

Since $x_1 < x_2 \Rightarrow f(x_1) < f(x_2)$ for all x_1 and x_2 in $[a,b]$.

Therefore, *f* is strictly increasing in $[a,b]$.

Theorem 7.9.3: If *f* is continuous on $[a,b]$ and $f'(x) \leq 0$ in $]a,b[$, then *f* is decreasing in $[a,b]$.

Proof: Let us define a function g by setting

$$g(x) = -f(x), \text{ for all x in } [a,b].$$

Then g is continuous on $[a,b]$, and

$$g'(x) = -f'(x) \geq 0 \text{ in }]a,b[.$$

Therefore by theorem 7.9.1 the function g is increasing in $[a,b]$.

This means that if x_1 and x_2 be any two distinct point of $[a,b]$ sluch that $x_1 < x_2$, then

$$g(x_1) \leq g(x_2)$$

i.e., $-f(x_1) \leq -f(x_2)$

i.e., $f(x_1) \geq f(x_2)$

Since $x_1 < x_2 \Rightarrow f(x_1) \geq f(x_2)$, for all x_1 and x_2 in $[a,b]$ therfore, f is decreasing in $[a,b]$

Theorem 7.9.4: If f is continuous on $[a,b]$ and $f'(x) < 0$ in $]a,b[$, then f is strictly decreasing in $[a,b]$.

Proof: Let us define a function g by setting

$$g(x) = -f(x), \text{ for all } x \text{ in } [a,b]$$

Then g is continuous on $[a,b]$ and $g'(x) - f'(x) > 0$, for all x in $[a,b]$.

Therefore by theorem 7.9.2 the function g is strictly increasing in $[a,b]$.

This means that if x_1 and x_2 be any two distinct points in $[a,b]$ such that

$x_1 < x_2$,

then $g(x_1) < g(x_2)$

$\Rightarrow \quad -f(x_1) < -f(x_2)$

$$\Rightarrow \quad f(x_1) > f(x_2)$$

Since $x_1 < x_2 \Rightarrow f(x_1) > f(x_2)$, for all x_1 and x_2 in $[a,b]$, therefore, f is strictly decreasing in $[a,b]$.

Examples

Example 1: If f'' be continuous on $[a,b]$ and derivable on $]a,b[$, then prove that

$$f(b)-f(a)-(b-a)\{f'(a)+f'(b)\} = -\frac{(b-a)^3}{12} f'''(d)$$

for some real number d between a and b.

Solution: Let g be the function defined on $[a, b]$ by setting

$$g(x) = f(x) - f(a) - \frac{1}{2}(x-a)\{b'(a) + f'(x)\} + A(x-a)^3,$$

for all x in $[a,b]$ where A is a constant to be suitably chosen.

Now g is continuous on $[a,b]$

and g is derivable on $]a,b[$

Let A be so chosen such that $g(a) = g(b)$

Since $g(a) = 0$, therefore

$$f(b) - f(a) - \frac{1}{2}(b-a)\{f'(a) + f'(b)\} + A(b-a)^3 = 0 \quad \text{...(i)}$$

Now, the function g satisfies all the conditions of Rolle's theorem in $[a,b]$ and therefore, there exists a real number c between a and b, such that

$$g'(c) = 0$$

$$\Rightarrow \frac{1}{2}\{f'(c) - f'(a)\} - \frac{1}{2}(c-a)f''(c) + 3A(c-a)^2 = 0 \quad \text{...(ii)}$$

Let h be a function defined on $[a,c]$ by setting

$$h(x) = \frac{1}{2}\{f'(x) - f'(a)\} - \frac{1}{2}(x-a)f''(x) + 3A(x-a)^2 = 0 \quad \ldots\text{(iii)}$$

for all x in $[a,c]$.

Now, h is continuous on $[a,c]$

h is derivable on $]a,c[$

$h(c) = 0$ by (ii), so that

$h(c) = h(a)$.

$g(x) = -f(x)$, for all x in $[a, b]$

Now, the fucntion h satisfies all the conditions of Rolle's theorem in $[a, c]$ and therefore, there exists a real number d $(a < d < c)$ such that

$$h'(d) = 0$$

$$\Rightarrow \frac{1}{2}f'(d) - \frac{1}{2}f''(a) - \frac{1}{2}(d-a)f'''(d) + 6A(d-a) = 0$$

$$\Rightarrow A = f'''(d)/12. \text{ since } d - a \neq 0$$

From (i) and (iv), we have

$$f(b) - f(a) - \frac{1}{2}(b-a)\{f'(a) + f'(b)\} = -(b-a)^3 f'''(d)/12.$$

Example 2: If in the cauchy's Mean value theorem, $f(x) = e^x$ and $g(x) = e^{-x}$, show that c is the arithmetic mean between a and b.

Solution: $\dfrac{f(b) - f(a)}{g(b) - g(a)} = \dfrac{e^b - e^a}{\bar{e}^b - \bar{e}^a}$

$$= -e^a e^b$$

$$= -e^{a+b}$$

$$\frac{f'(x)}{g'(x)} = \frac{e^x}{-\bar{e}^x}$$

$$\Rightarrow \quad \frac{f'(c)}{g'(c)} = -e^{2c}$$

$$\therefore \quad -e^{a+b} = -e^{2c}$$

$$\Rightarrow \quad c = \frac{a+b}{2}.$$

Example-3: Show that the number θ which occurs in Lagrange's remainder' after n terms in Taylor's expansion of a function f in $[a, a+h]$, the interval $[a, a+h]$ approaches $\frac{1}{n+1}$ as $h \to 0$ provided

(i) f^n is continuous in $[a, a+h]$

(ii) f^{n+1} exists in $[a, a+h[$;

(iii) f^{n+1} is continuous at $x = a$,

(iv) $f^{n+1}(a) \neq 0$

Solution: Since f^n is continuous in $[a, a+h]$ therefore, by applying Taylor's theorem with Lagrange's form of remainder after n terms to the function f in $[a, a+h]$, we have

$$f(a+h) = f(a) + \cdots\cdots + \frac{h^n}{n!} f^n(a + \theta h) \qquad \ldots(1)$$

where $0 < \theta < 1$.

Since f^n is continuous in $[a, a+h]$ and f^{n+1} exists in $]a, a+h[$, therefore, by Taylor's theorem with remainder after $n+1$ terms, we have

$$f(a+h) = f(a) + hf'(a) + \ldots\ldots + \frac{h^n f^n(a)}{n!} + \frac{h^{n+1}}{(n+1)!} f^{n+1}(a+1,h) \qquad \ldots(2)$$

for some θ, such that $0 < \theta_1 < 1$

From (1) & (2), we have

$$f^n(a+\theta h)-f^n(a)=\frac{h}{n+1}f^{n+1}(a+\theta_1 h) \qquad \ldots(3)$$

Since f^n is continuous in $[a,a+h]$ and f^{n+1} exists in $]a,a+h[$, therefore, the fucntion f^n satisfies the hypothesis of the Mean value theorem in $[a,a+\theta h]$. Consequently,

$$f^n(a+\theta h)-f^n(a)=h\theta_1 f^{n+1}(a+\theta_2\theta h) \qquad \ldots(4)$$

for some θ_2 such that $0<\theta_2<1$

Comparing (3) and (4), we have

$$\theta f^{n+1}(a+\theta_2\theta h)=\frac{1}{n+1}f^{n+1}(a+\theta_1,h) \qquad \ldots(5)$$

As $h\to 0$, $\theta_2\theta h\to 0$, $\theta,h\to 0$.

Since f^{n+1} is continuous at $x = a$ therefore

$$\lim_{h\to 0} f^{n+1}(a+\theta_2\theta h)=\lim_{h\to 0}(a+\theta,\, h)=f^{n+1}(a) \qquad \ldots(6)$$

Taking limits of both sides of (5) as $h\to 0$ and using (6), we have

$$\left(\lim_{h\to 0}\theta\right)f^{n+1}(a)=f^{n+1}(a)\frac{1}{n+1} \qquad \ldots(7)$$

Since $f^{n+1}(a)\neq 0$, therefore, (7) implies that

$$\lim_{h\to 0}\theta=\frac{1}{n+1}$$

Example 4: Show that sin*x* lies between

$$x-\frac{x^3}{6} \text{ and } x-\frac{x^3}{6}+\frac{x^5}{120}.$$

Solution: If $x=0$, then each of the expressions sin*x*, $x-\frac{x^3}{6}$ and $x-\frac{x^3}{6}+\frac{x^5}{120}$ has the value 0.

Case–I

Let $x > 0$

By applying Taylor's theorem to the function f defined by

$$f(x) = \sin x \text{ in } [0, x]$$

and writing the remainder after three terms, we have

$$\sin x = x - \frac{x^3}{6}\cos(\theta x), \text{ where } 0 < \theta < 1$$

Since $-1 \le -\cos\theta x$, and since, $x > 0$

therefore,

$$x - \frac{x^3}{6} \le x - \frac{x^3}{6}\cos(\theta x)$$

$$= x - \frac{x^3}{6} \le \sin x \quad \ldots(1)$$

Again, by applying Taylor's theorem to the function in $[0, x]$ and writing remainder after five terms, we have

$$\sin x = x - \frac{x^3}{6} + \frac{x^5}{120}\sin(\theta_1 x), \text{ where } 0 < \theta_1 < 1.$$

Since $\sin(\theta_1 x) \le 1$ and since $\sin x > 0$

therefore,

$$x - \frac{x^3}{6} + \frac{x^5}{120}\sin(\theta_1 x) \le x - \frac{x^3}{6} + \frac{x^5}{120}$$

$$\Rightarrow \sin x \le x - \frac{x^3}{6} + \frac{x^5}{120} \quad \ldots(2)$$

From (1) & (2), we find that

$$x - \frac{x^3}{6} \le \sin x \le x - \frac{x^3}{6} + \frac{x^5}{120}, \quad \ldots(3)$$

Case–II

Let $x < 0$

If we take $y=-x$, then $y > 0$.

From (3), we have

$$y-\frac{y^3}{6} \le \sin y \le y-\frac{y^3}{6}+\frac{y^5}{120} \qquad \text{...(4)}$$

Putting, $y = -x$ in (4) we have

$$x-\frac{x^3}{6} \ge \sin x \ge x-\frac{x^3}{6}+\frac{x^5}{120}$$

Thus, we find that the given statement is true for all values of x.

Example 5: Prove that

$$\tan^{-1} x = \tan^{-1}\frac{\pi}{4}+\frac{x-\pi/4}{1+\pi^2/16}-\frac{\pi(x-\pi/4)^2}{4(1+\pi^2/16)}+.....$$

Solution: We have,

$$f(x) = \tan^{-1} x$$

$$= \tan^{-1}\left(\frac{\pi}{4}+x-\frac{\pi}{4}\right)$$

Let $a=\frac{\pi}{4}$ and $h = x-\pi/4$

Now,

$$f(a+h) = f(a)+hf'(a)+\frac{h^2}{2}f''(a)+.......$$

$$\Rightarrow f(x) = f\left(\frac{\pi}{4}\right)+\left(x-\frac{\pi}{4}\right)f'\left(\frac{\pi}{4}\right)+\frac{(x-\pi/4)^2}{2!}f''(\pi/4)+......$$

$$\Rightarrow f(x) = \tan^{-1}\pi/4+\frac{(x-\pi/4)}{1+\pi^2/16}-\frac{\pi(x-\pi/4)^2}{4(1+\pi^2/16)^2}+......$$

Where $f(x) = \tan^{-1} x$

$$f(\pi/4) = \tan^{-1}\frac{\pi}{4}$$

$$f'(x) = \frac{1}{1+x^2}$$

$$f'\left(\frac{\pi}{4}\right) = \frac{1}{1+\pi^2/16}$$

$$f''(x) = \frac{-2x}{\left(1+x^2\right)^2}$$

$$\Rightarrow \quad f''(\pi/4) = \frac{-2\times\frac{\pi}{4}}{\left(1+\pi^2/16\right)^2}$$

Example 6: If $f''(x) > 0$ for all $x \in R$, then show that

$$f\left(\frac{1}{2}(x_1+x_2)\right) \le \frac{1}{2}\left[f(x_1)+f(x_2)\right]$$

for every pair of real numbers x_1 and x_2.

Solution:

If $x_1 = x_2$, the result is obvious.

If $x_1 \ne x_2$, let us suppose that $x_1 < x_2$.

Applying Lagrange's Mean value theorem to the function *f* in the intervals

$\left[x_1, \frac{1}{2}(x_1+x_2)\right]$ and $\left[\frac{1}{2}(x_1+x_2), x_2\right]$, we have

$$f((x_1+x_2)/2) - f(x_1) = \left[\frac{1}{2}(x_1+x_2) - x_1\right]f'(c_1) \quad \ldots(1)$$

$$f(x_2) - f((x_1+x_2)/2) = \left[x_2 - \frac{1}{2}(x_1+x_2)\right]f'(c_2) \quad \ldots(2)$$

where c_1 and c_2 are some real numbers such that

$$x_1 < c_1 < \frac{1}{2}(x_1 + x_2) < c_2 < x_2$$

Subtracting bothsides of (2) from the corresponding sides of (1), we have

$$2f\left\{\frac{1}{2}(x_1 + x_2)\right\} - f(x_1) - f(x_2) = \frac{1}{2}(x_2 - x_1)[f'(c_1) - f(c_2)]$$

Again applying Lagrange's mean value theorem to the function f' in the interval $[c_1, c_2]$, we find that

$$f'(c_2) - f'(c_1) = (c_2 - c_1)f''(d)$$

where d is a suitable real number in $[c_1, c_2]$

Since $c_2 - c_1 > 0$ and $f''(d) > 0$, therefore

$$f'(c_2) - f(c_1) > 0$$

Since $x_2 > x_1$, therefore from (3) and (5), we have

$$2f\left(\frac{1}{2}(x_1 + x_2)\right) - f(x_1) - f(x_2) < 0$$

$$f\left(\frac{1}{2}(x_1 + x_2)\right) < \frac{1}{2}[f(x_1) + f(x_2)]$$

Example 7: Assuming the possibility of expansion, expand $y = e^x \log(1+x)$ in ascending powers of x.

Solution: We have,

$$y = e^x \log(1+x) \qquad \text{...(i)}$$

Differentiating w.r.t. x, we get

$$y_1 = \frac{e^x}{1+x} + e^x \log(1+x) \qquad \text{...(ii)}$$

$$\Rightarrow \quad (1+x)y_1 = e^x + (1+x)y$$

Differentiating w.r.t. x, we have

$$(1+x)y_2 + y_1 = e^x + (1+x)y_1 + y$$

$$(1+x)y_2 - xy_1 - e^x - y = 0 \quad \text{...(iii)}$$

Applying Leibnitz's theorem, we get

$$(1+x)y_{n+2} + (n-x)y_{n+1} - (n+1)y_n - e^x = 0 \quad \text{...(iv)}$$

Putting $x = 0$ in (i), (ii), (iii) and (iv), we get.

$$y(0) = 0$$

$$y_1(0) = 1$$

$$y_2(0) = 1$$

$$y_{n+1}(0) = -ny_{n+1}(0) + (n+1)y_n(0) + 1$$

Putting $n = 1,2,3,4,......$, we get

$$y_3(0) = -y_2(0) + 2y_1(0) + 1 = 2$$

$$y_4(0) = -2y_3(0) + 3y_2(0) + 1 = 0$$

$$y_5(0) = -3y_4(0) + 4y_3(0) + 1 = 9$$

$$y_6(0) = -4y_5(0) + 5y_4(0) + 1 = -35.$$

Substituting these values in the expansion

$$f(x) = f(0) + xf'(0) + \frac{x^2}{2!}f''(0) + \frac{x^3}{3!}f'''(0) + ...$$

We have

$$f(x) = x + \frac{x^2}{2!} + \frac{2x^3}{3!} + \frac{9x^5}{5!} - \frac{35x^6}{6!} + ...$$

where $y = f(x)$.

Exercise – 7.2

1. Verify Cauchy's Mean value theorem for the functions x^2 and x^4 in the interval $[a,b]$; being positive.

2. Prove that

$$\cos x = 1 - \frac{x^2}{2!} + \frac{x^4}{4!} - .... + \frac{(-1)^n x^{2n}}{(2n)!} - .....$$

3. Prove that if f^{n-1} and g^{n-1} both exist in $[a, a+h]$ and are derivable in $]a, a+h[$, then there exists a number θ between 0 and 1 such that

$$\frac{f(a+b) - f(a) - hf'(a) - ..... \frac{h^{n-1}}{(n-1)!} f^{n-1}(a)}{g(a+h) - g(a) - hg'(a) - ...... \frac{h^{n-1}}{(n-1)!} g^{n-1}(a)} = \frac{f^n(a+\theta h)}{g^n(a+\theta h)}$$

4. Let the function *f* be continuous in $[a,b]$ and derivable in $]a,b[$. Show that there exists a number $c \in]a,b[$ such that

$$2c[f(a) - f(b)] = f'(c)[a^2 - b^2]$$

5. If, in the cauchy's Mean value theorem, we write for *f*(*x*) and *g*(*x*), $\sqrt{x}$ and $\frac{1}{\sqrt{x}}$ respectively, then, c, the geometric mean between *a* and *b*, and if we write $\frac{1}{x^2}$ and $\frac{1}{x}$ then, *c* is the harmonic mean between *a* and *b*. It is understood that *a* and *b* are both positive.

6. Show that

$$e^x = 1 + x + \frac{x^2}{2!} + ... + \frac{x^{n-1}}{(n-1)!} + \frac{x^n}{n!} e^{\theta x} \text{ for all } x \in R.$$

7. Find by Maclaurin's theorem, the first four terms and the remainder after *n* terms of the Maclaurin's expression of $e^{ax} \cos bx$ in terms of the ascending powers of *x*.

8. Show that

$$\sin x = x - \frac{x^3}{3!} + \frac{x^5}{5!} \ldots + (-1)^{n-1} \frac{x^{2n-1}}{(2n-1)!} + (-1)^n \frac{x^{2n}}{(2n)!} \sin \theta x$$

for all $x \in R$.

9. Express $\sin^2 x$ and $\cos^2 x$ as power series, valid for all $x \in R$.

10. Show that for all $x \in]-1, 1[$,

$$\frac{1}{1+x} = 1 - x + x^2 - x^3 + x^4 - \ldots\ldots$$

11. Show that

$$e^{ax} \sin bx = bx + abx^2 + \frac{3a^2b - b^3}{3!} x^3 + \ldots\ldots$$

$$+ \ldots\ldots + \frac{(a^2+b^2)^{\frac{1}{2}n}}{n!} x^n \left(n \tan^{-1} \frac{b}{a} \right) + \ldots\ldots$$

for all $x \in R$.

12. Show that for all $x \in R$

$$\sinh x = x + \frac{x^3}{3!} + \frac{x^5}{5!} + \frac{x^7}{7!} \ldots\ldots$$

13. Show that for all $x \in R$,

$$e^{ax} \cos bx = 1 + ax + \frac{a^2 - b^2}{2!} x^2 + \frac{a(a^2 - 3b^2)}{3!} x^3 + \ldots\ldots$$

$$+ \ldots\ldots + \frac{(a^2+b^2)^{\frac{1}{2}n}}{n!} x^n \left(n \tan^{-1} \frac{b}{a} \right) + \ldots\ldots$$

14. Show that for all $x \in]-1, 1[$

$$\log(1-x) = -x - \frac{x^2}{2} - \frac{x^3}{3} - \frac{x^4}{4} - \frac{x^5}{5} \ldots$$

15. Show that

$$\cosh x = 1 + \frac{x^2}{2!} + \frac{x^4}{4!} = \frac{x^6}{6!} + \ldots\ldots$$

for all $x \in R$.

16. Assuming the possibility of expansion, expand $f(x) = e^{m\sin^{-1}x}$ in ascending integral powers of x.

17. Show that $\cos^2 x = 1 - x^2 + \frac{1}{3}x^4 - \frac{2}{45}x^6 \ldots\ldots$

18. If $y = \log\left[x + \sqrt{(1+x^2)}\right]$, prove that

$$\left(1+x^2\right)\frac{d^2y}{dx^2} + x\frac{dy}{dx} = 0.$$

19. Prove that $e^x \sin^2 x = x^2 + x^3 + \frac{1}{6}x^4 \ldots\ldots$

20. If $y = \sin\log\left(x^2 + 2x + 1\right)$, prove that

$$(x+1)^2 y_{n+2} + (2n+1)(x+1)y_{n+1} + \left(n^2+4\right)y_n = 0.$$

21. Prove that

$$\sin\left(\frac{\pi}{4} + x\right) = \frac{1}{\sqrt{2}}\left(1 + x - \frac{x^2}{2!} - \frac{x^3}{3!} + \ldots\ldots\right)$$

22. Prove that

$$e^{\sin x} = 1 + x + \frac{x^2}{2} - \frac{x^4}{8} + \ldots\ldots$$

23. Prove that

$$y = \frac{x^2}{2} + \frac{x^4}{12} + \frac{x^6}{45} + \ldots\ldots$$

where $y = \log\sec x$.

24. Show that $e^x - 1$ greater than $(1+x)\log(1+x)$, if x is positive.

25. Show that $(1+x)[\log(1+x)]^2 < x^2$, whenever $x > 0$.

26. Use Taylor's theorem to establish the following inequatities :

(i) $\cos x \geq 1 - \frac{x^2}{2}$, for all $x \in R$

(ii) $x - \frac{x^3}{6} < \sin x < x$, if $x > 0$

27. Prove that

$$1 + x + \frac{x^2}{2} \leq e^x \leq 1 + x + \frac{x^2}{2} e^x$$

for all $x \geq 0$.

28. Prove that $x - \frac{x^2}{2} \leq \sin x \leq x$

for all $x > 0$.

29. Obtain the first three terms in the expansion of $\log(1 + \tan x)$ in powers of x.

30. Obtain the following expressions :

(i) $\frac{x}{2}\frac{e^x + 1}{e^x - 1} = 1 + \frac{1}{6}.\frac{x^2}{2!} - \frac{1}{30}.\frac{x^4}{4!} + \frac{1}{42}.\frac{x^6}{6!}$

(ii) $\log\frac{\tan x}{x} = \frac{x^2}{3} + \frac{7}{90}x^4 + ...$

(iii) $\sin^{-1} x = x + \frac{1}{2}.\frac{x^3}{3} + \frac{1}{2}.\frac{3}{4}.\frac{x^5}{5} + ...$

(iv) $\log\sec x = \frac{1}{2}x^2 + \frac{1}{12}x^4 + \frac{1}{45}x^6 + ...$

(v) $\frac{x}{e^x - 1} = 1 - \frac{x}{2} + \frac{1}{6}.\frac{x^2}{2!} - \frac{1}{30}.\frac{x^4}{4!} + ...$

8

Maxima and Minima

8.1. MAXIMUM AND MINIMUM VALUES OF A FUNCTION

Let f be a function defined on an interval I and let c be any interior point of I.

(a) f is said to have a **maximum value** at $x = c$ if there exists a positive number δ such that

$$0 < |x - c| < \delta$$

$$\Rightarrow f(x) < f(c)$$

(b) f is said to have a **minimum value** at $x = c$ if there exists a positive number δ such that

$$0 < |x - c| < \delta$$

$$\Rightarrow f(x) > f(c)$$

(c) f is said to have **extreme value** at $x = c$ if it has either a maximum or a minimum at $x = c$.

8.2. A NECESSARY CONDITION FOR EXTREME VALUES

A necessary condition for $f(c)$ to be an extreme value of f is that $f'(c) = 0$ so that $f(c)$ is an extreme value $\Rightarrow f'(c) = 0$

Proof: Assume that f is derivable at c.

Let $f(c)$ be a maximum value of f.

There exists an interval $]c-\delta, c+\delta[$, around c, such that, if $c + h$ is a number other than c, belonging to this interval, we have

$$f(c+h) < f(c)$$

Here, h may be positive or negative. Thus,

$$h > 0 \Rightarrow \frac{f(c+h)-f(c)}{h} < 0 \quad \text{...(i)}$$

$$h < 0 \Rightarrow \frac{f(c+h)-f(c)}{h} > 0 \quad \text{...(ii)}$$

From (i) and (ii), we have

$$\lim_{h \to (0+0)} \Rightarrow \frac{f(c+h)-f(c)}{h} \leq 0 \text{ and}$$

$$\lim_{h \to (0-0)} \Rightarrow \frac{f(c+h)-f(c)}{h} \geq 0 \quad \text{...(iii)}$$

The relations (iii) will simultaneously be true, if and only if $f'(c) = 0$

It can similarly be shown that $f'(c) = 0$, if $f(c)$ is a minimum value of f.

Note1: The vanishing of $f'(c)$ is only a necessary but not a sufficient condition for $f(c)$ to be an extreme value.

For example, $f(x) = x^3$, for all $x \in R$

Here $f'(0) = 0$, but f does not have an extreme value at $x = 0$.

Note 2: $f(0) = 0$ is a minimum value of $f(x) = |x|$ even though f is not derivable at 0.

Note 3: If $f'(x) = 0$ at $x = c$, then we say that $x = c$ is a stationary point for f. Also, $f(c)$ is then said to be a stationary value of f.

Greatest and least values of a function in an interval

The greatest and least values of a function f is an interval $[a, b]$ are $f(a)$ or $f(b)$ or are given by the values of x for which $f'(x) = 0$.

If $c_1, c_2, \ldots\ldots, c_k$, be the roots of the equation $f'(x) = 0$ which belong to $]a, b[$, then the greatest and least numbers respectively of the finite set.

$$\{f(a), f(c_1), f(c_2), \ldots\ldots, f(c_x), f(b)\}$$

Change of sign

A function is said to change of sign from positive to negative as x passes through a number c, if there exists some left-handed neighbourhood $]c-h, c[$ of c for every point of which the function is positive and also there exits some right handed neighbourhood $]c, c+h[$ of c for every point of which the function is negative.

If a continuous function $f(x)$ changes sign as x passes through c we must have $f(c)=0$.

8.3. SUFFICIENT CONDITION FOR EXTREME VALUES

Theorem: $f(c)$ is an extreme value of f if and only if $f'(x)$ changes sign as x passes through c.

Case I: Let $f'(x)$ changes sign form positive to negative as x passes through c.

In some left-handed neighbourhood $]c-\delta, c[$ of c, $f'(x)$ is positive and so f is strictly increasing implying that $f(c)$ is the greatest of all the values of f.

In some right-handed neighbourhood $]c, c+\delta[$ of c, $f'(x)$ is negative and so f is strictly decreasing implying that $f(c)$ is the greatest of all the values of f.

Thus, $f(c)$ is the greatest of all the values of $f(x)$ when $f(x) \in]c-\delta, c+\delta[$

Case II: Let $f'(x)$ changes sign from negative to positive as x passes through c. It may be shown that $f(c)$ is the least of all the values of f, in a certain complete neighbourhood of $c \Rightarrow f(c)$ is a minimum value of f.

Case III: If $f'(x)$ does not change sign, then $f(x)$ is either strictly increasing or decreasing throughout this neighborhood implying that $f(c)$ is not an extreme value of f.

8.4. USE OF SECOND ORDER DERIVATIVES

Theorem 1: $f(c)$ is a minimum value of the function f if

$f'(c)=0$ and $f''(c)>0$.

As $f''(c)>0$ is +ve, there exists an interval $[c-\delta, c+\delta]$ around, c for every point x of which the second derivative $f''(x)$ is positive. This implies that $f'(x)$ is strictly increasing in $[c-\delta, c+\delta]$. Also since $f'(c)=0$, it follows that

$$f'(x)<0 \;\forall\; x \in [c-\delta, c[$$

and $f'(x)>0 \;\forall\; x \in]c, c+\delta]$.

Thus, f is strictly decreasing in $[c-\delta, c[$ and strictly increasing in $]c, c+\delta]$ so that $f(c)$ is a minimum value of $f(x)$.

Theorem 2: $f(c)$ is a maximum value of the function f, if $f'(c)=0$ and $f''(c)<0$

Since $f''(c)$ is negative, there exists an interval $[c-\delta, c+\delta]$ around c for every point x of which the second derivative is negative.

Thus, $f'(x)$ is strictly decreasing in $[c-\delta, c+\delta]$.

Also $f'(c)=0$

It follows that

$$f'(x)>0 \forall x \in [c-\delta, c[$$

$$f'(x)<0 \forall x \in]c, c+\delta]$$

Thus f is strictly increasing in $[c-\delta, c]$ and strictly decreasing in $[c, c+\delta]$ so that $f(c)$ is a maximum value of $f(x)$.

Examples

Example 1: Find the greatest and least values of the function $x^4-4x^3-2x^2+12x+1$ in the interval [–2, 5].

Solution: We have,

$$f(x) = x^4 - 4x^3 - 2x^2 + 12x + 1$$

$$\Rightarrow f'(x) = 4x^3 - 12x^2 - 4x + 12$$

$$= 4(x^3 - 3x^2 - x + 3)$$

$$= 4(x+1)(x-1)(x-3)$$

$$\Rightarrow f'(x) = 0 \text{ for } x = 1, -1, 3.$$

Now, $f(1) = 8$

$f(-1) = -8$

$f(3) = -8$

Thus, the least value is –8 and the greatest value is 8.

Example 2: Show that the function *f*, defined by

$f(x) = x^5 - 5x^4 + 5x^3 - 1$, for all $x \in R$

has a maximum value when $x = 1$, a minimum value when $x = 3$ and neither when $x = 0$.

Solution: We have,

$$f(x) = x^5 - 5x^4 + 5x^3 - 1, \; \forall \; x \in R$$

$$\Rightarrow f'(x) = 5x^4 - 20x^3 + 15x^2, \; \forall \; x \in R$$

$$= 5x^2(x-1)(x-3)$$

Since *f* is derivable for all $x \in R$ and $f'(x) = 0$ when $x = 0, 1, 3$, therefore, the critical values are 0, 1 and 3.

Now, $f'(x) > 0$ when $x < 0$

$f'(x) = 0$ when $x = 0$

$f'(x) > 0$ when $0 < x < 1$

$f'(x) = 0$ when $x = 1$

$f'(x) < 0$ when $1 < x < 1$

$f'(x) = 0$ when $x = 3$

$f'(x) > 0$ when $x > 3$

Since $f'(x)$ changes sign from positive to negative as x passes through 1, therefore, f has a maximum value at $x = 1$.

Again, since $f'(x)$ changes sign from negative to positive as x passes through 3, therefore, f has a minimum value at $x = 3$.

Example 3: If $f'(x) = (x-a)^{2n}(x-b)^{2m+1}$ where m and n are positive integers is the derivative of a function f. Then show that $x = b$ gives a minimum but $x = a$ is neither a maximum nor a minimum.

Solution: We have

$$f'(x) = (x-a)^{2n}(x-b)^{2m+1}$$

Now, $f'(x) = 0$

$\Rightarrow x = a, b$

For $x < b, (x-a)^{2n}(x-b)^{2m+1} < 0$

For $x > b, (x-a)^{2n}(x-b)^{2m+1} > 0$

Thus, f' changes sign from negative to positive as x passes through b as such has maximum at b.

Again, since $2n$ is an even integer $(x-a)^{2n}(x-b)^{2m+1}$ does not change as x passes through a. This function has neither maximum nor minimum for $x = a$.

Example 4: Show that $\sin^p\theta.\cos^q\theta$ attains a maximum, when $\theta = \tan^{-1}\sqrt{\frac{p}{q}}$

Solution: Let $y = \sin^p\theta.\cos^q\theta$

$$\therefore \quad \frac{dy}{dx} = \sin^{p-1}\theta.\cos\theta.\cos^q\theta + \sin^p\theta.q\cos^{q-1}\theta(-\sin\theta)$$

$$= p\sin^{p-1}\theta.\cos^{q+1}\theta - q\sin^{p+1}\theta.\cos^{q-1}\theta$$

$$= \sin^{p-1}\theta.\cos^{q+1}\theta(p\cos^2\theta - q\sin^2\theta)$$

$$= \sin^p\theta.\cos^q\theta\left(\frac{p\cos^2\theta - q\sin^2\theta}{\sin\theta.\cos\theta}\right)$$

$$= \sin^p\theta.\cos^q\theta(p\cot\theta - q\tan\theta)$$

For maximum or minimum, we must have

$$\frac{dy}{d\theta} = 0$$

$$\Rightarrow \quad \sin^p\theta.\cos^q\theta(p\cot\theta - q\tan\theta) = 0$$

$$\Rightarrow \quad \sin^p\theta = 0 \text{ or } \cos^q\theta = 0 \text{ or } p\cot\theta - q\tan\theta = 0$$

$$\Rightarrow \quad \theta = 0 \text{ or } \theta = \pi/2 \text{ or } \theta = \tan^{-1}\sqrt{p/q}$$

Now, $\frac{dy}{d\theta} = \sin^p\theta.\cos^q\theta(p\cot 0 - q\tan\theta)$

$$= y(p\cot\theta - q\tan\theta)$$

$$\Rightarrow \frac{d^2y}{d\theta^2} = \frac{dy}{d\theta}(p\cot\theta - q\tan\theta) + y(-p\operatorname{cosec}^2\theta - q\sec^2\theta)$$

$$\Rightarrow \left(\frac{d^2y}{d\theta^2}\right)_{\theta=\tan^{-1}\sqrt{p/q}} = \left(\frac{dy}{d\theta}\right)_{\theta=\tan^{-1}\sqrt{p/q}}\left(p\sqrt{\frac{q}{p}} - q\sqrt{\frac{p}{q}}\right)$$

$$+\sin^p\theta.\cos^q\theta\left[-p\operatorname{cosec}^2\theta - q\sec^2\theta\right]$$

$$= 0 - \sin^p\theta.\cos^q\theta(p\operatorname{cosec}^2\theta + q\sec^2\theta) < 0$$

Hence, y is maximum, when $\theta = \tan^{-1}\sqrt{\frac{p}{q}}$

Exmaple 5: Investigate for maximum and minimum values the function given by $y = \sin x + \cos 2x$.

Solution: We have

$$y = \sin x + \cos 2x$$

$$\therefore \quad \frac{dy}{dx} = \cos x - 2\sin 2x$$

$$= \cos x - 4\sin x.\cos x$$

$$\Rightarrow \quad \frac{dy}{dx} = 0 \text{ when } \cos x = 0 \text{ or } \sin x = 1/4$$

Now, $\cos x = 0 \Rightarrow x = \pi/2, 3\pi/2$

and $\sin x = 1/4 \Rightarrow x = \sin^{-1}\left(\frac{1}{4}\right), \pi - \sin^{-1} 1/4$

Now, $\frac{d^2y}{dx^2} = -\sin x - 4\cos 2x$

For, $x = \frac{\pi}{2}$

$$\Rightarrow \frac{d^2y}{dx^2} = 3 > 0$$

For $x = 3\pi/2$, $\frac{d^2y}{dx^2} = 5 > 0$.

For $x = \sin^{-1} 1/4$, $\frac{d^2y}{dx^2} = -\sin x - 4(1 - 2\sin^2 x) = -15/4 < 0$.

For $x = \pi - \sin^{-1} 1/4$, $\frac{d^2y}{dx^2} = -15/4 < 0$.

Therefore, y is maximum for $x = \sin^{-1} \frac{1}{4}, \pi - \sin^{-1} \frac{1}{4}$ and is a maximum for $x = \frac{\pi}{2}, \frac{3\pi}{2}$.

Example 6: Show that the maximum value of $(1/x)^x$ is $(e)^{1/e}$.

Solution: Let $y = (1/x)^x$

$$\Rightarrow \log y = -x\log x$$

$$\Rightarrow \frac{1}{y}.\frac{dy}{dx} = -(1+\log x)$$

$$\Rightarrow \frac{dy}{dx} = -(1+\log x)(1/x)^x$$

$$\frac{dy}{dx} = 0$$

$$\Rightarrow \log x + 1 = 0$$

$$\Rightarrow x = e^{-1}$$

Again, $\dfrac{d^2y}{dx^2} = -\dfrac{1}{x}\left(\dfrac{1}{x}\right)^x + (1+\log x)^x (1/x)^x$

At $x = e^{-1}$, $\dfrac{d^2y}{dx^2} = -e(e)^{1/e} < 0$

$\Rightarrow$ y has maximum value for $x = e^{-1}$ and the maximum value of y is $e^{\frac{1}{e}}$.

Example7: Show that the function f defined by

$$f(x) = |x|^p\,|x-1|^q \quad \forall\ x \in R$$

has a maximum value $p^q q^p/(p+q)^{p+q}$, p,q being positive.

Solution: We have,

$$f(x) = \begin{cases} (-1)^p(1-x)^q, & x < 0 \\ 0, & x = 0 \\ x^p(1-x)^q, & 0 < x < 1 \\ x^p(x-1)^q, & x > 1 \end{cases}$$

Consider $f(x) - x^p(1-x)^q, \quad 0 < x < 1$

$$\Rightarrow f'(x) = px^{p-1}(1-x)^q - qx^p(1-x)^{q-1}$$

$$= x^{p-1}(1-x)^{q-1}\left[p - x(p+q)\right]$$

$$f'(x) = 0 \Rightarrow x = 0, 1, p/(p+q)$$

Again, $f''(x) = (p-1)x^{p-2}(1-x)^{q-1}[p-x(p+q)]$

$$-(q-1)x^{p-1}(1-x)^{q-2}[p-x(p+q)]-(p+q)x^{p-1}(1-x)^{q-1}$$

$$f''(p/p+q) = -(p+q)\{p/(p+q)\}^{p-1}\{q/(p+q)\}^{q-1}$$

< 0 where p & q are integers.

Thus, f has a maximum value at $x = \dfrac{p}{p+q}, 0 < \dfrac{p}{p+q} < 1$. The max. value

is $\dfrac{p^p q^p}{(p+q)^{p+q}}$.

Example 8: If A denotes the arithmetic mean of the real numbers

$a_1, a_2, a_3, \ldots\ldots a_n$, show that $\sum_{i=1}^{n}(x-a_i)^2$ has a maximum value at A.

Solution: Let $f(x) = \sum_{i=1}^{n}(x-a_i)^2$

$$\Rightarrow f'(x) = \sum_{i=1}^{n} 2(x-a_i)$$

$$= 2nx - (a_1 + a_2 + \ldots\ldots + a_n)$$

$$= 2nx - 2na \text{, where } A = \frac{a_1 + a_2 + \ldots\ldots + a_n}{n}$$

$$f'(x) = 0 \Rightarrow x = A$$

Again, $f''(x) = 2n > 0$

Thus, the given sum has a minimum value for $x = A$.

Exercise – 8.1

1. Prove that the function f defined by

   $f(x) = 3|x| + 4|x-1| \ \forall \ x \in R$

   has a minimum value 3 at $x = 1$.
2. Prove that the function f defined by

   $f(x) = 2|x-2| + 5|x-3| \ \forall \ x \in R$

   has a minimum value 2 at $x = 3$.
3. Find the greatest and least values of the following functions.

   (i) $3x^4 - 2x^3 - 6x^2 + 6x + 1$ in the interval $[0, 2]$

   (ii) $2x^3 - 15x^2 + 36x + 1$ in the interval $[2, 3]$

   (iii) $(x-2)\sqrt[3]{x-1}$ in $[1, 9]$

   (iv) $(x-1)^2 + 3$ in $[-3, 1]$
4. If $y = \dfrac{ax - b}{(x-1)(x-4)}$ has a turning point P (2, -1), find the values of a and b and show that y is maximum at P.
5. Find the maximum profit that a company can make, if the profit function is given by

   $p(x) = 41 + 24x - 18x^2$
6. Show that $\dfrac{\log x}{x}$ has a maximum value at $x = e$.
7. Show that the function $f(x) = (x+2)(x-1)^2(2x-1)(x-3)$ changes sign form positive to negative as x passes through ½ and from negative to positive as x passes through –2 or 3.
8. Show that the function $f(x) = (2x+3)(x+4)(x-2)(x-1)^3$ changes sign from positive to negative as x passes through –4 and 1 from negative to positive as x passes through –3/2 and 2.
9. Find the extreme values of $5x^6 + 18x^5 - 10$.
10. Show that $x^5 - 5x^4 + 3x^2 - 1$ has a maximum value when $x = 1$ and a

minimum value when $x = 3$ and neither when $x = 0$.

11. Find the minimum and maximum values of

(i) $\sin x.\cos 2x$ (ii) $e^x \cos(x-a)$

(iii) $a \sec x + b \operatorname{cosec} x$ (iv) $\sin x \cos^2 x$

12. Find the maximum value of $\log x / x$ in $[0, \infty]$.

13. Find the values of x for which $\sin x - x\cos x$ is a maximum or a minimum.

14. Find the extreme values of the expression $x^3/(x^4+1)$.

15. Find the maxima and minima as well as the greatest and the least values of the function $f(x) = x^3 - 12x^2 + 45x$ in the interval [0, 7].

8.5. APPLICATION TO PROBLEMS

Different types of problem can be solved by applying the theory of maxima and minima.

If y is 0 for $x = a$ as well as for $x = b$ and positive otherwise and has only one stationary value, then the stationary value is necessarily the maximum and the greatest.

If $y \to \infty$ as $x \to a$ and $x \to b$ and has only one stationary value, then the stationary value is necessarily the minimum and the least.

The following results will be useful for solving the applied problems in maxima and minima.

1. ***Square of side x :***

   $Area = x^2$

   $Perimeter = 4x$

2. ***Rectangle of sides x and y :***

   Area $= xy$

   Perimeter $= 2(x+y)$

3. ***Circle of radius, r :***

   Area $= \pi r^2$

   Circumference $= 2\pi r$

4. ***Sphere of radius, r :***

   Volume $= \dfrac{4}{3}\pi r^3$

   Surface Area $= 4\pi r^2$

5. ***Right circular cylinder of base radius r and height h :***

   Volume $= \pi r^2 h$

   Surface Area $= 2\pi rh + 2\pi r^2$

   Curved surface Area $= 2\pi rh$

6. ***Right circular cone of height h, slant height l and radius of the base r:***

   Volume $= (1/3)\pi r^2 h$

   Curved Surface Area $= \pi rl$

   Total surface Area $= \pi r^2 + \pi rl$

7. ***Cuboid of edges of lengths x, y and z :***

Volume $= xyz$

Surface Area $= 2(xy + yz + zx)$

8. ***Cube of edge length x :***

Volume $= x^3$

Surface Area $= 6x^2$

9. ***Equilateral triangle of side x :***

Area $= \frac{\sqrt{3}}{4}x^2$

Perimeter $= 3x$

Examples

Example 1: Find two positive numbers whose sum is 14 and the sum of whose squares is minimum.

Solution: Let the numbers be x and y. Then,

$$x + y = 14 \qquad \text{...(i)}$$

Let $S = x^2 + y^2$. Then,

$$S = x^2 + (14 - x)^2$$

$$\Rightarrow S = 2x^2 - 28x + 196$$

$$\Rightarrow \frac{dS}{dx} = 4x - 28 \text{ and } \frac{dS}{dx} = 0$$

$$\therefore \frac{dS}{dx} = 0 \Rightarrow 4x - 28 = 0 \Rightarrow x = 7$$

Now, $\frac{d^2S}{dx^2} = 4 > 0$

Thus, S is minimum when $x = 7$. putting $x = 7$ in (i), we get $y = 7$.

Hence, the required numbers are 7 and 7.

Example 2: Show that the rectangle of maximum perimeter which can be inscribed in a circle of radius a is a square of side $\sqrt{2}a$.

Solution: Let $ABCD$ be a rectangle in a given circle of radius a with centre at O.

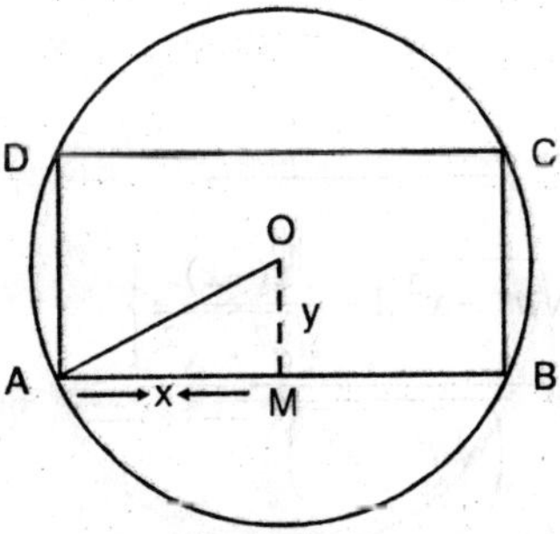

Figure 8.1

Let $AB = 2x$ and $AD = 2y$ be the sides of the rectangle. Then

$$AM^2 + OM^2 = OA^2$$

$$\Rightarrow x^2 + y^2 = a^2$$

$$\Rightarrow y = \sqrt{a^2 - x^2} \quad \ldots(i)$$

Let P be the perimeter of the rectangle $ABCD$. Then

$$P = 4x + 4y$$

$$\Rightarrow P = 4x + 4\sqrt{a^2 - x^2} \quad \text{[Using (i)]}$$

$$\Rightarrow \frac{dP}{dx} = 4 - \frac{4x}{\sqrt{a^2 - x^2}}$$

For maximum or minimum values of P, we have

$$\frac{dP}{dx} = 0$$

$$\Rightarrow 4 - \frac{4x}{\sqrt{a^2 - x^2}} = 0$$

$$\Rightarrow 4 = \frac{4x}{\sqrt{a^2 - x^2}}$$

$$\Rightarrow a^2 - x^2 = x^2$$

$$\Rightarrow 2x^2 = a^2$$

$$\Rightarrow x = \frac{a}{\sqrt{2}}$$

Now, $$\frac{d^2P}{dx^2} = \frac{-4\left\{\sqrt{a^2-x^2}.1 - \frac{x(-x)}{\sqrt{a^2-x^2}}\right\}}{\left(\sqrt{a^2-x^2}\right)^2}$$

$$= \frac{-4a^2}{\left(a^2-x^2\right)^{3/2}}$$

$$\therefore \left(\frac{d^2P}{dx^2}\right)_{x=a/\sqrt{2}} = \frac{-4a^2}{\left(a^2-\frac{a^2}{2}\right)^{3/2}} = \frac{-8\sqrt{2}}{a} < 0$$

Thus, P is maximum when $x = \frac{a}{\sqrt{2}}$

Putting $x = \frac{a}{\sqrt{2}}$ in (i), we obtain $y = \frac{a}{\sqrt{2}}$.

Therefore, $x = y = a\sqrt{2} \quad \Rightarrow 2x = 2y$

Hence, P is maximum when the rectangle is square of side $2x = \frac{2a}{\sqrt{2}} = \sqrt{2}a$.

Example 3: Show that the triangle of maximum area that can be inscribed in a given circle is an equilateral triangle.

***Solution*:** Let ABC be a triangle inscribed in a given circle with center O and radius r.

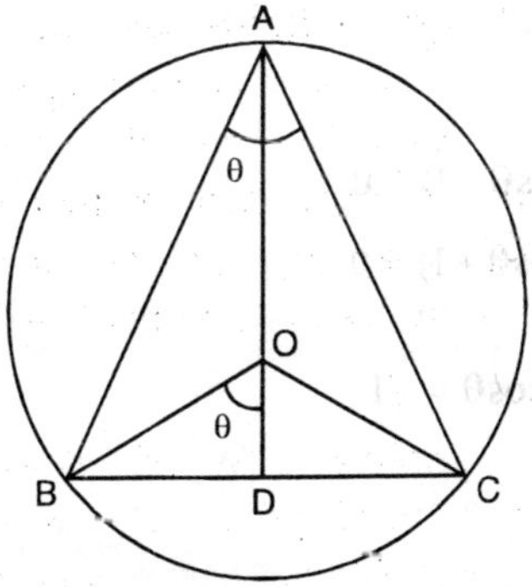

Figure 8.2

The area of the triangle will be maximum if its vertex *A* opposite to the base *BC* is at a maximum distance from the base *BC*. This is only possible only when *A* lies on the diameter perpendicular to *BC*. Thus, $AD \perp BC$. So, triangle *ABC* must be isosceles triangle.

Let $\angle CAB = \theta$. Then $\angle COB = 2\theta$. So $\angle COD = \theta$. Now, in triangle COD, we have $CD = r\sin\theta$ and $OD = r\cos\theta$

$\therefore \quad BC = 2CD = 2r\sin\theta$

and $AD = OA + OD = r + r\cos\theta$

Let *A* be the area of the triangle *ABC*. Then,

$$A = \frac{1}{2} BC.AD$$

$$\Rightarrow A = \frac{1}{2}(2r\sin\theta)(r + r\cos\theta)$$

$$= r^2(\sin\theta + \sin\theta.\cos\theta)$$

$$\Rightarrow \frac{dA}{d\theta} = r^2(\cos\theta + \cos^2\theta - \sin^2\theta)$$

$$= r^2(2\cos^2\theta + \cos\theta - 1)$$

and, $$\frac{d^2A}{d\theta^2} = r^2[-4\cos\theta.\sin\theta - \sin\theta]$$

For maximum or minimum, we have

$$\frac{dA}{d\theta} = 0$$

$$\Rightarrow r(2\cos^2\theta + \cos\theta - 1) = 0$$

$$\Rightarrow (2\cos\theta - 1)(\cos\theta + 1) = 0$$

$$\Rightarrow \cos\theta = \frac{1}{2} \text{ or } \cos\theta = -1$$

$$\Rightarrow \cos\theta = \frac{1}{2}$$

$$\Rightarrow \theta = \pi/3$$

And, $$\left(\frac{d^2A}{d\theta^2}\right)_{\theta=\pi/3} = r^2\left[-4\cos\frac{\pi}{3}.\sin\frac{\pi}{3} - \sin\frac{\pi}{3}\right]$$

$$= \frac{-3\sqrt{3}}{2}r^2 < 0$$

Thus, A is maximum when $\theta = \pi/3$, i.e., $\angle CAB = \pi/3$ but the triangle ABC is isosceles. Therefore, the triangle is equilateral. Hence A is maximum when the triangle is equilateral.

Example 4: Find the volume of the largest cylinder that can be inscribed in a sphere of radius r.

Solution: Let h be the height and R be the radius of the base of the inscribed cylinder.

Let V be the volume of the cylinder. Then,

$$V = \pi R^2 h \qquad \ldots\text{(i)}$$

From ΔOCA, we have $r^2 = \left(\frac{h}{2}\right)^2 + R^2$

$$\Rightarrow \quad R^2 = r^2 - h^2/4$$

$$\therefore \quad V = \pi\left(r^2 - h^2/4\right)h$$

$$\Rightarrow \quad V = \pi r^2 h - \frac{\pi}{4}h^3$$

$$\Rightarrow \quad \frac{dV}{dh} = \pi r^2 - \frac{3\pi h^2}{4} = 0$$

and $\dfrac{d^2V}{dh^2} = \dfrac{-3\pi h}{2}$

For maximum or minimum,

$$\frac{dV}{dh} = 0$$

$$\Rightarrow \quad \pi r^2 - \frac{3\pi h^2}{4} = 0$$

$$\Rightarrow \quad h^2 = \frac{4r^2}{3}$$

$$\Rightarrow \quad h = \frac{2}{\sqrt{3}} r$$

Now, $\left(\dfrac{d^2V}{dh^2}\right)_{h=\frac{2r}{\sqrt{3}}} = \dfrac{-\pi r}{\sqrt{3}} < 0$

Thus, V is maximum when $h = \dfrac{2r}{\sqrt{3}}$

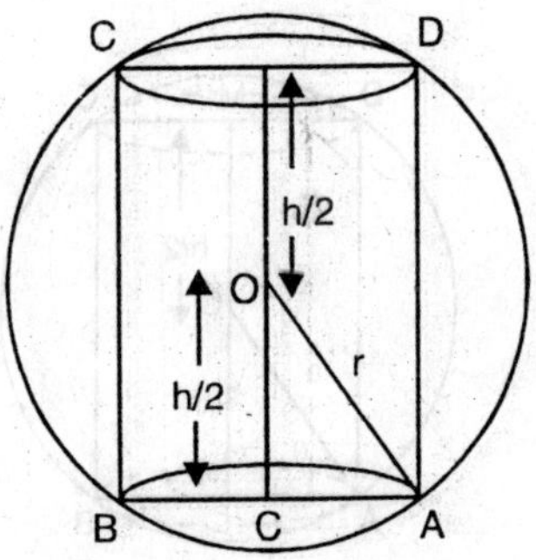

Figure 8.3

Putting $h = \frac{2r}{\sqrt{3}}$ in $R^2 = r^2 - \frac{h^2}{4}$, we obtain $R = \sqrt{\frac{2}{3}}r$

The maximum volume of the cylinder is given by

$$V = \pi R^2 h$$

$$= \pi\left(\frac{2}{3}\pi^2\right)\left(\frac{2r}{\sqrt{3}}\right)$$

$$= \frac{4\pi r^3}{3\sqrt{3}}$$

Example 5: Show that the height of the cylinder of maximum volume that can be inscribed in a sphere of radius a is $\frac{2a}{\sqrt{3}}$

Solution: Let r be the radius of the base and h be the height of the cylinder $ABCD$ which is inscribed in a sphere of radius a.

Let O be the centre of the sphere such that $\angle OAB = \theta$. Then

$r = AL = a\cos\theta$ and $h = (OL) = 2a\sin\theta$

Let V be the volume of the cylinder. Then

$$V = \pi r^2 h = 2\pi a^3 \cos^2\theta \sin\theta$$

$$\Rightarrow \frac{dV}{d\theta} = 2\pi a^3\left[-2\sin^2\theta.\cos\theta + \cos^3\theta\right]$$

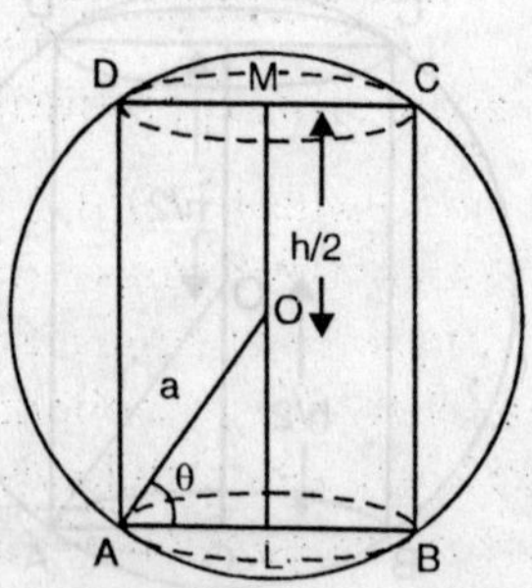

Figure 8.4

For maximum or minimum V, we have, $\frac{dV}{d\theta} = 0$

$$\therefore \quad 2\pi a^3\left(-2\sin^2\theta.\cos\theta + \cos^3\theta\right) = 0$$

$$\Rightarrow \quad \cos^3\theta\left(1 - 2\tan^2\theta\right) = 0$$

$$\Rightarrow \quad \cos\theta = 0 \text{ or } \tan\theta = \frac{1}{\sqrt{2}}$$

But $\cos\theta = 0$ is not possible as it gives $\theta = \pi/2$.

$$\tan\theta = \frac{1}{\sqrt{2}}$$

Now, $\frac{d^2V}{d\theta^2} = 2\pi a^3\left[-7\cos^2 0.\sin\theta + 2\sin^3\theta\right]$

Putting $\tan\theta = 1/\sqrt{2}$ i.e. $\sin\theta = 1/\sqrt{3}$ and $\cos\theta = \sqrt{2/3}$, we have

$$\frac{d^2V}{d\theta^2} = 2\pi a^3\left[\frac{14}{-3\sqrt{3}} + \frac{2}{3\sqrt{3}}\right]$$

$$= \frac{-8\pi a^3}{\sqrt{3}} < 0$$

Hence, the volume V is maximum when $\tan\theta = 1/\sqrt{2}$. Also, the height of the cylinder is $h = 2a\sin\theta = \frac{2a}{\sqrt{3}}$

Example 6: Show that the volume of the greatest cylinder which can be inscribed in a cone of height h and semi-vertical angle α is $\frac{4}{27}\pi h^3 + \tan^2\alpha$.

Solution: Let VAB be a given cone of height h; semi-vertical angle α and let x be the radius of the base of the cylinder $A'B'DC$ which is inscribed in the cone VAB. Then,

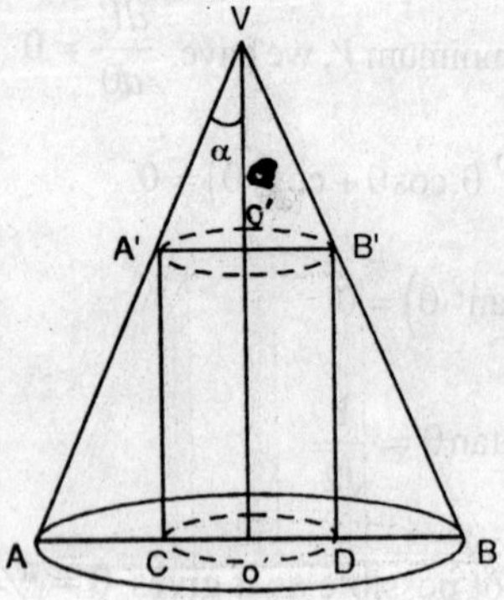

Figure 8.5

$$OO' = \text{height of the cylinder}$$
$$= VO - VO'$$
$$= h - x\cot\alpha$$

Let V be the volume of the cylinder, then,

$$V = \pi x^2(h - x\cot\alpha) \qquad \text{...(i)}$$

$$\Rightarrow \frac{dV}{dx} = 2\pi xh - 3\pi x^2 \cot\alpha$$

For maximum or minimum V,

$$\frac{dV}{dx} = 0$$

$$\Rightarrow 2\pi xh - 3\pi x^2 \cot\alpha = 0$$

$$\Rightarrow x = \frac{2h}{3}\tan\alpha \quad [\because x \neq 0]$$

Now, $\dfrac{d^2V}{dx^2} = 2\pi h - 6\pi x\cot\alpha$

When $x = \dfrac{2h}{3}\tan\alpha$, we have

$$\frac{d^2V}{dx} = \pi(2h - 4h) = -2\pi h < 0$$

Hence, V is maximum when $x = \frac{2h}{3}\tan\alpha$,

The maximum volume of the cylinder is

$$V = \left(\frac{2h}{3}\tan\alpha\right)^2\left(h - \frac{2h}{3}\right) \qquad \text{[From (i)]}$$

$$= \frac{4}{27}\pi h^3 \tan^2\alpha$$

Example 7: The combined resistance R of two resistors R_1 and $R_2 (R_1, R_2 > 0)$ is given by

$$\frac{1}{R} = \frac{1}{R_1} + \frac{1}{R_2}$$

If $R_1 + R_2 = C$ (a constant), show that the maximum resistance R is obtained by choosing $R_1 = R_2$.

Solution: We have,

$$\frac{1}{R} = \frac{1}{R_1} + \frac{1}{R_2} \text{ and } R_1 + R_2 = C$$

$$\therefore \quad \frac{1}{R} = \frac{R_1 + R_2}{R_1 R_2} = \frac{C}{R_1 R_2} = \frac{C}{R_1(C - R_1)}$$

$$\Rightarrow R = \frac{R_1 C - R_1^2}{C} = R_1 - \frac{R_1^2}{C}$$

$$\Rightarrow \frac{dR}{dR_1} = 1 - \frac{2R_1}{C}$$

$$\Rightarrow \frac{d^2R}{dR_1^2} = \frac{-2}{C}$$

For maximum or minimum, we must have

$$\frac{dR}{dR_1} = 0$$

$$\Rightarrow 1-\frac{2R_1}{C}=0$$

$$\Rightarrow R_1=\frac{C}{2}$$

Now, $\dfrac{d^2R}{dR_1^{\,2}}=\dfrac{-2}{C}<0$ for all values of R_1.

Thus, R is maximum when $R_1=C/2$

When $R_1=C/2$, we have, $R_2=C-\dfrac{C}{2}=\dfrac{C}{2}$

Hence, R is maximum when $R_1=R_2=C/2$

Example 8: Normal is drawn at a variable point p of an ellipse $x^2/a^2+y^2/b^2=1$; find the maximum distance of the normal from the center of the ellipse.

Solution: Take any point $p(a\cos\theta, a\sin\theta)$ on the ellipse; θ being the eccentric angle of the point. Because of the symmetry of the ellipse about the two co-ordinate axes, it is enough to consider only those of the values of θ which lie between 0 and $\pi/2$.

The equation of tangent at p is

$$\frac{x\cos\theta}{a}+\frac{y\sin\theta}{b}=1$$

Hence the slope of the normal at $P=\dfrac{a\sin\theta}{b\cos\theta}$.

Therefore, the equation of the normal at P is

$$ax\sin\theta-by\cos\theta=(a^2-b^2)\sin\theta.\cos\theta$$

If, p is perpendicular distance from the centre (0, 0), we obtain

$$p=\frac{(a^2-b^2)\sin\theta.\cos\theta}{\sqrt{a^2\sin^2\theta+b^2\cos^2\theta}} \qquad \text{...(i)}$$

$$\Rightarrow \frac{dp}{d\theta} = (a^2 - b^2)\frac{b^2\cos^4\theta - a^2\sin^4\theta}{(a^2\sin^2\theta + b^2\cos^2\theta)^{3/2}}$$

$$\Rightarrow \frac{dp}{d\theta} = 0 \text{ for } \tan^4\theta = \frac{b^2}{a^2}$$

i.e., for $\tan\theta = \pm\sqrt{\frac{b}{a}}$

Now, $p = 0$ when $\theta = 0$ or $\pi/2$

and p is +ve when θ lies between 0 and $\pi/2$

Therefore p is maximum, when $\tan\theta = \sqrt{(b/a)}$,

Substituting this value in (i), we see that the maximum value of p is $a - b$.

Example 9: Assuming that the petrol burnt (per hour) in driving a motor boat varies as the cube of its velocity, show that the most economical speed when going against a current of c miles per hour is $\frac{3}{2}c$ miles per hour.

Solution: Let v miles per hour be the velocity of the boat so that $v - c$ miles per hour is the velocity relative to water when going against the current.

Therefore the time required to cover a distance of d miles $= \dfrac{d}{v-c}$ hours.

The petrol burnt per hour $= kv^3$, where k is constant. Thus the total amount, y, of petrol burnt is given by

$$y = k.\frac{v^3 d}{v-c} = kd\frac{v^3}{v-c}$$

$$\Rightarrow \frac{dy}{dv} = kd\frac{3v^2(v-c) - 1.v^3}{(v-c)^2}$$

$$\Rightarrow \frac{dy}{dv} = 0 \text{ for } v = 0 \text{ and } \frac{3}{2}c\text{, of these } v = 0 \text{ is inadmissible.}$$

Also, $v \to c \Rightarrow y \to \infty, v \to \infty \Rightarrow y \to \infty$.

Thus, $v = \frac{3}{2}c$ gives the least value of C.

Exercise – 8.2

1. Divide 64 into two parts such that the sum of the cubes of two parts is minimum.
2. Of all closed cylindrical cans (right circular), which enclose a given volume of $100 cm^3$, which has the minimum surface area?
3. A beam is supported at the two ends and is uniformly loaded. The bending moment M at a distance x from one end is given by

   (i) $M = \frac{WL}{2}x - \frac{W}{2}x^2$ (ii) $M = \frac{Wx}{3} - \frac{W}{3}\frac{x^3}{L^2}$

   Find the point at which M is maximum in each case.
4. Show that the height of the cylinder of maximum volume that can be inscribed in a sphere of radius R is $\frac{2R}{\sqrt{3}}$.
5. Show that among all positive numbers x and y with $x^2 + y^2 = r^2$ the sum $(x + y)$ is largest when $x = y = r/\sqrt{2}$
6. Find the point on the curve $y^2 = 4x$ which is nearest to the point $(2, -8)$.
7. A rectangle is inscribed in a semi-circle of radius r with one of its sides a diameter of semi-circle. Find the dimensions of the rectangle so that its area is maximum. Find also the area.
8. The total cost of producing x radio sets per day is Rs $\left(\frac{x^2}{4} + 35x + 25\right)$ and the price per set at which they may be sold is Rs. $\left(50 - \frac{x}{2}\right)$. Find the daily output to maximize the total profit.
9. Find the area of greatest isosceles triangle that can be inscribed in a given ellipse having its vertices incident with one extremely of major axis.
10. Show that the cone of the greatest volume which can be inscribed in a given sphere has an attitude equal to 2/3 of the diameter of the sphere.

11. If $(x) = x^3 + ax^2 + bx + c$ has a maximum at $x = 1$ and minimum at $x = 3$, determine a, b and c.

12. A box of constant volume C is to be twice as long as it is wide. The material on the top and four sides cost three times as much per square meter as that in the bottom. What are the most economic dimensions?

13. Find the surface of the right circular cylinder of greatest surface which can be inscribe in a sphere of radius r.

14. A cone is circumscribed to a sphere of radius r; show that when the volume of the cone is least its altitude is $4r$ and its semi-vertical angle is

$$\sin^{-1}\frac{1}{3}.$$

15. The sum of the surfaces of a cube and a sphere is given; show that when the sum of their volumes is least the distance of the sphere is equal to the edge of the cube.

16. Show that the semi-vertical angle of the cone of maximum volume and of given slant height is $\tan^{-1}\sqrt{2}$.

17. Show that the height of a closed cylinder of given volume and least surface is equal to its diameter.

18. A cone is inscribed in a sphere of radius r; prove that its volume as well as its curved surface is greatest when its altitude is 4r/3.

19. Prove that the area of the triangle formed by the tangent at any point of the ellipse $x^2/a^2 + y^2/b^2 = 1$ and its axes is a minimum for the point

$$\left(a\sqrt{2}, b\sqrt{2}\right)$$

20 A rectangular sheet of metal has four equal square portions removed at the corners and the sides are then turned up so as to form an open rectangular box. Show that when volume contained in the box is a maximum, the depth will be

$$\frac{1}{4}\left[(a+b) - (a^2 - ab + b^2)^{1/2}\right]$$

where, a, b are the sides of the original rectangle.

21. Show that the maximum rectangle inscribable in a circle is a square.

9

Indeterminate Form

9.1. INTRODUCTION

Let $f(x)$ and $g(x)$ be two functions such that $\lim_{x\to a} f(x)$ and $\lim_{x\to a} g(x)$ both exist and $\lim_{x\to a} f(x) \neq 0$.

Then, $\lim_{x\to a} \dfrac{f(x)}{g(x)} = \dfrac{\lim_{x\to a} f(x)}{\lim_{x\to a} g(x)}$

Now, suppose that the denominator tends to zero as $x \to a$, then a necessary condition for $\lim_{x\to a}[f(x)/g(x)]$ to exist and be finite is that $\lim_{x\to a} f(x) = 0$

In fact, if $\lim_{x\to a} \dfrac{f(x)}{g(x)} = l$

then, $\lim_{x\to a} f(x) = \lim_{x\to a} \dfrac{f(x)}{g(x)} . g(x)$

$$= \lim_{x \to a} \frac{f(x)}{g(x)} . \lim_{x \to a} g(x)$$

$$= l.0 = 0$$

Thus, we have a contradiction.

The limit of the denominator in each of the following three cases is zero when $x \to 0$:

(i) $\lim(1/x^2) = +\infty$

(ii) $\lim(1/-x^2) = -\infty$

(iii) $\lim(1/x)$ does not exist.

The method which is necessary to evaluate $\lim_{x \to a}[f(x)/g(x)]$ when $\lim_{x \to a} f(x) = \lim_{x \to a} g(x) = 0$ is generally called **L'Hospital's rule.**

9.2. THE INDETERMINATE FORM: 0/0

If $\lim_{x \to a} f(x) = 0$, $\lim_{x \to a} g(x) = 0$, then

$$\lim_{x \to a} \frac{f'(x)}{g'(x)} = l \qquad \Rightarrow \lim_{x \to a} \frac{f(x)}{g(x)} = l$$

Suppose $f(a) = 0$, $g(a) = 0$ and thus render f and g continuous at a.

Also, since we have

$$\lim_{x \to a} \frac{f'(x)}{g'(x)} = l$$

it follows that there exits a neighbourhood of a for every point x of which $f'(x)/g'(x)$ is defined so that $g'(x) \neq 0$.

We have by Cauchy's Mean value theorem

$$\frac{f(a+h)}{g(a+h)} = \frac{f(a+h) - f(a)}{g(a+h) - g(a)}$$

$$= \frac{f'(a+\theta h)}{g'(a+\theta h)} \qquad 0<\theta<1$$

Since, $\dfrac{f'(a+\theta h)}{g'(a+\theta h)}$ tends to l as $h \to 0$,

it follows that

$$\lim_{h\to 0} \frac{f(a+h)}{g(a+h)} = l \Leftrightarrow \lim_{x\to 0} \frac{f(x)}{g(x)} = l$$

In general, if $\lim_{x\to a} g'(x) = 0$ and $\lim_{x\to a} f'(x) = 0$, then

$$\lim_{x\to a} \frac{g''(x)}{f''(x)} = l \Rightarrow \lim_{x\to a} \frac{g'(x)}{f'(x)} = l$$

and so on.

9.2.1. L'Hospital's Rule when $x \to a+0$ and when $x \to a-0$

if $f'(x)$ and $g'(x)$ exist for all $x \in]a, a+\delta]$, if $g'(x) \neq 0$ for any x in $]a, a+\delta]$, if $\lim_{x\to a+0} f(x) = \lim_{x\to a+0} g(x) = 0$ and if $\lim_{x\to a+0} \dfrac{f'(x)}{g'(x)} = l$, then

$$\lim_{x\to a+0} \frac{f(x)}{g(x)} = l.$$

Similarly, if $f'(x)$ and $g'(x)$ exist for all $x \in [a-\delta, a[$, if $g'(x) \neq 0$ for any x in $[a-\delta, a[$, if $\lim_{x\to a-0} f(x) = \lim_{x\to a-0} g(x) = 0$ and if $\lim_{x\to a-0} \dfrac{f'(x)}{g'(x)} = l$,

then $\lim_{x\to a-0} \dfrac{f(x)}{g(x)} = l$

9.2.2. L'Hospital's Rule when $x \to \infty$

If $f'(x)$ and $g'(x)$ exist for all $x \in [k, \infty[$, if $g'(x) \neq 0$ for any $x \in [k, \infty[$, if $\lim_{x\to\infty} f(x) = \lim_{x\to\infty} g(x) = 0$, and if $\lim_{x\to\infty} \dfrac{f'(x)}{g'(x)} = l$, then

$$\lim_{x\to\infty} \frac{f(x)}{g(x)} = l$$

where k is some positive real number.

9.2.3. L'Hospital's Rule for Infinite Limits

If $f'(x)$ and $g'(x)$ exist for all $x \in]a-\delta, a+\delta[$ except possibly at $x = a$, if $g'(x) \neq 0$ for any $x \in]a-\delta, a+\delta[$ except possibly at $x = a$, if $\lim_{x\to a} f(x) = 0 = \lim_{x\to a} g(x)$,and if $\lim_{x\to a} f'(x) = \infty$, then $\lim_{x\to a} \frac{f(x)}{g(x)} = \infty$

9.2.4. Preliminary Transformation

Sometimes use of certain known limits, such as,

$$\lim_{x\to 0} \frac{\sin x}{x} = 1, \lim_{x\to 0} \frac{\tan x}{x} = 1,$$

simplifies the process a good deal.

These limits may also be used to shorten the process at an intermediate stage.

Examples

Example1: Evaluate $\lim_{x\to\pi/2} \frac{\cos x}{\pi/2 - x}$

Solution: We can write, $\frac{\cos x}{\pi/2 - x} = \frac{f(x)}{g(x)}$

Where $f(x) = \cos x$ and $g(x) = \pi/2 - x$

$$\lim_{x\to\pi/2} \cos x = 0 \quad , \quad \lim_{x\to\pi/2}\left(\frac{\pi}{2} - x\right) = 0$$

Therefore, $\frac{f(x)}{g(x)}$ is of the form 0/0 as $x \to \pi/2$.

By L'Hospital's rule,

$$\lim_{x\to\pi/2} \frac{\cos x}{\pi/2 - x} = \lim_{x\to\pi/2} \frac{-\sin x}{-1} = 1$$

Example 2: Evaluate $\lim_{x\to 0} \frac{1-\cos 2x}{x^2}$

Solution: We can write, $\frac{1-\cos 2x}{x^2}$

Where $f(x) = 1-\cos 2x$ and $g(x) = x^2$

$\lim_{x\to 0}(1-\cos 2x) = 0$, $\lim_{x\to 0} x^2 = 0.$

Therefore, $\frac{f(x)}{g(x)}$ is the form 0/0 as $x \to 0$

Applying L'Hospital's rule,

$$\lim_{x\to 0} \frac{1-\cos 2x}{x^2} = \lim_{x\to 0} \frac{2\sin 2x}{2x}$$

Since $\frac{\sin 2x}{2x}$ is again of the form 0/0, therefore, again applying L'Hospital's rule,

$$\lim_{x\to 0} \frac{\sin 2x}{x} = \lim_{x\to 0} \frac{2\cos 2x}{1} = 2$$

$$\therefore \lim_{x\to 0} \frac{1-\cos 2x}{x^2} = 2$$

Example 3: Find the values of a and b in order that $\lim_{x\to 0} \frac{x(1+a\cos x) - b\sin x}{x^3}$ may be equal to 1.

Solution: The function is of the form (0/0).

$$\therefore \quad \lim_{x\to 0} \frac{x(1+a\cos x) - b\sin x}{x^3}$$

$$= \lim_{x\to 0} \frac{1 + a\cos x - ax\sin x - b\cos x}{3x^2}$$

The denominator being 0 for $x=0$, the fraction will not tend to a finite limit unless the numerator is also zero for $x = 0$. This requires

$$1+a-b=0 \qquad \ldots\text{(i)}$$

Again, supposing this relation is satisfied, we have

$$\lim_{x\to 0}\frac{1+a\cos x-ax\sin x-b\cos x}{3x^2}$$

$$=\lim_{x\to 0}\frac{-2a\sin x-ax\cos x+b\sin x}{6x}$$

$$=\lim_{x\to 0}\frac{-3a\cos x+ax\sin x+b\cos x}{6}$$

$$=\frac{b-3a}{6}$$

As given $\dfrac{b-3a}{6}=1$

$$\Rightarrow b-3a=6 \qquad \ldots\text{(ii)}$$

From (i) and (ii), we have

$$a=-5/2,\quad b=-3/2$$

Example 4: Evaluate :

$$\lim_{x\to 0}\frac{\sinh x-\sin x}{x\sin^2 x}$$

Solution: We have

$$\frac{\sinh x-\sin x}{x\sin^2 x}=\frac{\sinh x-\sin x}{x^3}.\left(\frac{x}{\sin x}\right)^2$$

$$\therefore \quad \lim_{x\to 0}\frac{\sinh x-\sin x}{x\sin^2 x}=\lim_{x\to 0}\frac{\sinh x-\sin x}{x^3}.\lim_{x\to 0}\left(\frac{x}{\sin x}\right)^2$$

$$=\lim_{x\to 0}\frac{\sinh x-\sin x}{x^3}$$

Now, the function is of the form 0/0.

Then the given limit will be equal to

$$= \lim_{x\to 0} \frac{\cosh x - \cos x}{3x^2} \quad (0/0)$$

$$= \lim_{x\to 0} \frac{\sinh x + \sin x}{6x} \quad (0/0)$$

$$= \lim_{x\to 0} \frac{\cosh x + \cos x}{6}$$

$$= \frac{1}{3}$$

Example 5: Evaluate : $\lim_{x\to 0} \frac{\tan x - x}{x^2 \tan x}$

Solution: Writing,

$$\frac{\tan x - x}{x^2 \tan x} = \frac{\tan x - x}{x^3} . \frac{x}{\tan x}$$

We find that

$$\lim_{x\to 0} \frac{\tan x - x}{\tan x} = \lim_{x\to 0} \frac{\tan x - x}{x^3} . \lim_{x\to 0} \frac{x}{\tan x}$$

$$= \lim_{x\to 0} \frac{\tan x - x}{x^3} \quad (0/0)$$

$$= \lim_{x\to 0} \frac{\sec^2 x - 1}{3x^2} \quad (0/0)$$

$$= \lim_{x\to 0} \frac{2\sec^2 x . \tan x}{6x}$$

$$= \frac{1}{3} \lim_{x\to 0} \sec^2 x . \lim_{x\to 0} \frac{\tan x}{x}$$

$$= \frac{1}{3}$$

Example 6: Evaluate $\lim_{x\to 0}\dfrac{(1+x)^{1/x}-e+\frac{1}{2}ex}{x^2}$

Solution: $\lim_{x\to 0}(1+x)^{1/x}=e$ is a 0/0 form

Let $y=(1+x)^{1/x}$

$$\Rightarrow \log y=\frac{1}{x}\log(1+x)$$

$$=\frac{1}{x}\left(x-\frac{x^2}{2}+\frac{x^3}{3}-\frac{x^4}{4}+\cdots\cdots\right)$$

$$=1-\frac{x}{2}+\frac{x^2}{3}-\frac{x^3}{4}+\cdots\cdots$$

$$\Rightarrow y=e^{1-\frac{x}{2}+\frac{x^2}{3}-\frac{x^3}{4}+\cdots\cdots}$$

$$=e.e^{-x/2+x^2/3-x^3/4+\cdots\cdots}$$

$$=e\left[1+\left(-\frac{x}{2}+\frac{x^2}{3}-\frac{x^3}{4}+\cdots\cdots\right)+\frac{1}{2}\left(-\frac{x}{2}+\cdots\cdots\right)^2\right]$$

$$=e\left[1-\frac{x}{2}+\frac{11}{24}x^2\cdots\cdots\cdots\right]$$

$$\therefore\quad \lim_{x\to 0}\frac{(1+x)^{1/x}-e+\frac{1}{2}ex}{x^2}=\lim_{x\to 0}\frac{e\left[1-\frac{x}{2}+\frac{11}{24}x^2\cdots\cdots\cdots\right]-e.\frac{ex}{2}}{x^2}$$

$$=\lim_{x\to 0}\left(\frac{11e}{24}+\text{terms containing } x\right)$$

$$=\frac{11e}{24}$$

Exercise – 9.1

Evaluate the following limits:

1. $\lim_{x\to 0} \dfrac{\tan x - x}{\sin x - x}$

2. $\lim_{x\to 0} \dfrac{\sinh x - x}{\sin x - x\cos x}$

3. $\lim_{x\to 1} \dfrac{1+\log x - x}{1-2x+x^2}$

4. $\lim_{x\to 0} \dfrac{xe^x - \log(x+1)}{\cosh x - \cos x}$

5. $\lim_{x\to 0} \dfrac{\log(1-x^2)}{\log\cos x}$

6. $\lim_{x\to 0} \dfrac{e^{ax} - e^{-ax}}{\log(1+bx)}$

7. $\lim_{x\to 0} \dfrac{\log(1+x) - x}{1-\cos x}$

8. $\lim_{x\to 0} \dfrac{x\cos x - \log(1+x)}{x^2}$

9. $\lim_{x\to 0} \dfrac{a^x - 1}{b^x - 1}$

10. $\lim_{x\to 0} \dfrac{\sin ax}{\sin bx}$

11. $\lim_{x\to 0} \dfrac{e^x \sin x - x - x^2}{x^2 + x\log(1-x)}$

12. $\lim_{x\to 0} \dfrac{x^2 + 2\cos x - 2}{x\sin^3 x}$

13. $\lim_{x\to 0} \dfrac{\sin 2x + 2\sin^2 x - 2\sin x}{\cos x - \cos^2 x}$

14. If the limit of $\lim_{x\to 0} \dfrac{\sin 2x + a\sin x}{x^3}$ be finite, find the value of a and the limit.

15. Determine the values of p and q for which $\lim_{x\to 0} \dfrac{x(1+p\cos x) - q\sin x}{x^3}$ exists and equals 1.

9.4. THE INDETERMINATE FORM ∞/∞

If $\lim_{x \to a} f(x) = \infty$, $\lim_{x \to a} g(x) = \infty$, then $f(x)/g(x)$ is said to assume the indeterminate form ∞/∞ as $x \to a$.

Theorem: If $f'(x)$ and $g'(x)$ exists for all $x \in]0,\delta]$, if $g'(x) \neq 0$ for any $x \in]0,\delta]$, if $f(x) \to \infty$ and $g(x) \to \infty$ as $x \to 0+0$, and if $\lim_{x\to 0+0} \frac{f'(x)}{g'(x)} = l$, then

$$\lim_{x\to 0+0} \frac{f(x)}{g(x)} = l$$

Proof: Let h be the function defined on $]0,\delta]$ and $h(x) = f(x) - g(x)$, for all $x \in]0,\delta]$,

Then $h'(x) = f'(x) - lg'(x)$, for all $x \in]0,\delta]$ ...(1)

Since $g'(x) \neq 0$ for any $x \in]0,\delta]$ and $\lim_{x\to 0+0} \frac{f'(x)}{g'(x)} = l$

Therefore, it follows from (1) that

$$\lim_{x\to 0+0} \frac{h'(x)}{g'(x)} = 0 \qquad ...(2)$$

Now (2) implies that given any $\in > 0$ we can find $\delta_1 > 0$, such that

$$0 < x < \delta_1 \quad \Rightarrow \quad |h'(x)/g'(x)| < \in/2 \qquad ...(3)$$

Also, since $g(x) \to \infty$ as $x \to 0+0$, therefore, we can find $\delta_2 > 0$ such that

$$0 < x < \delta_2 \quad \Rightarrow \quad g(x) > 0 \qquad ...(4)$$

Let $\delta_3 = \frac{1}{2} Min.\{\delta_1.\delta_2\}$, so that from (3) and (4), we have

$|h'(x)/g'(x)| < \in/2$ and

$$g(x) > 0 \text{, wherever } 0 < x \le \delta_3 \qquad \ldots(5)$$

Let x be any number in $]0, \delta_3]$, then the functions g and h satisfy the conditions of Cauchy's Mean value theorem in $[x, \delta_3]$, and consequently, there exists $c \in]x, \delta_3[$, such that

$$\frac{h(\delta_3) - h(x)}{g(\delta_3) - g(x)} = \frac{h'(c)}{g'(c)} \qquad \ldots(6)$$

Since $0 < c < \delta_3 < \delta_1$, therefore, from (3),

$$|h'(c)/g'(c)| < \in/2 \qquad \ldots(7)$$

Also, since $g(x) \to \infty$ as $x \to 0+0$, therefore, we can choose a positive number $\delta_4 < \delta_3$ such that

$$g(x) > g(\delta_3) \text{, for all } x \in]0, \delta_4[\qquad \ldots(8)$$

From (4) and (8), we have

$$0 < g(x) - g(\delta_2) < g(x) \text{, for all } x \in]0, \delta_4[\qquad \ldots(9)$$

From (6), (7) and (9), we have

$$\frac{|h(\delta_3) - h(x)|}{g(x)} < \frac{\in}{2} \text{, for all } x \in]0, \delta_4[\qquad \ldots(10)$$

Again since $g(x) \to \infty$ as $x \to 0+0$, therefore, we can find a positive number $\delta_3 < \delta_4$, such that

$$g(x) > \frac{2}{\in}|h(\delta_3)| \text{, for all } x \in]0, \delta_5[\qquad \ldots(11)$$

Since $\dfrac{h(x)}{g(x)} = \dfrac{h(x) - h(\delta_3)}{g(x)} + \dfrac{h(\delta_3)}{g(x)}$

therefore, from (10) & (11), we have

$$\left|\frac{h(x)}{g(x)}\right| \le \left|\frac{h(x) - h(\delta_3)}{g(x)}\right| + \left|\frac{h(\delta_3)}{g(x)}\right|$$

$$< \in \text{ for all } x \in]0, \delta_5[$$

Hence, $\lim_{x \to 0+0} \dfrac{h(x)}{g(x)} = 0$

Since, $h(x) = f(x) - lg(x)$, for all $x \in]0, \delta[$, therefore, it follows that

$$\lim_{x \to 0+0} \frac{f(x)}{g(x)} = l$$

9.5. THE INDETERMINATE FORM 0.∞

If $\lim_{x \to a} f(x) = 0$ and $\lim_{x \to a} g(x) = \infty$, then we write $f(x).g(x) = \dfrac{f(x)}{1/g(x)}$

Now, $\dfrac{f(x)}{1/g(x)}$ is of the form 0/0 and can be determined.

9.6. THE INDETERMINATE FORM ∞ − ∞

If $\lim_{x \to a} f(x) = \infty = \lim_{x \to a} g(x)$, then we write

$$f(x) - g(x) = \left[\frac{1}{f(x)} - \frac{1}{g(x)}\right] \div \frac{1}{f(x).g(x)}$$

so that the numerator and denominator both tend to 0 as x tends to a. Now, the limit may be determined.

9.7. THE INDETERMINATE FORM 0^0

Consider $\lim_{x \to a} [f(x)]^{g(x)}$

when $\lim_{x \to a} f(x) = 0$, $\lim_{x \to a} g(x) = 0$

We write, $y = [f(x)]^{g(x)}$

so that $\log y = g(x) \log f(x)$

9.8. THE INDETERMINATE FORM ∞^0

Consider $\lim_{x \to a} [f(x)]^{g(x)}$

when $\lim_{x \to a} f(x) = \infty$, $\lim_{x \to a} g(x) = 0$

We write, $y = [f(x)]^{g(x)}$

so that $\log y = g(x) \log f(x)$

9.9. THE INDETERMINATE FORM 1^∞

Consider $\lim_{x \to a} [f(x)]^{g(x)}$

when $\lim_{x \to a} f(x) = 1$, $\lim_{x \to a} g(x) = \infty$

We write $y = [f(x)]^{g(x)}$

so that $\log y = g(x) \log f(x)$

In each of the three cases ,(i.e., 9.7, 9.8 & 9.9), we see that the right hand side assumes the intermediate form $0.\infty$ and its limit, therefore be determined.

Let $\lim_{x \to a} \{g(x).\log f(x)\} = 1$

$\therefore \quad \lim \log y = 1$

$\Rightarrow \log \lim y = 1$

$\Rightarrow \quad \lim y = e^1$

$\Rightarrow \quad \lim [f(x)]^{g(x)} = e^1$

Examples

Example 1: Determine $\lim_{x \to a} \dfrac{\log(x-a)}{\log(e^x - e^a)}$

Solution: The function is of the form ∞ / ∞

$$\therefore \lim_{x\to a} \frac{\log(x-a)}{\log(e^x - e^a)} = \lim_{x\to a}\left(\frac{1}{x-a}\right)\Big/\left(\frac{e^x}{e^x - e^a}\right)$$

$$= \lim_{x\to a} \frac{e^x - e^a}{e^x(x-a)}\left(\frac{0}{0}\right)$$

$$= \lim_{x\to a} \frac{e^x}{e^x(x-a) + e^x}$$

$$= \frac{e^a}{e^a} = 1$$

Example 2: Evaluate $\lim\limits_{x\to\pi-0} \dfrac{\log(\sin x)}{\log(\sin 2x)}$

Solution: The given expression is of the form ∞/∞

$$\therefore \lim_{x\to\pi-0} \frac{\log(\sin x)}{\log(\sin 2x)} = \lim_{x\to\pi-0} \frac{\cot x}{2\cot 2x}(\infty/\infty)$$

$$= \lim_{x\to\pi-0} \frac{-\operatorname{cosec}^2 x}{-4\operatorname{cosec}^2 2x}$$

$$= \lim_{x\to\pi-0} \cos^2 x$$

$$= 1$$

Example 3: Evaluate:

$$\lim_{x\to 0}(\cot x)^{1/\log x}$$

Solution: Let $y = (\cot x)^{1/\log x}$

$$\Rightarrow \log y = \frac{1}{\log x}\log(\cot x)$$

$$\Rightarrow \lim_{x\to 0}\log y = \lim_{x\to 0}\frac{\log(\cot x)}{\log x}(\infty/\infty \text{ form})$$

$$= \lim_{x\to 0} \frac{\dfrac{-\operatorname{cosec}^2 x}{\cot x}}{1/x}$$

$$= \lim_{x\to 0} \frac{-x}{\sin x} \cdot \frac{1}{\cos x} = -1$$

$$\Rightarrow \log \lim_{x\to 0} y = -1$$

$$\Rightarrow \lim_{x\to 0} y = e^{-1} = 1/e$$

Example 4: Evaluate:

$$\lim_{x\to 0}\left(\frac{1}{x^2} - \frac{1}{\sin^2 x}\right)$$

Solution: We have,

$$\lim_{x\to 0}\left(\frac{1}{x^2} - \frac{1}{\sin^2 x}\right) = \lim_{x\to 0} \frac{\sin^2 x - x^2}{x^2 \sin^2 x} \quad \left(\frac{0}{0} \text{ form}\right)$$

$$= \lim_{x\to 0}\left(\frac{\sin^2 x - x^2}{x^4}\right)\left(\frac{x}{\sin x}\right)^2$$

$$= \lim_{x\to 0}\left(\frac{\sin^2 x - x^2}{x^4}\right)$$

$$= \lim_{x\to 0} \frac{\sin 2x - 2x}{4x^3}$$

$$= \lim_{x\to 0} \frac{2\cos 2x - 2}{12x^2}$$

$$= \lim_{x\to 0} \frac{\cos 2x - 1}{6x^2}$$

$$= \lim_{x\to 0} \frac{-2\sin^2 x}{6x^2}$$

$$= -\frac{1}{3}\lim_{x\to 0}\left(\frac{\sin x}{x}\right)^2$$

$$= -\frac{1}{3}$$

Example 5: Determine

$$\lim_{x\to 0}(1+x)^{1/x}$$

Solution: Let $y = (1+x)^{1/x}$

$$\Rightarrow \log y = \frac{\log(1+x)}{x} \qquad \left(\frac{0}{0} \text{ form}\right)$$

Now, $\lim_{x\to 0}\frac{\log(1+x)}{x} = \lim_{x\to 0}\frac{(1+x)^{-1}}{1}$

So that $\lim_{x\to 0}(\log y) = 1$

$$\Rightarrow \log\left(\lim_{x\to 0} y\right) = 1$$

$$\Rightarrow \lim_{x\to 0} y = e$$

Example 6: Evaluate, $\lim_{x\to 0} x\log x$

Solution: We have,

$$x\log x = \frac{\log x}{1/x}$$

$$\therefore \lim_{x\to 0}(x\log x) = \lim_{x\to 0}\frac{\log x}{1/x} \qquad \left(\frac{\infty}{\infty}\right)$$

$$= \lim_{x\to 0}\left(\frac{1}{x} \div \frac{-1}{x^2}\right)$$

$$= \lim_{x\to 0}(-x) = 0$$

Example 7: Evaluate. $\lim\limits_{x \to a} x^x$

Solution: Let $y = x^x$

$$\Rightarrow \log y = x \log x$$

$$= \frac{\log x}{1/x} \quad \left(\frac{\infty}{\infty} \text{ form}\right)$$

Now, $\lim\limits_{x \to 0} \dfrac{\log x}{1/x} = \lim\limits_{x \to 0} \dfrac{1/x}{-1/x^2} = 0$

Since, $\lim\limits_{x \to 0} \log y = 0$

$$\therefore \log\left(\lim_{x \to 0} y\right) = 0$$

$$\Rightarrow \lim_{x \to 0} y = e^0 = 1$$

Exercise – 9.2

1. *Determine the following limits:*

(i) $\lim\limits_{x\to\infty} \dfrac{\log x}{x}$ (ii) $\lim\limits_{x\to 0} \dfrac{\log \sin x}{\cot x}$

(iii) $\lim\limits_{x\to 0} \dfrac{\log \tan 2x}{\log \tan x}$ (iv) $\lim\limits_{x\to 0} \log_{\tan x} \tan 2x$

(v) $\lim\limits_{x\to 1} \dfrac{\log(1-x)}{\tan(x\pi/2)}$ (vi) $\lim\limits_{x\to\infty} \dfrac{\log x}{x^n}$, where $n > 0$.

2. *Evaluate the following limits:*

(i) $\lim\limits_{x\to 0} x \tan(\pi/2 - x)$ (ii) $\lim\limits_{x\to 0} x \log \tan x$

(iii) $\lim\limits_{x\to a} (a-x) \tan \dfrac{\pi x}{2a}$ (iv) $\lim\limits_{x\to \pi/2} (1-\sin x) \tan x$

(v) $\lim\limits_{x\to 1} (1-x) \tan \dfrac{\pi}{2} x$ (vi) $\lim\limits_{x\to 0} \sin x . \ln x^2$

3. *Determine the following limits:*

(i) $\lim\limits_{x\to 1} \left(\dfrac{x}{x-1} - \dfrac{1}{\ln x} \right)$ (ii) $\lim\limits_{x\to 0} \left(\dfrac{1}{x} - \dfrac{1}{e^x - 1} \right)$

(iii) $\lim\limits_{x\to 0} \left(\dfrac{1}{x} - \cot x \right)$ (iv) $\lim\limits_{x\to 0} \left(\dfrac{a}{x} - \cot \dfrac{x}{a} \right)$

(v) $\lim\limits_{x\to 0} \left(\cot^2 x - \dfrac{1}{x^2} \right)$ (vi) $\lim\limits_{x\to \pi/2} \sec x - \tan x$

(vii) $\lim\limits_{x\to 0} \left\{ \dfrac{\pi}{4x} - \dfrac{\pi}{2x(e^{\pi x} + 1)} \right\}$ (viii) $\lim\limits_{x\to \pi/2} \left(\sec x - \dfrac{1}{1-\sin x} \right)$

(ix) $\lim\limits_{x\to 0} \left\{ \dfrac{1}{x} - \dfrac{1}{x^2} \ln(1+x) \right\}$ (x) $\lim\limits_{x\to 4} \left\{ \dfrac{1}{\ln(x-3)} - \dfrac{1}{x-4} \right\}$

(xi) $\lim_{x\to 1}\left\{\frac{x}{x-1}-\frac{1}{\ln x}\right\}$ (xii) $\lim_{x\to 0}(\operatorname{cosec} x-\cot x)$

4. *Evaluate the following limits:*

(i) $\lim_{x\to a}(x-a)^{x-a}$ (ii) $\lim_{x\to a} x^{x}$

(iii) $\lim_{x\to\pi/2}(\cos x)^{\cos x}$ (iv) $\lim_{x\to 0}(\tan x)^{\sin 2x}$

(v) $\lim_{x\to 0}(\cot x)^{\sin 2x}$ (vi) $\lim_{x\to\pi/2}(\sin x)^{\tan x}$

(vii) $\lim_{x\to 0}(1+\sin x)^{\cot x}$ (viii) $\lim_{x\to\pi/2}(\sec x)^{\cot x}$

(ix) $\lim_{x\to 0}\left(\ln\frac{1}{x}\right)^{x}$ (x) $\lim_{x\to 1}\frac{1}{x^{x}-1}$

(xi) $\lim_{x\to 0}\left(\frac{\tan x}{x}\right)^{1/x^2}$ (xii) $\lim_{x\to 1}(2-x)^{\tan\frac{\pi x}{2}}$

(xiii) $\lim_{x\to 1} x^{1/(x-1)}$ (xiv) $\lim_{x\to 0}\left(\frac{\sin x}{x}\right)^{1/x^2}$

(xv) $\lim_{x\to 0}(\cos x)^{1/x^3}$ (xvi) $\lim_{a\to b}\frac{a^b-b^a}{a^a-b^b}$

(xvii) $\lim_{x\to 1}\frac{1-x+\log x}{1-\sqrt{(2x-x^2)}}$ (xviii) $\lim_{x\to 0}\frac{\log_{\sec\frac{x}{2}}\cos x}{\log_{\sec x}\cos\frac{x}{2}}$

(xix) $\lim_{x\to 0}\frac{x^2\sin 1/x}{\tan x}$ (xx) $\lim_{x\to 0}\frac{(1+x)e^{-x}-(1-x)e^{x}}{x\left(e^{x}-e^{-x}\right)-2x^2e^{-x}}$

5. If $f(x) = e^{-1/x^2}, x \neq 0$

$f(0) = 0$

show that the derivative of every order of f vanishes for $x = 0$, i.e.,

$f^n(0) = 0$, for all n.

6. $\lim_{x \to 0} (1+x)^{a/x}$, a being any non-zero real number.

7. $\lim_{x \to \infty} \left(x + \frac{1}{x} \right)^{ax}$, a being any non-zero real number.

10

Partial Differentiation

10.1. INTRODUCTION

Till now, we have studied real valued functions of a single real variable. In this chapter, we shall introduce the concept of functions of two or more variables, their limits, continuity and partial derivatives.

10.2. FUNCTIONS OF TWO VARIABLES

Let D be a subset of R^2 so that D is a set of ordered pairs of real numbers and let f be a function from D into R, so that f assigns to each $(x, y) \in D$, a real number $f(x, y)$.

For example, $f(x, y) = 2x^2 + 3y^2$ for all $(x, y) \in R^2$. Then f is a real valued function defined on R^2. Similarly, $f(x, y) = \sqrt{1 - x^2 - y^2}$ is a function defined on a subset C of R consisting of points within and on the circle $x^2 + y^2 = 1$.

Note 1. If $z = f(x, y)$. x and y are called the **independent variables** and z is called the **dependent variable**.

2. If a function of two variables, the set of points $\{(x, y) : f(x, y) = 0\}$ represents a curve in the xy-plane.

10.3. FUNCTIONS OF THREE OR MORE VARIABLES

Let $f(x,y,z)=3x+8y-z$ defines a function $u=f(x,y,z)$, where $f(x,y,z)=3x+8y-z$, of three independent variables.

The domain of this function is R^3 and its range is a subset of R.

Similarly, let $f(x_1,x_2,\cdots\cdots x_n)=x_1{}^3+x_2{}^3+\cdots\cdots+x^n$ defines a function of n independent variables with domain R^n and range a subset of R.

10.4. NEIGHBOURHOOD OF A POINT (a, b)

Let δ be any positive number. the points (x, y) such that

$$a-\delta \le x \le a+\delta,$$
$$b-\delta \le y \le b+\delta$$

determine a square bounded by the lines

$$x=a-\delta, \quad x=a+\delta,$$
$$y=b-\delta, \quad y=b+\delta$$

Its centre is at the point (a, b).

This square is called a *Neighbourhood* of the point (a, b).

For every value of δ, we will get a neighbourhood. Thus the set

$$(x,y):\{a-\delta \le x \le a+\delta, b-\delta \le y \le b+\delta\}$$

is a neighborhood of the point (a, b).

10.5. CONTINUITY OF A FUNCTION OF TWO VARIABLES

Let (a, b) be a point of the domain of

$$z=f(x,y)$$

f is continuous at (a, b) if for points (x, y) near (a, b), the value $f(x, y)$ of the function is near $f(a, b)$.

Thus for continuity at (a, b), there exists a square bounded by the lines

$$x=a-\delta, \quad x=a+\delta,$$
$$y=b-\delta, \quad y=b+\delta$$

such that, for any point (x, y) of this square, $f(x, y)$ lies between

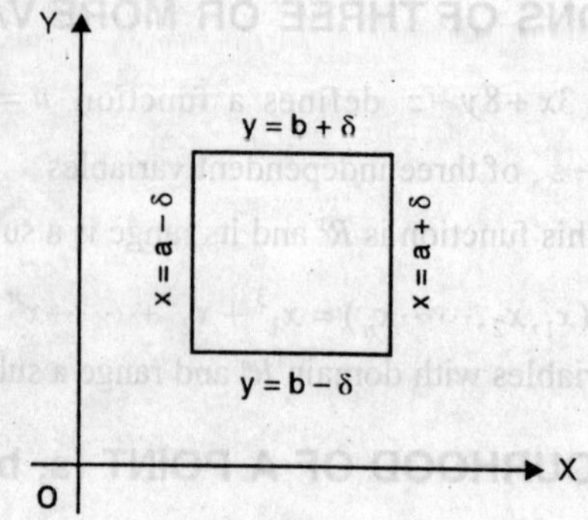

Figure 10.1

$f(a,b) - \in$ and

$f(a,b) + \in$

where $\in$ is any positive number, however small.

A function f is said to be continuous if it is continuous at every point of its domain.

10.6. LIMIT OF A FUNCTION OF TWO VARIABLES

Definition. A function is said to tend to the limit l, as (x, y) tends to (a, b), if corresponding to any pre-assigned positive number $\in$, there exists a positive number δ such that

$$|f(x,y) - l|| < \in$$

for all points (x, y), other than possibly (a, b), such that

$$a - \delta \le x \le a + \delta,$$
$$b - \delta \le y \le b + \delta$$

In symbols, we then write

$$\lim_{(x,y)\to(a,b)} f(x,y) = l$$

Limit of a continuous function.

Definition. Let f be a function of x and y defined on some neighbourhood of (a, b). f is said to be continuous at a point (a, b) if given $\in > 0$, there exits $a\delta > 0$ such that

$$|f(x, y) - f(a, b)| < \in$$

whenever $|x - a| < \delta,\ |y - b| < \delta$

Continuity of f at (a, b) is equivalent to requiring that exists and is equal to $f(a, b)$. i.e., the limit of the function = value of the function.

10.7. PARTIAL DERIVATIVES

Let $z = f(x, y)$, then

$$\lim_{h \to 0} \frac{f(a + h, b) - f(a, b)}{h}$$

if it exists, is said to be the **partial derivative** of f w.r.t. x at (a, b) and is denoted by

$$\left(\frac{\partial z}{\partial x}\right)_{(a,b)} \text{ or } f_x(a, b)$$

Similarly, the partial derivative of f w.r.t. x at (a, b) is the derivative of the function $f(x) = f(x, b)$ of a single real variable at $x = a$.

Again,

$$\lim_{k \to 0} \frac{f(a, b + k) - f(a, b)}{k}$$

if it exists, is called the **partial derivative** of $f(x, y)$ w.r.t. y at (a, b) and is denoted by

$$\left[\frac{\partial z}{\partial y}\right]_{(a,b)} \text{ or } f_y(a, b)$$

10.7.1. Partial Derivative of Higher Orders

If $z = f(x, y)$ is a function of two independent variables x and y, then the process of taking partial derivatives gives rise to two new functions

$$f_1 = \frac{\partial z}{\partial x} \text{ and } f_2 = \frac{\partial z}{\partial y}$$

which are again functions of the two variables x and y.

These are called **second order partial derivatives** of z and are denoted as follows:

$$\frac{\partial}{\partial z}\left(\frac{\partial z}{\partial x}\right) = \frac{\partial^2 z}{\partial x^2} = \frac{\partial^2 f}{\partial x^2} = fx^2 = fxx$$

$$\frac{\partial}{\partial y}\left(\frac{\partial z}{\partial y}\right) = \frac{\partial^2 z}{\partial y^2} = \frac{\partial^2 f}{\partial y^2} = fy^2 = fyy$$

$$\frac{\partial}{\partial x}\left(\frac{\partial z}{\partial y}\right) = \frac{\partial^2 z}{\partial x.\partial y} = \frac{\partial^2 f}{\partial x.\partial y} = fxy$$

$$\frac{\partial}{\partial y}\left(\frac{\partial z}{\partial x}\right) = \frac{\partial^2 z}{\partial y.\partial x} = \frac{\partial^2 f}{\partial y.\partial x} = fyx$$

10.8. GEOMETRICAL REPRESENTATION OF A FUNCTION OF TWO VARIABLES

Consider a pair of perpendicular lines *OX* and *Y*. draw a line $Z'OZ'$ perpendicular to the *XY* plane through the point *O* and call it *z*-axis.

The three co-ordinate axes, taken in pairs determine three planes, viz., *XY*, *YZ* and *ZX* which are taken as the co-ordinate planes.

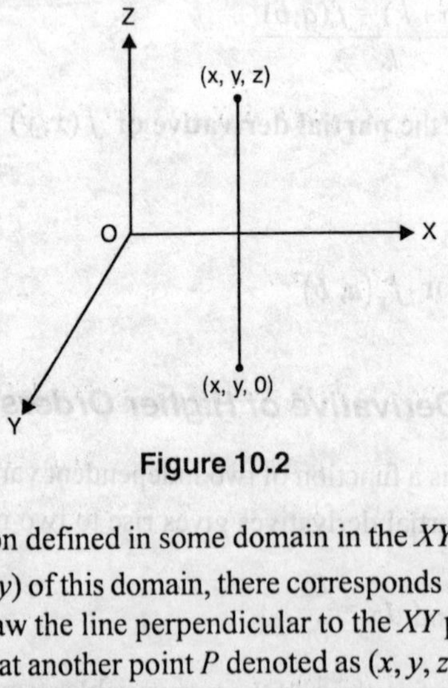

Figure 10.2

Let f be a function defined in some domain in the *XY* plane.

To each point (x, y) of this domain, there corresponds a value of z. Through the point (x, y) we draw the line perpendicular to the *XY* plane equal in length to z, so that we arrive at another point *P* denoted as (x, y, z); lying on one or the other side of the plane according as z is positive or negative.

The set of the points P determines a surface which is said to represent the function geometrically.

10.8.1 Geometrical Interpretation of Partial Derivatives of the First Order

Let $z = f(x,y)$ ...(i)

Equation (i) represents a surface geometrically.

Now, we seek the geometrical interpretation of the partial derivatives.

$$\left[\frac{\partial z}{\partial x}\right]_{(a,b)} \text{ and } \left[\frac{\partial z}{\partial y}\right]_{(a,b)}$$

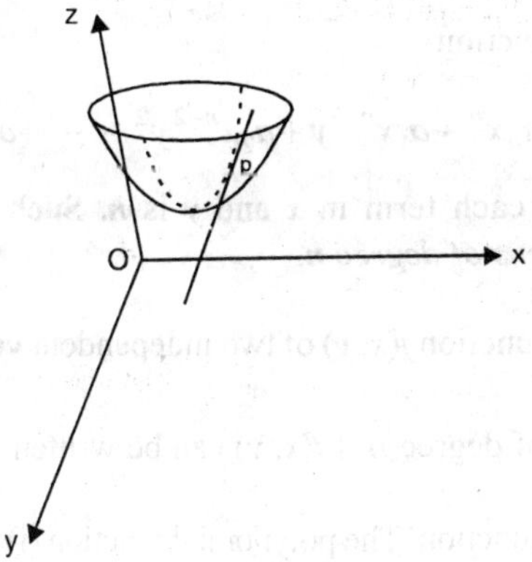

Figure 10.3

The point $P[(a,b), f(a,b)]$ on the surface corresponds to the point (a, b) of the domain of the function.

If a variable point, starting from P, changes its position on the surface such that y remains constantly equal to b, then it is clear that the locus of the point is the curve of intersection of the surface and the plane $y = b$.

On this curve x and z vary according to the relation

$$z = f(x,b)$$

Also, $\left(\frac{\partial z}{\partial x}\right)_{(a,b)}$ is the ordinary derivative of $f(x, b)$ w.r.t. x for $x = a$.

Hence, $\left[\dfrac{\partial z}{\partial x}\right]_{(a,b)}$ denotes the tangent of the angle which the tangent to the curve, in which the plane $y = b$ parallel to the ZX plane cuts the surface at $P[(a,b), f(a,b)]$ makes with x-axis.

Similarly, it may be seen that $\left[\dfrac{\partial z}{\partial y}\right]$ is the tangent of the angle which the tangent at $P[(a,b), f(a,b)]$ to the curve of intersection of the surface and the plane $x = a$ parallel to the ZY plane, makes with y-axis.

10.9. HOMOGENOUS FUNCTION

Consider the function

$$f(x,y) = a_o x^n + a_1 x^{n-1} y + a_2 x^{n-2} y^2 + \cdots\cdots + a_n y^n \qquad \ldots\text{(i)}$$

The degree of each term in **x** and **y** is ***n***. **Such** functions are called *homogenous functions of degree n.*

Definition-10.9 A function $f(x, y)$ of two independent variables x and y is said to be homogenous of degree n if $f(x, y)$ can be written in the form $x^n \phi\left(\dfrac{y}{x}\right)$, where ϕ can be any function. The polynomial function (i) which can be rewritten as

$$x^n\left[a_0 + a_1\frac{y}{x} + a_2\left(\frac{y}{x}\right)^2 + \cdots\cdots + a_n\left(\frac{y}{x}\right)^n\right]$$

is a homogenous expression of order n according to the definition.

Some examples of homogenous expression are

(i) $f(x,y) = x^n \sin(y/x)$

(ii) $f(x,y) = x^3 - 3xy^2 + y^3$

(iii) $f(x,y) = \left(\sqrt{y} + \sqrt{x}\right) / (y + x)$

10.9.1. Euler Theorem on Homogenous Functions

If $z = f(x,y)$ be a homogenous function of x, y of degree n, then

$$x\frac{\partial z}{\partial x}+y\frac{\partial z}{\partial y}=nz\text{, for all } x, y \in \text{ the domain of the function.}$$

Proof. We have,

z is a homogenous function of degree n.

$\therefore z = x^n f(y/x)$

$$\Rightarrow \frac{\partial z}{\partial x}=nx^{n-1}f\left[\frac{y}{x}\right]+x^n f'\left(\frac{y}{x}\right)\left(\frac{-y}{x^2}\right)$$

$$=nx^{n-1}f\left[\frac{y}{x}\right]-yx^{n-2}f'\left(\frac{y}{x}\right)$$

Similarly, $\frac{\partial z}{\partial y}=x^n f'\left(\frac{y}{x}\right)\left(\frac{1}{x}\right)=x^{n-1}f'\left(\frac{y}{x}\right)$

Thus, we have

$$x\frac{\partial z}{\partial x}+y\frac{\partial z}{\partial y}=nx^n f\left(\frac{y}{x}\right)=nz$$

Hence the result.

Corollary 1: If $z = f(x,y)$ is a homogenous of x and y of degree n, then

$$x^2\frac{\partial^2 z}{\partial x^2}+2xy\frac{\partial^2 z}{\partial x.\partial y}+y^2\frac{\partial^2 z}{\partial y^2}=n(n-1)z$$

By Euler's theorem, we have

$$x\frac{\partial z}{\partial x}+y\frac{\partial z}{\partial y}=nz \quad \text{...(1)}$$

Differentiating (1) partially w.r.t. x, we get

$$\frac{\partial z}{\partial x}+x\frac{\partial^2 z}{\partial x^2}+y\frac{\partial^2 z}{\partial x.\partial y}=n\frac{\partial z}{\partial x} \quad \text{...(2)}$$

Again, differentiating (1) partially w.r.t. y, we get

$$x\frac{\partial^2 z}{\partial y.\partial x}+\frac{\partial z}{\partial y}+y\frac{\partial^2 z}{\partial y^2}=n\frac{\partial z}{\partial y} \quad \ldots(3)$$

Multiplying (2) and (3) by x and y respectively and adding, we get

$$x\frac{\partial z}{\partial x}+x^2\frac{\partial^2 z}{\partial x^2}+xy\frac{\partial^2 z}{\partial x.\partial y}+y\frac{\partial z}{\partial y}+y^2\frac{\partial^2 z}{\partial y^2}=n\left(x\frac{\partial z}{\partial x}+y\frac{\partial z}{\partial y}\right)$$

$$\Rightarrow x^2\frac{\partial^2 z}{\partial x^2}+2xy\frac{\partial^2 z}{\partial x.\partial y}+y^2\frac{\partial^2 z}{\partial y^2}=(n-1)\left(x\frac{\partial z}{\partial x}+y\frac{\partial z}{\partial y}\right)$$

$$=n(n-1)z$$

Corollary 2: We can define a homogenous function of any number of variables as follows:

A function $z=f(x_1,x_2\cdots\cdots x_k)$ of k independent variables is called a homogenous function of degree *n*, if it can be written in the form

$$z=x_1{}^n\phi\left(\frac{x_2}{x_1},\frac{x_3}{x_1},\cdots\cdots\frac{x_k}{x_1}\right)$$

Generalizing Euler's theorem for functions of two independent variables, we can similarly prove the following:

If $z=f(x_1,\cdots\cdots,x_k)$ is a homogenous function of k independent variables $x_1,\cdots\cdots,x_k$, then

$$x_1\frac{\partial z}{\partial x_1}+x_2\frac{\partial z}{\partial x_2}+\cdots\cdots+x_k\frac{\partial z}{\partial x_k}=nz$$

We have, $z=x_1{}^n\phi\left(\frac{x_2}{x_1},\cdots\cdots\frac{x_k}{x_1}\right)$

$$=x_1{}^n\phi(y_2,y_3\cdots\cdots y_k)$$

Where $y_2=\frac{x_2}{x_1},y_3=\frac{x_3}{x_1}\cdots\cdots,y_k=\frac{x_k}{x_1}$

Now, $\frac{\partial z}{\partial x_1}=nx_1{}^{n-1}\phi(y_2,y_3\cdots\cdots,y_k)$

$$+\left(\frac{\partial\phi}{\partial y_2}\frac{\partial y_2}{\partial x_1}+\frac{\partial\phi}{\partial y_3}\frac{\partial y_3}{\partial x_1}+\cdots\cdots+\frac{\partial\phi}{\partial y_k}\frac{\partial y_k}{\partial x_1}\right)$$

Also, $\dfrac{\partial y_2}{\partial x_1}=\dfrac{-x_2}{{x_1}^2},\dfrac{\partial y_3}{\partial x_1}=\dfrac{-x_3}{{x_1}^2},\cdots\cdots$

$$\therefore \quad \frac{\partial z}{\partial x_1}=nx^{n-1}\phi-{x_1}^{n-2}\left\{x_2\frac{\partial\phi}{\partial y_2}+x_3\frac{\partial\phi}{\partial y_3}+\cdots\cdots+x_k\frac{\partial\phi}{\partial y_k}\right\} \qquad \ldots(1)$$

Again, $\dfrac{\partial z}{\partial x_2}={x_1}^n\dfrac{\partial\phi}{\partial y_2}.\dfrac{\partial y_2}{\partial x_2}={x_1}^n\dfrac{\partial\phi}{\partial y_2}.\dfrac{1}{x_1}={x_1}^n\dfrac{\partial\phi}{\partial y_2}$ ...(2)

Similarly, $\dfrac{\partial z}{\partial x_3}={x_1}^n\dfrac{\partial\phi}{\partial y_3}.\dfrac{\partial y_3}{\partial x_3}={x_1}^n\dfrac{\partial\phi}{\partial y_3}.\dfrac{1}{x_1}={x_1}^{n-1}\dfrac{\partial\phi}{\partial y_3}$ ...(3)

and so on till we get

$$\frac{\partial z}{\partial x_k}={x_1}^{n-1}\frac{\partial\phi}{\partial y_k} \qquad \ldots(k)$$

Multiplying (1)..........(k) by $x_1,x_2\cdots\cdots,x_k$ respectively and adding, we get

$$x_1\frac{\partial z}{\partial x_1}+x_2\frac{\partial z}{\partial x_2}+\cdots\cdots+x_k\frac{\partial z}{\partial x_k}=n{x_1}^n\phi\left(\frac{x_2}{x_1},\cdots\cdots\frac{x_k}{x_1}\right)=nz$$

Examples

Example 1: Let f be a function defined on R^2 by setting $f(x,y)=|x|+|y||$, for all $x,y\in R$. Show that f is continuous at the origin.

Solution: Let $\in>0$

Now,

$$|f(x,y)-f(0,0)|=|x|+|y| \qquad (\because f(0,0)=0)$$

$$<\in,\text{ whenever } |x|<\frac{1}{2}\in,\ |y|<\frac{1}{2}\in$$

Setting $\delta = \frac{1}{2} \in$, we find that

$$|f(x, y) - f(0, 0)| < \in \text{ whenever}$$

$$|x| < \delta, \ |y| < \delta$$

Hence f is continuous at (0, 0).

Example 2: If $z = x^2 \tan^{-1}(y/x) - y^2 \tan^{-1}(x/y)$, prove that

$$\frac{\partial^2 z}{\partial x \partial y} = \frac{x^2 - y^2}{x^2 + y^2}$$

Solution: We have,

$$z = x^2 \tan^{-1}(y/x) - y^2 \tan^{-1}(x/y)$$

$$\therefore \quad \frac{\partial z}{\partial y} = \frac{x^2}{1 + y^2/x^2} \cdot \frac{1}{x} - 2y \tan^{-1}\frac{x}{y} - \frac{y^2}{1 + x^2/y^2} \cdot \frac{-x}{y^2}$$

$$= \frac{x^3}{x^2 + y^2} + \frac{xy^2}{x^2 + y^2} - 2y \tan^{-1}\frac{x}{y}$$

$$= x - 2y \tan^{-1}\frac{x}{y}$$

$$\therefore \quad \frac{\partial^2 z}{\partial x \partial y} = \frac{\partial}{\partial x}\left(x - 2y \tan^{-1}\frac{x}{y}\right)$$

$$= 1 - \frac{2y}{1 + x^2/y^2} \cdot \frac{1}{y}$$

$$= 1 - \frac{2y^2}{x^2 + y^2}$$

$$= \frac{x^2 - y^2}{x^2 + y^2}$$

Example 3: Verify Euler's theorem on homogenous functions for the function

$$z = x^n \ln \frac{y}{x}$$

Solution: z is a homogenous function of x and y of degree n.

$$\therefore \quad x\frac{\partial z}{\partial x} + y\frac{\partial z}{\partial y} = nz$$

Now, $\dfrac{\partial z}{\partial x} = nx^{n-1} \ln \dfrac{y}{x} + x^n \left(\dfrac{x}{y} \cdot \dfrac{y}{x^2} \right)$

$$= nx^{n-1} \ln \frac{y}{x} - x^{n-1}$$

and $\dfrac{\partial z}{\partial y} = x^n \dfrac{x}{y} . \dfrac{1}{x}$

$$= \frac{x^n}{y}$$

$$\therefore x\frac{\partial z}{\partial x} + y\frac{\partial z}{\partial y} = x\left(nx^{n-1} \ln \frac{y}{x} - x^{n-1} \right) + y.\frac{x^n}{y}$$

$$= nx^n \ln \frac{y}{x} - x^n + x^n$$

$$= nx^n \ln \frac{y}{x}$$

$$= nz$$

Example 4: If $u = \dfrac{1}{\sqrt{\left(x^2 + y^2 + z^2\right)}}; x^2 + y^2 + z^2 \neq 0$, then show that

$$\frac{\partial^2 u}{\partial x^2} + \frac{\partial^2 u}{\partial y^2} + \frac{\partial^2 u}{\partial z^2} = 0$$

Solution: We have,

$$u = \frac{1}{\sqrt{(x^2+y^2+z^2)}}$$

$$\therefore \frac{\partial u}{\partial x} = -\frac{1}{2}(x^2+y^2+z^2)^{-3/2}\,2x$$

$$= -x(x^2+y^2+z^2)^{-3/2}$$

$$\frac{\partial^2 u}{\partial x^2} = -(x^2+y^2+z^2)^{-3/2} + 3x^2(x^2+y^2+z^2)^{-5/2}$$

Similarly, $\frac{\partial^2 u}{\partial y^2} = -(x^2+y^2+z^2)^{-3/2} + 3y^2(x^2+y^2+z^2)^{-5/2}$

and $\frac{\partial^2 u}{\partial z^2} = -(x^2+y^2+z^2)^{-3/2} + 3z^2(x^2+y^2+z^2)^{-5/2}$

Adding, we get

$$\frac{\partial^2 u}{\partial x^2} + \frac{\partial^2 u}{\partial y^2} + \frac{\partial^2 u}{\partial z^2} = 0$$

Example 5: If $u = \sin^{-1}\left(\frac{x^2+y^2}{x+y}\right)$, prove that

$$x\frac{\partial u}{\partial x} + y\frac{\partial u}{\partial y} = \tan u$$

Solution: Let $\frac{x^2+y^2}{x+y} = z$. Then

$$\sin u = z$$

where $z = \frac{x^2+y^2}{x+y} = x\frac{1+y^2/x^2}{1+y/x}$ is a homogenous function of x and y of degree 1.

$\therefore$ By Euler's theorem, we have

$$x\frac{\partial z}{\partial x}+y\frac{\partial z}{\partial y}=z$$

But $\frac{\partial z}{\partial x}=\frac{\partial}{\partial x}(\sin u)=\cos u.\frac{\partial u}{\partial x}$

and $\frac{\partial z}{\partial y}=\frac{\partial}{\partial y}(\sin u)=\cos u.\frac{\partial u}{\partial y}$

Thus, $z=x\frac{\partial z}{\partial x}+y\frac{\partial z}{\partial y}$

$$=\cos u\left(x\frac{\partial u}{\partial y}+y\frac{\partial u}{\partial y}\right)$$

Since $z=\sin u$

$$\therefore \sin u=\cos u\left(x\frac{\partial u}{\partial x}+y\frac{\partial u}{\partial y}\right)$$

$$\Rightarrow x\frac{\partial u}{\partial x}+y\frac{\partial u}{\partial y}=\tan u.$$

Example 6: If $u=\cot^{-1}\frac{x+y}{\sqrt{x}+\sqrt{y}}$, show that

$$x\frac{\partial u}{\partial x}+y\frac{\partial u}{\partial y}+\frac{1}{4}\sin 2u=0$$

Solution: Let $z=\cot u=\frac{x+y}{\sqrt{x}+\sqrt{y}}$ ...(1)

z is a homogenous function of x and y of degree ½. Therefore by Euler's theorem

$$x\frac{\partial z}{\partial x}+y\frac{\partial z}{\partial y}+\frac{1}{2}z \quad ...(2)$$

From (1),

$$\frac{\partial z}{\partial x}=-\operatorname{cosec}^2 u.\frac{\partial u}{\partial x}$$

and $\frac{\partial z}{\partial y} = -\operatorname{cosec}^2 u.\frac{\partial u}{\partial x}$

From (2),

$$-x\operatorname{cosec}^2 u.\frac{\partial u}{\partial x} - y\operatorname{cosec}^2 u.\frac{\partial u}{\partial y} = \frac{1}{2}\cot u$$

$$\Rightarrow x\frac{\partial u}{\partial x} + y\frac{\partial u}{\partial y} = \frac{-1}{2}\frac{\cot u}{\operatorname{cosec}^2 u}$$

$$\Rightarrow x\frac{\partial u}{\partial x} + y\frac{\partial u}{\partial y} + \frac{1}{4}\sin^2 u = 0$$

Example 7: If $u = \tan^{-1}\frac{x^3+y^3}{x-y}, x \neq y$, show that

$$x^2\frac{\partial^2 u}{\partial x^2} + 2xy\frac{\partial^2 u}{\partial x.\partial y} + y^2\frac{\partial^2 u}{\partial y^2} = (1-4\sin^2 u)\sin 2u$$

Solution: Let, $z = \tan u = \frac{x^3+y^3}{x-y}$

$$= x^2\frac{1+(y/x)^3}{1-(y/x)}$$

is a homogenous function of x, y of degree 2.

$\therefore$ By Euler's theorem, we have

$$x\frac{\partial z}{\partial x} + y\frac{\partial z}{\partial y} = 2z$$

But

$$\frac{\partial z}{\partial x} = \sec^2 u\frac{\partial u}{\partial x}$$

and $\frac{\partial z}{\partial y} = \sec^2 u\frac{\partial u}{\partial y}$

Also, $$\frac{\partial^2 z}{\partial x^2} = \sec^2 u \frac{\partial^2 u}{\partial x^2} + 2\sec^2 u \tan u \left(\frac{\partial u}{\partial x}\right)^2$$

$$\frac{\partial^2 z}{\partial y^2} = \sec^2 u \frac{\partial^2 u}{\partial y^2} + 2\sec^2 u \tan u \left(\frac{\partial u}{\partial y}\right)^2$$

and $$\frac{\partial^2 z}{\partial x.\partial y} = \sec^2 u \frac{\partial^2 u}{\partial x.\partial y} + 2\sec^2 u \tan u \frac{\partial u}{\partial x}.\frac{\partial u}{\partial y}$$

Also by corollary of Euler's Theorem,

$$x^2 \frac{\partial^2 z}{\partial x^2} + 2xy \frac{\partial^2 z}{\partial x.\partial y} + y^2 \frac{\partial^2 z}{\partial y^2} = 2(2-1)z$$

$$\Rightarrow \sec^2 u \left(x^2 \frac{\partial^2 u}{\partial x^2} + 2xy \frac{\partial^2 u}{\partial x.\partial y} + y^2 \frac{\partial^2 u}{\partial y^2} \right)$$

$$+ 2\sec^2 u \tan u \left(x^2 \left(\frac{\partial u}{\partial x}\right)^2 + 2xy \frac{\partial u}{\partial x}.\frac{\partial u}{\partial y} + y^2 \left(\frac{\partial u}{\partial y}\right)^2 \right) = 2\tan u$$

$$\Rightarrow x^2 \frac{\partial^2 z}{\partial x^2} + 2xy \frac{\partial^2 u}{\partial x.\partial y} + y^2 \frac{\partial^2 u}{\partial y^2} + 2\tan u \left(x\frac{\partial u}{\partial x} + y\frac{\partial u}{\partial y} \right)^2$$

$$= 2\sin u \cos u$$

$$\Rightarrow x^2 \frac{\partial^2 u}{\partial x^2} + 2xy \frac{\partial^2 u}{\partial x.\partial y} + y^2 \frac{\partial^2 u}{\partial y^2} = \sin 2u - 2\tan u \sin^2 2u$$

$$= (1 - 2\tan u.\sin 2u)\sin 2u$$

$$= \left(1 - 4\sin^2 u\right)\sin 2u$$

Exorcise – 10.1

1. *Find the first order partial derivatives of*

   (i) $\cot^{-1}(x+y)$

   (ii) $e^{ax}\cos bx$

   (iii) $\sin(x^2y^2)$

   (iv) $\dfrac{x+y}{x-y}$

2. Find the second order partial derivatives of

   (i) $\tan\left(\tan^{-1}x+\tan^{-1}y\right)$

   (ii) $\dfrac{xy}{\sqrt{1+x^2+y^2}}$

   (iii) $\ln\left(x\tan^{-1}y\right)$

   (iv) $\sin\left(e^{ax}+e^{by}\right)$

   (v) e^{x^y}

3. Verify that $\dfrac{\partial^2 u}{\partial x.\partial y}=\dfrac{\partial^2 u}{\partial y.\partial x}$

   when u is $\log(y\sin x+x\sin y)$

4. If $z=\sin^{-1}\left(\dfrac{\sqrt{x}-\sqrt{y}}{\sqrt{x}+\sqrt{y}}\right)$, show that

   $\dfrac{\partial z}{\partial x}=-\dfrac{y}{x}.\dfrac{\partial z}{\partial y}$

5. Find the value of

   $\dfrac{1}{a^2}\dfrac{\partial^2 z}{\partial x^2}+\dfrac{1}{b^2}.\dfrac{\partial^2 z}{\partial y^2}$

when $a^2x^2+b^2y^2-c^2z^2=0$

6. If $z=f(x+ay)+\phi(x-ay)$, prove that

$$\frac{\partial^2 z}{\partial y^2}=\frac{a^2\partial^2 z}{\partial x^2}$$

7. If $z=\tan(y+ax)+(y-ax)^{3/2}$, find the value of

$$\frac{\partial^2 z}{\partial x^2}-a^2\frac{\partial^2 z}{\partial y^2}$$

8. If $v=\left(x^2+y^2+z^2\right)^{-1/2}$, show that

$$\frac{\partial^2 v}{\partial x^2}+\frac{\partial^2 v}{\partial y^2}+\frac{\partial^2 v}{\partial z^2}=0$$

9 If $z(x+y)=x^2+y^2$, show that

$$\left(\frac{\partial z}{\partial x}-\frac{\partial z}{\partial y}\right)^2=4\left(1-\frac{\partial z}{\partial x}-\frac{\partial z}{\partial y}\right)$$

10. If $z=3xy-y^3+\left(y^2-2x\right)^{3/2}$, verify that

$$\frac{\partial^2 z}{\partial x^2}\cdot\frac{\partial^2 z}{\partial y^2}=\left(\frac{\partial^2 z}{\partial x.\partial y}\right)^2$$

11. If $u=\log\dfrac{x^2+y^2}{x+y}$, prove that

$$x\frac{\partial u}{\partial x}+y\frac{\partial u}{\partial y}=1$$

12. If $u=\dfrac{1}{\sqrt{1-2xy+y^2}}$, show that

$$\frac{\partial}{\partial x}\left[(1-x)^2\frac{\partial u}{\partial x}\right]+\frac{\partial}{\partial y}\left(y^2\frac{\partial u}{\partial y}\right)=0$$

13. If $z=\tan^{-1}(y/x)$, verify that

$$\frac{\partial^2 z}{\partial x^2}+\frac{\partial^2 z}{\partial y^2}=0$$

14. If $u=\sin^{-1}\dfrac{x^2+y^2}{x+y}$, show that

$$x\frac{\partial u}{\partial x}+y\frac{\partial u}{\partial y}=\tan u$$

15. If $u=\cos^{-1}\dfrac{x+y}{\sqrt{x}+\sqrt{y}}$, show that

$$x\frac{\partial u}{\partial x}+y\frac{\partial u}{\partial y}+\frac{1}{2}\cot u=0$$

16. If $z=(x+y)+(x+y)\phi(y/x)$, prove that

$$x\left(\frac{\partial^2 z}{\partial x^2}-\frac{\partial^2 x}{\partial y\partial x}\right)=y\left(\frac{\partial^2 z}{\partial y^2}-\frac{\partial^2 z}{\partial x\partial y}\right)$$

17. If $z=\sec^{-1}\dfrac{x^3+y^3}{x+y}$, show that

$$x\frac{\partial z}{\partial x}+y\frac{\partial z}{\partial y}=2\cot z$$

18. If $u=\tan^{-1}\left(\dfrac{x+y}{\sqrt{x}+\sqrt{y}}\right)$, show that

$$x\frac{\partial u}{\partial x}+y\frac{\partial u}{\partial y}=\frac{1}{4}\sin 2u$$

19. If $V=r^m$ where $r^2=x^2+y^2+z^2$, show that

$$\frac{\partial^2 v}{\partial x^2}+\frac{\partial^2 v}{\partial y^2}+\frac{\partial^2 v}{\partial z^2}=m(m+1)r^{m-2}$$

20. If $u = f(r)$, where $r=\sqrt{(x^2+y^2)}$, prove that

$$\frac{\partial^2 u}{\partial x^2}+\frac{\partial^2 u}{\partial y^2}=f''(r)+\frac{1}{r}f'(r)$$

21. If $\theta = t^n e^{-r^2}/4t$, find the value of n which will make

$$\frac{1}{r^2}\frac{\partial}{\partial r}\left(r^2\frac{\partial\theta}{\partial r}\right)=\frac{\partial\theta}{\partial t}$$

22. If $u=\log\left(x^2+y^2+z^2\right)$, prove that

$$x\frac{\partial^2 u}{\partial y.\partial z}=y\frac{\partial^2 u}{\partial z.\partial x}=z\frac{\partial^2 u}{\partial x.\partial y}$$

23. If $u=f(ax^2+2hxy+by^2)$, $v=\phi(ax^2+2hxy+by^2)$ prove that

$$\frac{\partial}{\partial y}\left(u\frac{\partial v}{\partial x}\right)=\frac{\partial}{\partial x}\left(u\frac{\partial v}{\partial y}\right)$$

24. If $u=\sin^{-1}\left[\frac{x^{1/3}+y^{1/3}}{x^{1/2}+y^{1/2}}\right]^{1/2}$, show that

$$x^2\frac{\partial^2 u}{\partial x^2}+2xy\frac{\partial^2 u}{\partial x\partial y}+y^2\frac{\partial^2 u}{\partial y^2}=\frac{\tan u}{144}(13+\tan^2 u)$$

10.10. TOTAL DIFFERENTIALS

Let $z = f(x, y)$ be a function of two independent variables x and y.

Let $\delta x, \delta y$ denote small changes in the independent variables x and y and let δz be the corresponding change in z.

Then,

$$z + \delta z = f(x + \delta x, y + \delta y)$$

$$\therefore \delta z = f(x + \delta x, y + \delta y) - f(x, y)$$

Adding and subtracting $f(x + \delta x, y)$, we get

$$\delta z = f(x + \delta x, y + \delta y) - f(x + \delta x, y) + f(x + \delta x, y) - f(x, y)$$

$$= \{f(x + \delta x, y + \delta y) - f(x + \delta x, y)\} + \{f(x + \delta x, y) - f(x, y)\} \quad \text{...(i)}$$

Here the change δz has been expressed as the sum of two differences to each of which we shall apply Lagrange's Mean value theorem.

Consider the function $f(y) = f(x + \delta x, y)$, where $x + \delta x$ is constant, so that by the Mean value theorem

$$f(x + \delta x, y + \delta y) - f(x + \delta x, y) = \delta_y f_y(x + \delta x, y + \theta_1 \delta y)$$

We write,

$$f(x + \delta x, y + \theta_1 \delta y) - f_y(x, y) = \epsilon_2 \quad \text{...(ii)}$$

so that ϵ_2 depends on $\Delta x, \Delta y$, and because of the assumed continuity of f_y tends to zero as $\delta x, \delta y$ both tend to zero.

Again by considering $f(x) = f(x, y)$, keeping y constant, we have by the Mean value theorem,

$$f(x + \delta x, y) - f(x, y) = \delta_x f_x(x + \theta_2 \delta x, y)$$

Again, we can write

$$f_x(x + \theta_2 \delta x, y) - f_x(x, y) = \epsilon_1 \quad \text{...(iii)}$$

so that ϵ_1 depends upon δx and because of the assumed continuity of f_x tends to 0 as δx tends to zero.

From (i), (ii) & (iii), we get

$$\delta z = \delta x . f_x(x, y) + \delta y . f_y(x, y) + \epsilon_1 \, \delta x + \epsilon_2 \, \delta_y$$

$$= \frac{\partial z}{\partial x}\delta x + \frac{\partial z}{\partial y}\delta y + \epsilon_1\, \delta x + \epsilon_2\, \delta y$$

Now, when δx and $\delta y \to 0$, $\delta z \to 0$

Also, $\epsilon_1, \epsilon_2 \to 0$ when δx and $\delta y \to 0$ and $\frac{\partial z}{\partial x}, \frac{\partial z}{\partial y}$ are finite.

The part $\frac{\partial z}{\partial x}\delta x + \frac{\partial z}{\partial y}\delta y$ of ∂z is called **total internal differential** of z and is denoted by dz.

Thus, $\boldsymbol{dz = \frac{\partial z}{\partial x}dx + \frac{\partial z}{\partial y}dy}$

10.11. DIFFERENTIATION OF COMPOSITE FUNCTION

Let $z = f(x, y)$

possesses continuous partial derivatives and let

$x = \phi(t)$

$y = \Psi(t)$

possess continuous derivatives.

Then, $\frac{dz}{dt} = \frac{\partial z}{\partial x}.\frac{dx}{dt} + \frac{\partial z}{\partial y}.\frac{dy}{dt}$

Proof: Let $t, t + \Delta t$ be any two values.

Let $\Delta x, \Delta y, \Delta z$ be the changes in x, y, z consequent to the change Δt in t. We have

$$x + \Delta x = \phi(t + \Delta t)$$

$$y + \Delta y = \Psi(t + \Delta t)$$

and $z + \Delta z = f(x + \Delta x, y + \Delta y)$

$$\Rightarrow \quad \Delta z = f(x + \Delta x, y + \Delta y) - f(x, y)$$

$$= [f(x + \Delta x, y + \Delta y) - f(x, y + \Delta y)] + [f(x, y + \Delta y) - f(x, y)]$$

Applying Lagrange's Mean value theorem to the two differences on the right and obtain

$$\Delta z = \Delta x f_x(x + \theta_1 \Delta x, y + \Delta y) + \Delta y f_y(x, y + \theta_2 \Delta y)(0 < \theta_1, \theta_2 < 1)$$

$$\Rightarrow \frac{\Delta z}{\Delta t} = \frac{\Delta x}{\Delta t} f_x(x + \theta_1 \Delta x, y + \Delta y) + \frac{\Delta y}{\Delta t} f_y(x, y + \theta_2 \Delta y) \quad \ldots(i)$$

Let $\Delta t \to 0$ so that Δx and $\Delta y \to 0$

Because of the continuity of partial derivatives, we have

$$\lim_{(\Delta x, \Delta y) \to (0,0)} f_x(x + \theta_1 \Delta x, y + \Delta y) = f_x(x, y) = \frac{\partial z}{\partial x}$$

$$\lim_{\Delta y \to 0} f_y(x, y + \theta_2 \Delta y) = f_y(x, y) = \frac{\partial z}{\partial y}$$

Hence in the limit, (i) becomes

$$\frac{dz}{dt} = \frac{\partial z}{\partial x} \cdot \frac{dx}{dt} + \frac{\partial z}{\partial y} \cdot \frac{dy}{dt}$$

10.12. CHANGE OF VARIABLES

Let $z = f(x, y)$ ...(i)

possess continuous first order partial derivatives w.r.t. x, y.

Let $x = \phi(u, v)$

$y = \Psi(u, v)$

possesses continuous first order partial derivatives. We write,

$$\frac{\partial z}{\partial u} = \frac{\partial z}{\partial x} \cdot \frac{\partial x}{\partial u} + \frac{\partial z}{\partial y} \cdot \frac{\partial y}{\partial u} \quad \ldots(ii)$$

and $$\frac{\partial z}{\partial v} = \frac{\partial z}{\partial x} \cdot \frac{\partial x}{\partial v} + \frac{\partial z}{\partial y} \cdot \frac{\partial y}{\partial v} \quad \ldots(iii)$$

Now, solve equations (ii) and (iii) as simultaneously linear equations in $\frac{\partial z}{\partial x}$ and $\frac{\partial z}{\partial y}$. This will give us expressions for $\frac{\partial z}{\partial x}$ and $\frac{\partial z}{\partial y}$ in terms of $\frac{\partial z}{\partial u}, \frac{\partial z}{\partial v}$ and the easily determined quantities $\frac{\partial x}{\partial u}, \frac{\partial x}{\partial v}, \frac{\partial y}{\partial u}, \frac{\partial y}{\partial v}$.

These values of $\frac{\partial z}{\partial x}$ and $\frac{\partial z}{\partial y}$ together with values of x and y as given by (i) when substituted in any expression involving $z, x, y. \frac{\partial z}{\partial x}, \frac{\partial z}{\partial y}$ etc. will give the required expression in terms of $z, u, v. \frac{\partial z}{\partial u}$ and $\frac{\partial z}{\partial v}$ etc. Sometimes, it is easy to solve equations (i) for u and v in terms of x and y. That is if

$$u = U(x,y),\ v = V(x,y) \qquad \text{...(iv)}$$

then it will be simpler to make use of the formulae

$$\frac{\partial z}{\partial x} = \frac{\partial z}{\partial u}.\frac{\partial u}{\partial x} + \frac{\partial z}{\partial v}.\frac{\partial v}{\partial x} \qquad \text{...(v)}$$

and $$\frac{\partial z}{\partial y} = \frac{\partial z}{\partial u}.\frac{\partial u}{\partial y} + \frac{\partial z}{\partial v}.\frac{\partial v}{\partial y} \qquad \text{...(vi)}$$

10.13. DIFFERENTIATION OF IMPLICIT FUNCTIONS

Let $f(x,y) = 0$ defines y as a function of x implicitly.

We shall obtain the value of $\frac{dy}{dx}$ in terms of the partial derivatives $\frac{\partial f}{\partial x}$ and $\frac{\partial f}{\partial y}$.

Since $f(x,y)$ is a function of x and y and y is a function of x, therefore we can look upon $f(x,y)$ as a composite function of x.

$$\therefore \frac{df}{dx} = \frac{\partial f}{\partial x}.\frac{dx}{dx} + \frac{\partial f}{\partial y}.\frac{dy}{dx}$$

$$\Rightarrow 0 = \frac{\partial f}{\partial x} + \frac{\partial f}{\partial y}.\frac{dy}{dx} \qquad \text{...(i)}$$

$$\therefore \frac{dy}{dx} = -\frac{\partial f / \partial x}{\partial f / \partial y}$$

$$= -\frac{fx}{fy}, \text{ provided } fy \neq 0$$

Differentiating (i) again w.r.t x, we get

$$\frac{d}{dx}\left(\frac{\partial f}{\partial x}\right)+\frac{d}{dx}\left(\frac{\partial f}{\partial y}.\frac{dy}{dx}\right)=0$$

$$\Rightarrow \frac{d}{dx}\left(\frac{\partial f}{\partial x}\right)+\frac{d}{dx}\left(\frac{\partial f}{\partial y}\right)\frac{dy}{dx}+\frac{\partial f}{\partial y}\frac{d^2 y}{dx^2}=0$$

$$\Rightarrow \frac{\partial^2 f}{\partial x^2}+\frac{\partial^2 f}{\partial y \partial x}.\frac{dy}{dx}+\left(\frac{\partial^2 f}{\partial y \partial x}+\frac{\partial^2 f}{\partial y^2}.\frac{dy}{dx}\right)\frac{dy}{dx}+\frac{\partial f}{\partial y}\frac{d^2 y}{dx^2}=0$$

$$\Rightarrow \frac{\partial^2 f}{\partial x^2}+2\frac{\partial^2 f}{\partial x \partial y}.\frac{dy}{dx}+\frac{\partial^2 f}{\partial y^2}\left(\frac{dy}{dx}\right)^2+\frac{\partial f}{\partial y}\frac{d^2 y}{dx^2}=0$$

$$\Rightarrow f_{xx}+2f_{xy}\left(\frac{-f_x}{f_y}\right)+f_{yy}\left(\frac{-f_x}{f_y}\right)^2+f_y\frac{d^2 y}{dx^2}=0$$

On simplification, this gives,

if $f_y \neq 0$

Examples

Example 1: If $x = \tan^{-1}\left(\frac{y}{x}\right)$, prove that

$$dz = \frac{xdy - ydx}{x^2 + y^2}$$

Solution: We know,

$$dz = \frac{\partial z}{\partial x}dx + \frac{\partial z}{\partial y}dy$$

But,

$$\frac{\partial z}{\partial x} = \frac{1}{1+(y/x)^2}\cdot\left(\frac{-y}{x^2}\right)$$

$$= \frac{-y}{x^2+y^2}$$

and $$\frac{\partial z}{\partial y} = \frac{1/x}{1+(y/x)^2}$$

$$= \frac{x}{x^2+y^2}$$

$$\therefore \quad dz = \frac{-y}{x^2+y^2}dx + \frac{x}{x^2+y^2}dy$$

$$= \frac{xdy - ydx}{x^2+y^2}$$

Example 2: Find dz/dt when

$$z = xy^2 + x^2y, \quad x = at^2, \quad y = 2at$$

Verify by direct substitution.

Solution: We have

$$z = xy^2 + x^2y$$

$$\frac{\partial z}{\partial x} = y^2 + 2xy$$

$$\frac{\partial z}{\partial y} = 2xy + x^2$$

$$\frac{dx}{dt} = 2at \quad \text{and} \quad \frac{dy}{dt} = 2a$$

$$\therefore \frac{dz}{dt} = \frac{\partial z}{\partial x}.\frac{dx}{dt} + \frac{\partial z}{\partial y}.\frac{dy}{dt}$$

$$= \left(y^2 + 2xy\right)2at + \left(2xy + x^2\right)2a$$

$$= \left(4a^2t^2 + 4a^2t^3\right)2at + \left(4a^2t^3 + a^2t^4\right)2a$$

$$= a^3\left(16t^3 + 10t^4\right)$$

Again,

$$z = x^2y + xy^2$$

$$= 2a^3t^5 + 4a^3t^4$$

$$\Rightarrow \frac{dz}{dt} = 10a^3t^4 + 16a^3t^3$$

$$= a^3\left(16t^3 + 10t^4\right)$$

Hence the verification

Example 3: If $z = x^2 + y$ and $y = z^2 + x$, then find the differential co-efficients of the first order when x is the independent variable.

Solution: $dz = \frac{\partial z}{\partial x} dx + \frac{\partial z}{\partial y} dy$

Since $z = x^2 + y$

$$\therefore \quad \frac{\partial z}{\partial x} = 2x, \quad \frac{\partial z}{\partial y} = 1$$

Thus, $dz = 2xdx + dy$

Similarly, $dy = \frac{\partial y}{\partial x} dx + \frac{\partial y}{\partial z} dz,$

$$= dx + 2zdz$$

Thus, $dz = 2xdx + dx + 2zdz$

$$\therefore \quad dz(1 - 2z) = dx(2x + 1)$$

$$\therefore \quad \frac{dz}{dx} = \frac{2x + 1}{1 - 2z}$$

Also, $dy = dx + 2z(2xdx + dy)$

$$= dx(1 + 4xz) + 2zdy$$

$$\therefore \quad dz(1-2z) = dx(1+4xz)$$

$$\Rightarrow \quad \frac{dy}{dx} = \frac{1+4xz}{1-2z}$$

Example 4: z is a function of *x* and *y*, prove that if $x = e^{u} + e^{-v}$, $y = e^{-u} - e^{-v}$, then $\frac{\partial z}{\partial u} - \frac{\partial z}{\partial v} = x\frac{\partial z}{\partial x} - y\frac{\partial z}{\partial y}$

Solution: z is a composite function of u, v. We have

$$\frac{\partial z}{\partial u} = \frac{\partial z}{\partial x}.\frac{\partial x}{\partial u} + \frac{\partial z}{\partial y}.\frac{\partial y}{\partial u}$$

$$= \frac{\partial z}{\partial x}.e^{u} - \frac{\partial z}{\partial y}.e^{-u}$$

$$\frac{\partial z}{\partial v} = \frac{\partial z}{\partial x}.\frac{\partial x}{\partial v} + \frac{\partial z}{\partial y}.\frac{\partial y}{\partial v}$$

$$= -\frac{\partial z}{\partial x}.e^{-v} - \frac{\partial z}{\partial y}.e^{v}$$

Subtracting, we get

$$\frac{\partial z}{\partial u} - \frac{\partial z}{\partial v} = \frac{\partial z}{\partial x}\left(e^{u} + e^{-v}\right) - \frac{\partial z}{\partial y}\left(e^{-u} - e^{v}\right)$$

$$= x\frac{\partial z}{\partial x} - y\frac{\partial z}{\partial y}$$

Example 5: If $z = e^{x}\sin y$, where $x = \ln t$ and $y = t^{2}$, then find $\frac{dz}{dt}$.

Solution: We know,

$$\frac{dz}{dt} = \frac{\partial z}{\partial x}.\frac{dx}{dt} + \frac{\partial z}{\partial y}.\frac{dy}{dt}$$

But,

$$\frac{\partial z}{\partial x} = e^{x}\sin y,$$

$$\frac{\partial z}{\partial y} = e^x \cos y$$

$$\frac{dx}{dt} = \frac{1}{t} \text{ and } \frac{dy}{dt} = 2t$$

$$\therefore \quad \frac{dz}{dt} = e^x \sin y.\frac{1}{t} + \left(e^x \cos y\right)2t$$

$$= \frac{e^x}{t}\left(\sin y + 2t^2 \cos y\right)$$

Example 6: If $H = f(y-z, z-x, x-y)$, prove that

$$\frac{\partial H}{\partial x} + \frac{\partial H}{\partial y} + \frac{\partial H}{\partial z} = 0$$

Solution. Let, $u = y - z,\ v = z - x,\ w = x - y$

$$\Rightarrow H = f(u, v, w)$$

H is a composite function of x, y, z.

We have,

$$\frac{\partial H}{\partial x} = \frac{\partial H}{\partial u}.\frac{\partial u}{\partial x} + \frac{\partial H}{\partial v}.\frac{\partial v}{\partial x} + \frac{\partial H}{\partial w}.\frac{\partial w}{\partial x}$$

$$= \frac{\partial H}{\partial u}.0 + \frac{\partial H}{\partial v}(-1) + \frac{\partial H}{\partial w}.1$$

$$= -\frac{\partial H}{\partial v} + \frac{\partial H}{\partial w}$$

Similarly,

$$\frac{\partial H}{\partial y} = -\frac{\partial H}{\partial w} + \frac{\partial H}{\partial u}$$

$$\frac{\partial H}{\partial z} = -\frac{\partial H}{\partial u} + \frac{\partial H}{\partial v}$$

Adding all the above, we get

$$\frac{\partial H}{\partial x} + \frac{\partial H}{\partial y} + \frac{\partial H}{\partial z} = 0$$

Example 7: If $ax^2+2hxy+by^2=1$, then find $\frac{dy}{dx}$.

Solution: Here $f(x,y)=ax^2+2hxy+by^2-1=0$

$$\therefore \frac{dy}{dx}=-\frac{(\partial f/\partial x)}{(\partial f/\partial y)}$$

$$=-\frac{(2ax+2hy)}{2hx+2by}$$

$$=-\frac{(ax+hy)}{hx+by}$$

Example 8: H is a homogenous function of *x*, *y*, *z* of degree *n*, prove that

$$x\frac{\partial H}{\partial x}+y\frac{\partial H}{\partial y}+z\frac{\partial H}{\partial z}=nz$$

Solution: We have,

$$H=x^n f\left[\frac{y}{x},\frac{z}{x}\right]$$

$$=x^n f(u,v), \text{ where } y/x=u, z/x=v$$

$$\therefore \frac{\partial H}{\partial x}=nx^{n-1}f(u,v)+x^n\left[\frac{\partial f}{\partial u}.\frac{\partial u}{\partial x}+\frac{\partial f}{\partial v}.\frac{\partial v}{\partial x}\right]$$

But,

$$\frac{\partial u}{\partial x}=-\frac{y}{x},\ \frac{\partial v}{\partial x}=-\frac{z}{x^2}$$

Hence

$$\frac{\partial H}{\partial x}=nx^{n-1}f(u,v)-x^{n-2}\left[y\frac{\partial f}{\partial u}+z\frac{\partial f}{\partial v}\right]$$

Again $\frac{\partial H}{\partial y}=x^n\left[\frac{\partial f}{\partial u}.\frac{\partial u}{\partial y}+\frac{\partial f}{\partial v}.\frac{\partial v}{\partial y}\right]$

$$= x^{n-1}\frac{\partial f}{\partial u}, \text{ for } \frac{\partial u}{\partial y}=\frac{1}{x}, \quad \frac{\partial v}{\partial y}=0$$

Similarly

$$\frac{\partial H}{\partial z}=x^{n-1}\frac{\partial f}{\partial v}$$

$$\Rightarrow x\frac{\partial H}{\partial x}+y\frac{\partial H}{\partial y}+z\frac{\partial H}{\partial z}=nx^n f(u,v)=nz.$$

Example 9: If $f(x,y)=0$, $\phi(y,z)=0$, show that

$$\frac{\partial f}{\partial y}\left(\frac{\partial \phi}{\partial z}\frac{dz}{dx}\right)=\frac{\partial f}{\partial x}\frac{\partial \phi}{\partial y}$$

Solution: $f(x,y)=0$, we have

$$\frac{dy}{dx}=\frac{-(\partial f/\partial x)}{(\partial f/\partial y)} \quad \text{...(i)}$$

Again, $\phi(y,z)=0$, we have

$$\frac{dz}{dy}=\frac{-(\partial \phi/\partial y)}{(\partial \phi/\partial z)} \quad \text{...(ii)}$$

By multiplying (i) and (ii), we get

$$\frac{dy}{dx}\frac{dz}{dy}=\frac{\dfrac{\partial f}{\partial x}\cdot\dfrac{\partial \phi}{\partial y}}{\dfrac{\partial f}{\partial y}\cdot\dfrac{\partial \phi}{\partial z}}$$

$$\Rightarrow \quad \frac{dz}{dx}=\frac{\dfrac{\partial f}{\partial x}\cdot\dfrac{\partial \phi}{\partial y}}{\dfrac{\partial f}{\partial y}\cdot\dfrac{\partial \phi}{\partial z}}$$

$$\Rightarrow \quad \frac{\partial f}{\partial y}\left(\frac{\partial f}{\partial y}\cdot\frac{\partial \phi}{\partial z}\right)=\frac{\partial f}{\partial x}\cdot\frac{\partial \phi}{\partial y}.$$

Example 10: If $x = r\cos\theta, y = r\sin\theta$ and $V = f(x,y)$, then show that

$$\frac{\partial^2 V}{\partial x^2}+\frac{\partial^2 V}{\partial y^2}=\frac{\partial^2 V}{\partial r^2}+\frac{1}{r}.\frac{\partial V}{\partial r}+\frac{1}{r^2}\frac{\partial^2 V}{\partial\theta^2}$$

Solution: We have,

$$x = r\cos\theta, y = r\sin\theta$$

$$\therefore dx = \cos\theta dr - r\sin\theta.d\theta \qquad ...(i)$$

and $dy = \sin\theta.dr + r\cos\theta.d\theta$ ...(ii)

Solving (i) and (ii) as simultaneous linear equations in dr and $d\theta$, we get

$$dr = \cos\theta dx + \sin\theta.dy \qquad ...(iii)$$

$$d\theta = \frac{1}{r}(\cos\theta dy - \sin\theta dx) \qquad ...(iv)$$

Now, $dr = \frac{\partial r}{\partial x}dx + \frac{\partial r}{\partial y}dy$ ...(v)

and $d\theta = \frac{\partial\theta}{\partial x}dx + \frac{\partial\theta}{\partial y}dy$ ...(vi)

Comparing (iii) and (v), we get

$$\frac{\partial r}{\partial x} = \cos\theta, \frac{\partial r}{\partial y} = \sin\theta$$

Ccomparing (iv) and (vi), we get

$$\frac{\partial\theta}{\partial x} = -\frac{1}{r}\sin\theta, \frac{\partial\theta}{\partial y} = \frac{1}{r}\cos\theta$$

Thus, $\frac{\partial v}{\partial x} = \frac{\partial v}{\partial r}.\frac{\partial r}{\partial x} + \frac{\partial v}{\partial\theta}.\frac{\partial\theta}{\partial x}$

$$= \cos\theta\cdot\frac{\partial v}{\partial r} - \frac{1}{r}\sin\theta\frac{\partial v}{\partial\theta}$$

$$= \left(\cos\theta.\frac{\partial}{\partial r} - \frac{1}{r}\sin\theta\frac{\partial}{\partial\theta}\right)v \qquad ...(vii)$$

Similarly, $\dfrac{\partial v}{\partial y}=\dfrac{\partial v}{\partial r}\cdot\dfrac{\partial r}{\partial y}+\dfrac{\partial v}{\partial \theta}\cdot\dfrac{\partial \theta}{\partial y}$

$$=\cos\theta\cdot\frac{\partial v}{\partial r}-\frac{1}{r}\sin\theta\frac{\partial v}{\partial \theta}$$

$$=\left(\sin\theta.\frac{\partial}{\partial r}+\frac{1}{r}\cos\theta\frac{\partial}{\partial \theta}\right)v \qquad \text{...(viii)}$$

It follows that form (vii)

$$\frac{\partial^2 V}{\partial x^2}=\left(\cos\theta.\frac{\partial}{\partial r}-\frac{1}{r}\sin\theta\frac{\partial}{\partial \theta}\right)\left(\cos\theta.\frac{\partial v}{\partial r}-\frac{1}{r}\sin\theta\frac{\partial v}{\partial \theta}\right)$$

$$=\cos^2\theta\frac{\partial^2 V}{\partial r^2}-\frac{1}{r}\sin\theta.\cos\theta\frac{\partial^2 V}{\partial r\partial\theta}-\frac{1}{r}\sin\theta.\cos\theta\frac{\partial^2 V}{\partial r\partial\theta}+\frac{\sin^2\theta}{r}\frac{\partial V}{\partial r}$$

$$+\frac{1}{r^2}\sin^2\theta\frac{\partial^2 V}{\partial\theta^2}+\frac{1}{r^2}\sin\theta.\cos\theta.\frac{\partial V}{\partial\theta} \qquad \text{...(ix)}$$

Similarly from (viii) it follows that

$$\frac{\partial^2 V}{\partial y^2}=\left(\sin\theta.\frac{\partial}{\partial r}+\frac{1}{r}\cos\theta.\frac{\partial}{\partial \theta}\right)\left(\sin\theta.\frac{\partial v}{\partial r}+\frac{1}{r}\cos\theta\frac{\partial v}{\partial \theta}\right)$$

$$=\sin^2\theta\frac{\partial^2 V}{\partial r^2}+\frac{1}{r}\sin\theta.\cos\theta\frac{\partial^2 V}{\partial r\partial\theta}+\frac{1}{r}\sin\theta.\cos\theta\frac{\partial^2 V}{\partial\theta\partial r}+\frac{1}{r}\cos^2\theta\frac{\partial V}{\partial r}$$

$$+\frac{1}{r^2}\cos^2\theta\frac{\partial^2 V}{\partial\theta^2}-\frac{1}{r^2}\sin\theta.\cos\theta.\frac{\partial V}{\partial\theta} \qquad \text{...(x)}$$

Adding (ix) and (x), we get

$$\frac{\partial^2 V}{\partial x^2}+\frac{\partial^2 V}{\partial y^2}=\left(\cos^2\theta+\sin^2\theta\right)\frac{\partial^2 V}{\partial r^2}+\frac{1}{r^2}\left(\cos^2\theta+\sin^2\theta\right)\frac{\partial^2 V}{\partial\theta^2}$$

$$+\frac{1}{r}\left(\sin^2\theta+\cos^2\theta\right)\frac{\partial V}{\partial r}$$

$$=\frac{\partial^2 V}{\partial r^2}+\frac{1}{r^2}\frac{\partial^2 V}{\partial\theta^2}+\frac{1}{r}\frac{\partial v}{\partial r}$$

Example 11: If A, B, C are the angles of a triangle such that $\sin^2 A + \sin^2 B + \sin^2 C$ = constant, then prove that

$$\frac{dA}{dB} = \frac{\tan B - \tan C}{\tan C \quad \tan A}$$

Solution: We have,

$$\sin C = \sin\left(\pi - \overline{A + B}\right)$$

$$= \sin(A + B)$$

Let $f(A, B) = \sin^2 A + \sin^2 B + \sin^2 (A + B)$

$$f_A(A, B) - 2\sin A.\cos A + 2\sin(A + B).\cos(A + B)$$

$$f_B(A, B) = 2\sin B.\cos B + 2\sin(A + B).\cos(A + B)$$

$$\therefore \frac{dA}{dB} = \frac{-f_B}{f_A}$$

$$= -\frac{2\sin B.\cos B + 2\sin(A + B).\cos(A + B)}{2\sin A.\cos A + 2\sin(A + B).\cos(A + B)}$$

$$= -\frac{\sin 2B - \sin 2C}{\sin 2A - \sin 2C}$$

$$= \frac{-\cos A\{\sin C.\cos B - \cos C \sin B\}}{-\cos B\{\sin A \cos C - \cos A.\sin C\}}$$

$$= \frac{\cos A.\sin B \cos C - \cos A \cos B \sin C}{\cos A \cos B \sin C - \sin A \cos B \cos C}$$

$$= \frac{\tan B - \tan C}{\tan C - \tan A}$$

Example 12: Prove that if $y^3 - 3ax^2 + x^3 - 0$, then

$$\frac{d^2 y}{dx^2} + \frac{2a^2 x^2}{y^5} = 0$$

Solution: Let, $f(x, y) = y^3 - 3ax^2 + x^3 = 0$

We have,

$$f_x(x,y) = -6ax + 3x^2$$

$$f_y(x,y) = 3y^2$$

$$f_x{}^2(x,y) = -6a + 6x$$

$$f_{xy}(x,y) = 0$$

$$f_y{}^2(x,y) = 6y$$

Now, $$\frac{d^2y}{dx^2} = -\frac{f_x{}^2(f_y)^2 - 2f_{yx}f_xf_y + f_y{}^2(f_x)^2}{(f_y)^3}$$

$$= -\frac{6(x-a)9y^4 + (3x^2 - 6ax)^2 6y}{27y^6}$$

$$= -2\frac{(x-a)(3ax - x^3) + (x^2 - 2ax)^2}{y^5}$$

$$= -2\frac{a^2x^2}{y^5}$$

Thus, $$\frac{d^2y}{dx^2} + \frac{2a^2x^2}{y^5} = 0$$

Differentiating the given relation w.r.t. x

$$3y^2\frac{dy}{dx} = 6ax - 3x^2$$

$$\Rightarrow \frac{dy}{dx} = \frac{2ax - x^2}{y^2}$$

$$\Rightarrow \frac{d^2y}{dx^2} = \frac{(2a - 2x)y^2 - 2y(2ax - x^2)dy/dx}{y^4}$$

$$= \frac{(2a-2x)y^2 - 2y(2ax-x^2)(2ax-x^2)/y^2}{y^4}$$

$$= \frac{2(a-x)y^3 - 2(2ax-x^2)^2}{y^5}$$

$$= -\frac{2a^2x^2}{y^5}$$

Exercise –10.2

1. If $z = x^m y^n$, then prove that

$$\frac{dz}{z} = m\frac{dx}{x} + n\frac{dy}{y}.$$

2. If $u = x^2 - y^2$, $x = 2r - 3s + 4$, $y = -r + 8s - 5$, find $\partial u / \partial r$.

3. If $x = r\cos\theta, y = r\sin\theta$, then show that

   (i) $dx = \cos\theta dr - r\sin\theta . d\theta$.

   (ii) $dy = \sin\theta dr + r\cos\theta . d\theta$.

   Deduce that

   (i) $dx^2 + dy^2 = dr^2 + r^2 d\theta^2$.

   (ii) $xdy - ydx = r^2 d\theta$.

4. If $z = (\cos y)/x$ and $x = u^2 - v, y = e^V$, find $\partial z / \partial v$.

5. If $z = x^2 + y$ and $y = z^2 + x$, find differential co-efficients of the first order when

   (i) y is the independent variable.

   (ii) z is the independent variable.

6. If $z = \dfrac{\sin u}{\cos v}$, $u = \dfrac{\cos y}{\sin x}$, $v = \dfrac{\cos x}{\sin y}$ find $\partial z / \partial x$.

7. If $z = \tan^{-1}\dfrac{y}{x}$ where $x = \ln t, y = e^t$, find $\dfrac{dz}{dt}$.

8. If $u = (x + y)/(1 - xy)$, $x = \tan(2r - s^2)$, $y = \cot(r^2 s)$ then find $\partial u / \partial s$.

9. If $z = x^2 - y^2$, where□ $x = e^t \cos t, y = e^t \sin t$, find $\dfrac{dz}{dt}$.

10. If $z = xyf(x, y)$ and z is a constant, show that

$$\frac{f'(y/x)}{f(y/x)} = \frac{x[y + x(dy/dx)]}{y[y - x(dy/dx)]}.$$

11. Find $\frac{\partial z}{\partial x}$ and $\frac{\partial z}{\partial y}$ if $z = u^2 + v^2 + w^2$, where $u = ye^x$, $v = xe^{-y}$,

$w = y/x$.

12. If $f(x,y) = 0, \phi(y,z) = 0$,show that

$$\frac{\partial f}{\partial y} \cdot \frac{\partial \phi}{\partial z} \cdot \frac{dz}{dx} = \frac{\partial f}{\partial x} \cdot \frac{\partial \phi}{\partial y}.$$

13. If $z = e^{ax+by} f(ax \quad by)$, prove that

$$b\frac{\partial z}{\partial x} + a\frac{\partial z}{\partial y} = 2abz.$$

14. If $x\sqrt{(1-y^2)} + y\sqrt{(1-x^2)} = a$, show that

$$\frac{d^2y}{dx^2} = \frac{a}{(1-x^2)^{3/2}}.$$

15. Find $\frac{dy}{dx}$ if

(i) $x^4 + y^4 = 5a^2 axy$.

(ii) $x^y + y^x = (x+y)^{x+y}$.

16. If $x\sqrt{(1-y^2)} + y\sqrt{(1-x^2)} = a$, show that

$$\frac{d^2y}{dx^2} = \frac{a}{(1-x^2)^{3/2}}.$$

17. If $f(x,y,z) = 0$ be an equation is three variables x, y, z, prove that

$$\left(\frac{dy}{dx}\right)_z \left(\frac{dx}{dz}\right)_y \left(\frac{dz}{dy}\right)_r = -1.$$

18. If $\frac{x^2}{a^2}+\frac{y^2}{b^2}=1$, prove that

$$\frac{dy}{dx}=-\frac{b^2x}{a^2y} \text{ and } \frac{d^2y}{dx^2}=-\frac{b^4}{a^2y^3}.$$

19. If $z=\sin(x^2+y^2)$, where $a^2x^2+b^2y^2=c^2$, find $\frac{dz}{dx}$.

20. If $x\sqrt{1-y^2}+y\sqrt{1-x^2}=P$, show that

$$\frac{d^2y}{dx^2}=\frac{-P}{(1-x^2)^{3/2}}.$$

21. If $x^m+y^m=a^m$, show that

$$\frac{d^2y}{dx^2}=-(m-1)a^m.\frac{x^{m-2}}{y^{2m-1}}.$$

22. If $z=2(ax+by)^2-(x^2+y^2)$ and $a^2+b^2=1$, find the value of

$$\frac{\partial^2 z}{\partial x^2}+\frac{\partial^2 z}{\partial y^2}.$$

23. If u and v are functions of x and y defined by

$x=u+e^{-v}\sin u$, $y=v+e^{-v}\cos u$.

prove that

$$\frac{\partial u}{\partial y}=\frac{\partial v}{\partial x}.$$

24. Find partial derivatives of u, v with respect to x and y if

$x+y=u$, $y=uv$.

25. If $x=u+e^{-u}\sin u$, $y=v+e^{-v}\cos u$, prove that

$$\frac{\partial u}{\partial y}=\frac{\partial v}{\partial x}.$$

26. If $x = r\cos\theta$, $y = r\sin\theta$ show that

$$\frac{\partial^2 r}{\partial x^2}+\frac{\partial^2 r}{\partial y^2}=\frac{1}{r}\left\{\left(\frac{\partial r}{\partial y}\right)^2+\left(\frac{\partial r}{\partial y}\right)^2\right\}.$$

27. If $u = f(y-z, z-x, x-y)$, prove that

$$\frac{\partial u}{\partial x}+\frac{\partial u}{\partial y}+\frac{\partial u}{\partial z}=0.$$

28. If the curves $f(x,y)=0$ and $\phi(x,y)=0$ touch each other, show that at the point of contact.

$$\frac{\partial f}{\partial x}\cdot\frac{\partial \phi}{\partial y}-\frac{\partial f}{\partial y}\cdot\frac{\partial \phi}{\partial x}=0.$$

29. If $z = f(r)$, where $r^2 = x^2 + y^2$, prove that

$$\frac{\partial^2 z}{\partial x^2}+\frac{\partial^2 z}{\partial y^2}=f''(r)+\frac{1}{r}f'(r).$$

11

Curvature and Evolutes

Curvature is a subject of mathematical investigation.

11.1. THE CURVATURE OF A CURVE AT A GIVEN POINT

Consider a curve and a point P there on.

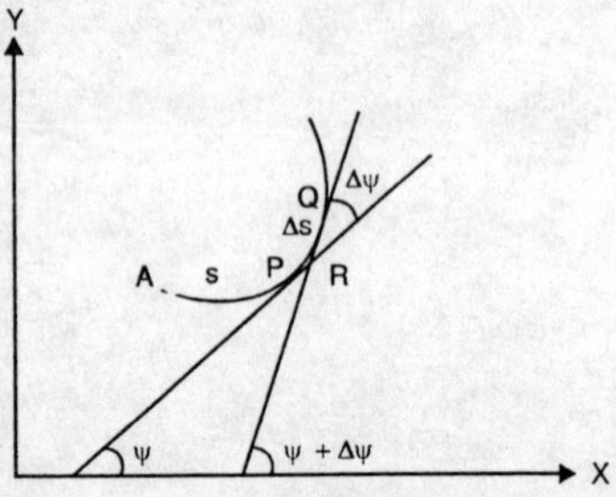

Figure 11.1

We take a point Q on the curve lying near P. Let A be a fixed point on the curve. Let arc $AP = s$, arc $AQ = s + \Delta s$. So that arc $PQ = \Delta s$.

Let tangents be drawn at P and Q making angles ψ and $\psi + \Delta\psi$ with a fixed straight line so that $\Delta\psi$ is the angle between these tangents.

It is clear that $\Delta\psi$ is greater or less according as the whole bending of the curve from P to Q is greater or less.

If Δs is the arc length *PQ*, then $\frac{\Delta\psi}{\Delta s}$ is the average bending or the average curvature of the arc *PQ*.

The limit of $\frac{\Delta\psi}{\Delta s}$ as *Q* approaches *P* (i.e., as $\Delta s \to 0$) is defined to be the curvature of the curve at the point P. Thus, the curvature of a curve at a point *P* is equal to

$$\lim_{\Delta x \to 0} \frac{\Delta\psi}{\Delta s} = \frac{d\psi}{ds}$$

where *s* denotes the arc length and ψ is the angle which the tangent at *P* makes with a fixed straight line.

11.2. CURVATURE OF A CIRCLE

The curvature of a circle is the same at every point and that the larger the radius of the circle, the smaller will be its curvature.

Consider a circle with radius, *r* and centre *O*.

Let *P*, *Q* be two points on the circle and let arc $PQ = \Delta s$. Let *L* be the point where the tangents *PT*, *QT* at *P* and *Q* meet. We have

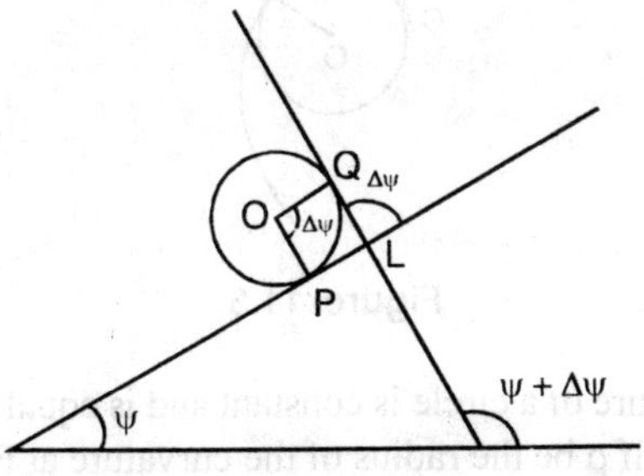

Figure 11.2

$$\angle POQ = \angle TLT' = \Delta\psi$$

From trigonometry

$$\frac{arcPQ}{OP} = \angle POQ$$

$$\Rightarrow \frac{\Delta s}{r} = \Delta\psi$$

$$\Rightarrow \frac{\Delta\psi}{\Delta s} = \frac{1}{r}$$

By definition, the curvature of the circle at P is

$$\lim_{\Delta s \to 0} \frac{\Delta\psi}{\Delta S} = \lim_{\Delta\psi \to 0} \frac{\Delta\psi}{r\Delta\psi}$$

$$= \frac{1}{r}$$

Thus, the curvature of a circle at any point is constant and is equal to the reciprocal of its radius.

11.3. RADIUS OF CURVATURE

The reciprocal of the curvature of a curve at any point is called its ***radius of curvature*** *at the point. It is generally denoted by f.*

Let P be any point on a curve. The radius of curvature of the curve at P is defined to be the radius of the circle C which touches the curve at the point P and has the same curvature at P as the given curve.

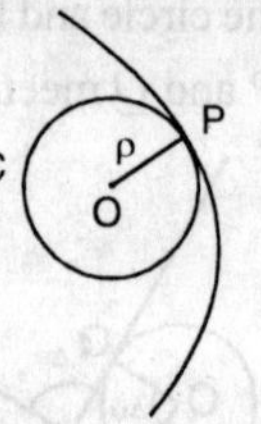

Figure. 11.3

Since the curvature of a circle is constant and is equal to the reciprocal of its radius, therefore, if ρ be the radius of the curvature at P, then we have,

$$\frac{d\psi}{ds} = \frac{1}{\rho}$$

$$\Rightarrow \rho = \frac{ds}{d\psi}$$

The circle C is called the **circle of curvature** at P and its centre O is called the **centre of curvature** at P.

Theorem: Let P be a given point on the curve and let Q be any other point on it. Let the normal on P and Q intersect in the point R. If R tends to the point N as Q tends to P, then PN is the radius of the curvature at the point P.

Proof: Let ψ and $\psi+\Delta\psi$ be the angles which the tangents at P and Q make with the axis of x respectively.

Let the arc length PQ be Δs.

By definition, the radius of curvature,

$$\rho = \lim_{\Delta\psi\to 0}\frac{\Delta s}{\Delta\psi}$$

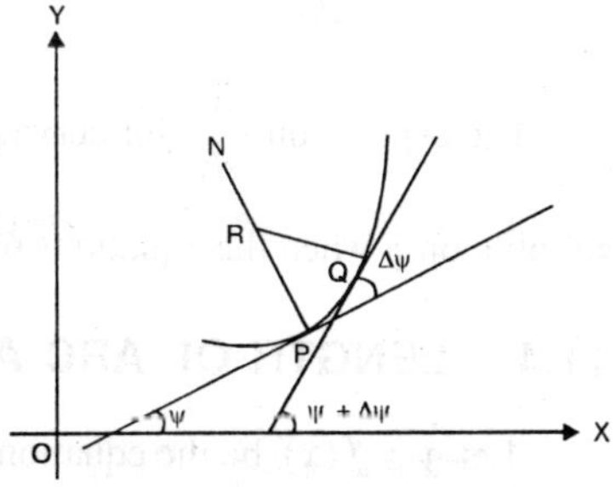

Figure 11.4

Now, from the ΔPQR,

$$\frac{PR}{chord\ PQ} = \frac{\sin\angle RQP}{\sin\angle PRQ}$$

$$= \frac{\sin\angle RQP}{\sin\Delta\psi}$$

Also, as $Q\to P, \Delta\psi\to 0$

Therefore, $PN = \lim\limits_{Q\to P} PR = \lim\limits_{\Delta\psi\to 0} PR$

$$= \lim_{\Delta\psi\to 0}\frac{chord\ PQ\sin\angle RQP}{\sin\Delta\psi}$$

$$= \lim_{\Delta\psi\to 0}\frac{chord\ PQ}{\Delta s}.\frac{\Delta s}{\Delta\psi}.\frac{\Delta\psi}{\sin\Delta\psi}.\sin\angle RQP$$

But $\lim\limits_{\Delta\psi\to 0}\dfrac{\Delta s}{\Delta\psi} = \dfrac{ds}{d\psi}$,

$$\lim_{\Delta\psi\to 0}\frac{chord\ PQ}{\Delta s} = 1$$

$$\lim_{\Delta\psi\to 0}\frac{\Delta\psi}{\sin\Delta\psi} = 1$$

As $Q\to P, \angle RQP\to\dfrac{\pi}{2}$

Therefore, $\lim_{\Delta\psi \to 0} \sin \angle RQP = \sin \frac{\pi}{2} = 1$

Thus, $\rho = \frac{ds}{d\psi} = PN$

The expression $\frac{d\psi}{ds}$ for curvature and $\frac{ds}{d\psi}$ for the radius of curvature are suitable only when the equations of the curve are given in terms of s and ψ.

11.4. LENGTH OF ARC AS A FUNCTION

Let $y = f(x)$ be the equation of a given curve on which we take a fixed point A.

To any given value of x corresponds a value of y viz., $f(x)$ and to this pair of numbers x and $f(x)$ corresponds a point $P(x, y)$ on the curve and this point P has some actual length 's' from A.

Thus, we have a function s of x for the curve

$y = f(x)$

Take a point $Q(x + \Delta x, y + \Delta y)$ on the curve near P.

Let arc $AQ = s + \Delta s$ so that arc $PQ = \Delta s$

From the right angled ΔPQN, we have

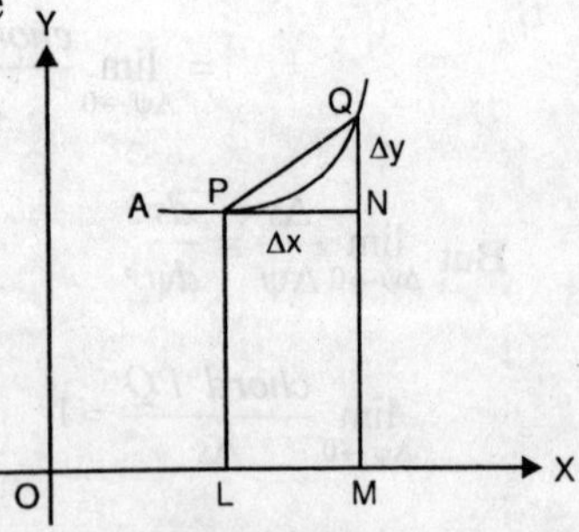

Figure 11.5

$$PQ^2 = PN^2 + NQ^2$$

$$= (\Delta x)^2 + (\Delta y)^2$$

$$\Rightarrow \left(\frac{PQ}{\Delta x}\right)^2 = 1 + \left(\frac{\Delta y}{\Delta x}\right)^2$$

$$\Rightarrow \left(\frac{\text{chord } PQ}{\text{arc } PQ}\right)\left(\frac{\Delta s}{\Delta x}\right)^2 = 1 + \left(\frac{\Delta y}{\Delta x}\right)^2$$

Assuming, on an initiative basis that

$$\lim_{Q \to P} \frac{\text{chord } PQ}{\text{arc } PQ} = 1$$

We obtain in the limit

$$\left(\frac{ds}{dx}\right)^2 = 1 + \left(\frac{dy}{dx}\right)^2$$

For the curve $y = f(x)$, 's' is measured positively in the direction of x increasing, so that, s, increases with x.

Hence $\frac{ds}{dx}$ is positive.

Thus, we have

$$\frac{ds}{dx} = \sqrt{\left[1 + \left(\frac{dy}{dx}\right)^2\right]}$$

Corollary 1: $\frac{dx}{ds} = \cos \psi$

$$\frac{dy}{ds} = \sin \psi$$

Corollary 2: For parametric Cartesian equations with parameter t,

$$\left(\frac{ds}{dt}\right)^2 = \left(\frac{dx}{dt}\right)^2 + \left(\frac{dy}{dt}\right)^2$$

11.5. RADIUS OF CURVATURE FOR CARTESIAN CURVES

11.5.1. Explicit form y = f(x)

Consider the curve $y = f(x)$

Differentiating,

$$\tan \psi = \frac{dy}{dx}$$

with respect to s, we get

$$\sec^2\psi\frac{d\psi}{ds}=\frac{d^2y}{dx^2}\cdot\frac{dx}{ds}$$

$$\Rightarrow\frac{ds}{d\phi}=\frac{\sec^2\psi ds/dx}{d^2y/dx^2}$$

$$=\frac{\left(1+\tan^2\phi\right)ds/dx}{d^2y/dx^2}$$

But $\frac{ds}{dx}=\sqrt{\left[1+\left(\frac{dy}{dx}\right)^2\right]}$

where the positive sign is to be taken before the radical.

So, $\frac{ds}{d\psi}=\frac{\left[1+(dy/dx)^2\right]\sqrt{\left[1+(dy/dx)^2\right]}}{d^2y/dx^2}$

Hence, $\rho=\frac{\left[1+(dy/dx)^2\right]^{3/2}}{d^2y/dx^2}$ ...(1)

$$=\frac{\left(1+y_1^2\right)^{3/2}}{y_2}$$

Note: From the definition of the radius of curvature ρ, that its value is independent of the choice of axes and depends only on the curve. Hence, interchanging the axes of x and y, the value of ρ is also given by

$$\rho=\frac{\left[1+(dx/dy)^2\right]^{3/2}}{d^2x/dy^2}$$

This formula is specially useful when the tangent is perpendicular to x-axis, in which case

$$\frac{dx}{dy}=0$$

11.5.2. Implicit form f(x,y)= 0

Let the equation of the curve is given implicitly in the form

$$f(x,y)=0$$

Now, $\dfrac{dy}{dx}=-\dfrac{f_x}{f_y}$ (if $f_y \neq 0$)

$$\Rightarrow \frac{d^2y}{dx^2}=-\frac{f_{xx}\cdot f_y^2-2f_{xx}\cdot f_x\cdot f_y+f_{yy}\cdot f_x^2}{f_y^3} \qquad \text{(if } f_y \neq 0\text{)}$$

By substituting the above values for $\dfrac{dy}{dx}$ and $\dfrac{d^2y}{dx^2}$ in

$$\rho=\frac{\left[1+(dy/dx)^2\right]^{3/2}}{d^2y/dx^2}, \text{ we get}$$

$$\rho=-\frac{\left(f_x^2+f_y^2\right)^{3/2}}{f_{xx}\cdot f_y^2-2f_{xy}\cdot f_x\cdot f_y+f_{yy}\cdot f_x^2}$$

11.5.3. Parametric form x= f(t),y= g(t)

Consider the equation of the curve is given in the parametric form. $x=f(t), y=g(t)$

Now, $\dfrac{dy}{dx}=\dfrac{dy}{dt}\bigg/\dfrac{dx}{dt}$

$$=\frac{g'(t)}{f'(t)}$$

and $$\frac{d^2y}{dx^2}=\frac{\dfrac{d}{dt}\left(\dfrac{dy}{dx}\right)-\dfrac{d^2x}{dt^2}\cdot\dfrac{dy}{dt}}{\left(\dfrac{dx}{dt}\right)^3}$$

$$= \frac{g''(t)f'(t) - f''(t)g'(t)}{(f'(t))^3}$$

Substituting the value of $\frac{dy}{dx}$ and $\frac{d^2y}{dx^2}$ in (1), we get

$$\rho = \frac{\left[(f'(t))^2 + (g'(t))^2\right]^{3/2}}{f'(t)g''(t) - g'(t)f''(t)}$$

11.5.4. Parametric form $x = f(s), y = g(s)$

Consider the equation of the curve is given in the form

$$x = f(s), \quad y = g(s)$$

where the parameter s is the arc length measured from a fixed point on the curve.

Now, $\frac{dx}{ds} = \cos\psi$ and $\frac{dy}{ds} = \sin\psi$ ...(i)

Differentiating with respect to s, we get

$$\frac{d^2x}{ds^2} = -\sin\psi \, , \frac{d\psi}{ds} = -\frac{\sin\psi}{\rho} \qquad \text{...(ii)}$$

and $\frac{d^2y}{ds^2} = \cos\psi \, , \frac{d\psi}{ds} = \frac{\cos\psi}{\rho}$ ...(iii)

Thus, using (i), (ii) and (iii), we get

$$\rho = -\frac{\sin\psi}{d^2x/ds^2}$$

$$= -\frac{dy/ds}{d^2x/ds^2}$$

and also, $\rho = \frac{\cos\psi}{d^2y/ds^2}$

$$= \frac{dx/ds}{d^2y/ds^2}$$

Squaring (ii) and (iii), an expression for ρ is also given by

$$\frac{1}{\rho}=\left(\frac{d^2x}{ds^2}\right)^2+\left(\frac{d^2y}{ds^2}\right)^2$$

11.6. RADIUS OF CURVATURE FOR POLAR CURVES

11.6.1. Radius of Curvature for Curves $r = f(\theta)$

Consider the equation of the curve is given in the polar form as $r = f(\theta)$

Let us take

$$\psi = \theta + \phi$$

where ψ is the angle which the tangent at any point $P(r,\theta)$ makes with the initial line and ϕ is the angle between the tangent at P and the radius vector through P.

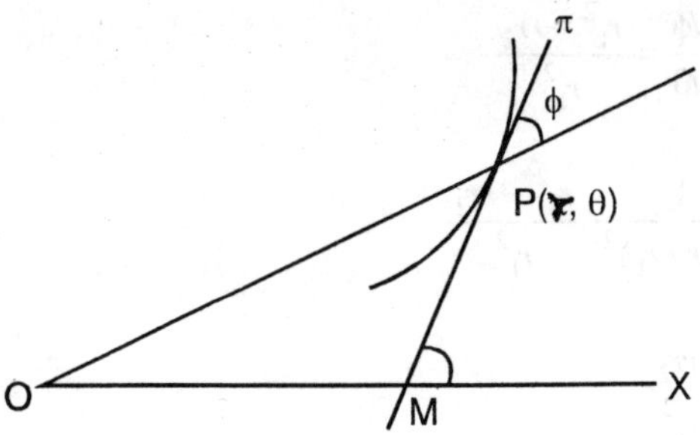

Figure 11.6

Now, the radius of curvature, ρ is given by

$$\frac{1}{\rho}=\frac{d\psi}{ds}$$

$$=\frac{d}{ds}(\theta+\phi)$$

$$=\left(\frac{d}{d\theta}(\theta+\phi)\right)\frac{d\theta}{ds}$$

$$= \left(1 + \frac{d\phi}{d\theta}\right)\frac{d\theta}{ds}$$

We know that,

$$\tan\phi = r\frac{d\theta}{dr} = \frac{r}{r_1}$$

where r_1 stands for $\dfrac{dr}{d\theta}$.

If r_2 stands for $\dfrac{d^2r}{d\theta^2}$, then by differentiating

$$\tan\phi = \frac{r}{r_1}$$

with respect to θ, we get

$$\sec^2\phi.\frac{d\phi}{d\theta} = \frac{r_1^{\ 2} - rr_2}{r_1^{\ 2}}$$

$$\therefore \frac{d\phi}{d\theta} = \frac{1}{1+(r/r_1)^2}.\frac{r_1^{\ 2} - rr_2}{r_1^{\ 2}}$$

$$= \frac{r_1^{\ 2} - rr_2}{r^2 + r_1^{\ 2}}$$

Also, $\dfrac{ds}{d\theta} = \sqrt{r^2 + r_1^{\ 2}}$

$$\Rightarrow \quad \frac{d\theta}{ds} = \frac{1}{\sqrt{r^2 + r_1^{\ 2}}}$$

Hence, $\dfrac{1}{\rho} = \left(1 + \dfrac{d\phi}{d\theta}\right)\dfrac{d\theta}{ds}$

$$= \left[1 + \frac{r_1^{\ 2} - rr_2}{r^2 + r_1^{\ 2}}\right].\frac{1}{\sqrt{r^2 + r_1^{\ 2}}}$$

$$= \frac{r^2 + 2r_1^2 - rr_2}{\left(r^2 + r_1^2\right)^{3/2}}$$

$$\therefore \quad \rho = \frac{\left(r^2 + r_1^2\right)^{\frac{3}{2}}}{r^2 + 2r_1^2 - rr_2} \qquad \ldots(1)$$

11.6.2. Radius of curvature for curves $u = f(\theta)$, where $u = \frac{1}{r}$

Consider the equation of the curve is given in the form

$$u = f(\theta)$$

We know that,

$$u = \frac{1}{r}$$

$$\therefore r_1 = -\frac{1}{u^2}\frac{du}{d\theta} = \frac{-u_1}{u^2}$$

$$\text{and } r_2 = -\frac{u^2 u_2 + u_1 . 2uu_1}{u^4}$$

$$= \frac{2u_1^2 - uu_2}{u^3}$$

Substituting these values of r_1 and r_2 in (1) above, we get

$$\rho = \frac{\left(u^2 + u_1^2\right)^{3/2}}{u^3\left(u + u_2\right)}$$

11.6.3. Radius of Curvature for Pedal Curves

Consider the equation of the curve is given in the pedal form as

$$r = f(p)$$

Now, $p = r\sin\phi$

$$\therefore \quad \frac{dp}{dr} = \sin\phi + r\cos\phi.\frac{d\phi}{dr}$$

$$= r\frac{d\theta}{ds} + r\frac{dr}{ds}.\frac{d\phi}{dr}$$

$$= r\left(\frac{d\theta}{ds} + \frac{d\phi}{ds}\right)$$

$$= r\frac{d}{ds}(\theta + \phi)$$

$$= r\frac{d\psi}{ds}$$

Since $\rho = \dfrac{ds}{d\psi}$

$$\therefore \quad \rho = r\frac{dr}{dp}$$

11.6.4. Radius of Curvature for Polar Tangent Curves

Consider the equation of the curve is given in the polar tangential form,

$$p = f(\psi)$$

Here ψ is the angle which the tangent at any point $P(x, y)$ makes with the x-axis and p is the length of the perpendicular from the origin on the tangent at P.

If ρ be the angle which this perpendicular makes with the x-axis, then

$$\alpha = \psi - \frac{\pi}{2}$$

Also, $p = x\cos\alpha + y\sin\alpha$

$$= x\cos\left(\psi - \frac{\pi}{2}\right) + y\sin\left(\psi - \frac{\pi}{2}\right)$$

$$= x\sin\psi - y\cos\psi$$

$$\therefore \quad \frac{dp}{d\psi} = x\cos\psi + \sin\psi.\frac{dx}{d\psi} + y\sin\psi - \cos\psi\frac{dy}{d\psi}$$

Now, $\frac{dx}{d\psi} = \frac{dx}{ds}.\frac{ds}{d\psi}$

$= \rho \cos\psi$

and $\frac{dy}{d\psi} = \frac{dy}{ds}.\frac{ds}{d\psi}$

$= \rho \sin\psi$

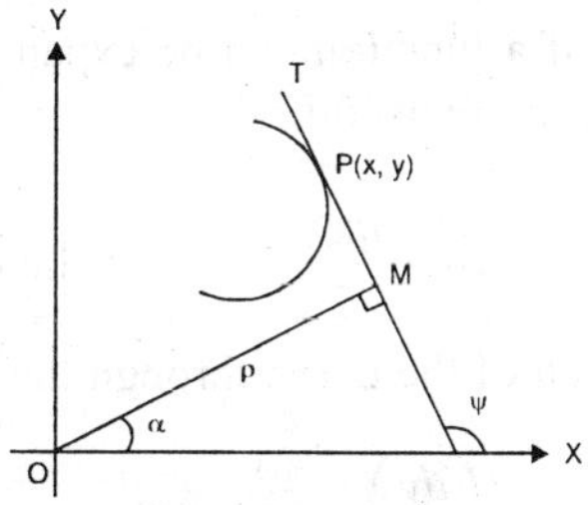

Figure 11.7

Thus, $\frac{dp}{d\psi} = x\cos\psi + y\sin\psi + \sin\psi.\rho\cos\psi - \cos\psi.\rho\sin\psi$

$\Rightarrow \frac{dp}{d\psi} = x\cos\psi + y\sin\psi$

Differentiating with respect to ψ

$$\frac{d^2p}{d\psi^2} = x\sin\psi + \cos\psi\frac{dx}{d\psi} + y\cos\psi + \sin\psi\frac{dy}{d\psi}$$

$$= -(x\sin\psi - y\cos\psi) + \cos\psi.\rho\cos\psi + \sin\psi.\rho\sin\psi$$

$$= -(x\sin\psi - y\cos\psi) + \rho\left(\cos^2\psi + \sin^2\psi\right)$$

$$= -p + \rho$$

Hence, $\rho = p + \frac{d^2p}{d\psi^2}$.

11.7. RADIUS OF CURVATURE AT THE ORIGIN

11.7.1. Radius of Curvature at the Origin by Maclaurin's Series

The value of the radius of curvature at the origin of a curve passes through the origin can be found by the formula

$$\rho = \frac{\left[1 + (dy/dx)^2\right]^{3/2}}{d^2y/dx^2}$$

by substituting $x = 0$ and $y = 0$ in the value of ρ as obtained by this formula.

But if a function can be expanded by Maclaurin's series, the following method is more useful.

Let $y = px + \frac{qx^2}{2} + \cdots\cdots$ is the expansion of y by Maclaurin's series for any branch of the curve through the origin, so that

$$p = \left(\frac{dy}{dx}\right)_{\substack{x=0\\y=0}} = y_1(0)$$

$$q = \left(\frac{d^2y}{dx^2}\right)_{\substack{x=0\\y=0}} = y_2(0)$$

Notice that $y(0) = 0$ as the curve passes through the origin.

Substitute $px + \frac{qx^2}{2} + \cdots\cdots$ for y in the equation of the given curve and determine the values of p and q by equating to zero the co-efficients of various powers of x in the identity obtained in this manner.

The required value of the radius of curvature at the origin is given by

$$\rho = \frac{\left(1 + p^2\right)^{3/2}}{q}$$

We have assumed in the above that the y-axis is not a tangent to the branch under consideration.

If y-axis happens to be a tangent, the corresponding value of ρ can be obtained by the formula

$$\rho = \frac{\left(1 + p_1{}^2\right)^{3/2}}{q_1}$$

By using the expansion in the form

$$x = p_1 y + \frac{q_1 y^2}{2} + \cdots\cdots$$

where $p_1 = \left(\dfrac{dx}{dy}\right)_{\substack{x=0\\y=0}}$

$$q_1 = \left(\frac{d^2x}{dy^2}\right)_{\substack{x=0\\y=0}}$$

If either x-axis or y-axis is a tangent to the curve at the origin, the radius of curvature is usually determined by a method due to Newton.

11.7.2. Radius of Curvature by Newton's Formula

(i) Newton Method. If a curve passes through the origin and the axis of x is tangent at the origin, then

$$\lim_{x\to 0} \frac{x^2}{2y}$$

gives the radius of curvature at the origin.

Proof: Here,

$$y_1(0) = \left(\frac{dy}{dx}\right)_{(0,0)} = 0$$

Now, $x^2/2y$ assumes the indeterminate form $0/0$ as $x \to 0$

$$\therefore \lim_{x\to 0} \frac{x^2}{2y} = \lim_{x\to 0} \frac{2x}{2y_1} \quad \left(\frac{0}{0}\right)$$

$$= \lim_{x\to 0} \frac{1}{y_2} = \frac{1}{y_2(0)}$$

But $\rho = \dfrac{\left(1+y_1^{\,2}\right)^{3/2}}{y_2}$

We have at the origin

$$\rho = \frac{(1+0)^{3/2}}{y_2(0)} = \frac{1}{y_2(0)}$$

Thus, at the origin where x-axis is a tangent

$$\rho = \lim_{x\to 0}\left(\frac{x^2}{2y}\right)$$

Similarly, it can be shown that at the origin where y-axis is a tangent

$$\rho = \lim_{x\to 0}\left(\frac{y^2}{2x}\right)$$

These formulae are known as *Newton's formula.*

(ii) Another proof of Newton's formula

Let $y = px + \dfrac{qx^2}{2!} + \ldots\ldots$ ...(i)

be the Maclaurin's expansion of y in ascending powers of x. The radius of curvature at the origin

$$= \frac{\left(1+p^2\right)^{3/2}}{q} = \frac{1}{q} \quad (\because p = 0)$$

Dividing (i) by y and taking limits as $x \to 0$, we get

$$\frac{1}{q} = \lim_{x\to 0}\left(\frac{x^2}{2y}\right)$$

But 1/q is the radius of curvature at the origin.

Hence the result.

(iii) Generalized Newtonian formula

Let the axes be chosen so that x-axis is a tangent at the origin P.

Let $P(x, y)$ be a point adjacent to O in which the curve and the curve intersect.

If PM be the perpendicular on x-axis, then

$$x^2 + y^2 = OP^2$$

is the square of the distance of any point $P(x, y)$ on the curve form the origin O and y is the distance of the point P form the tangent x-axis at O.

If OT be the tangent at any given point O of a curve, and PM, be the length of the perpendicular drawn form any point P to the tangent at O, then the radius of curvature at O,

$$\rho = \lim \frac{OP^2}{2PM}$$

when the point P tends to O as its limit.

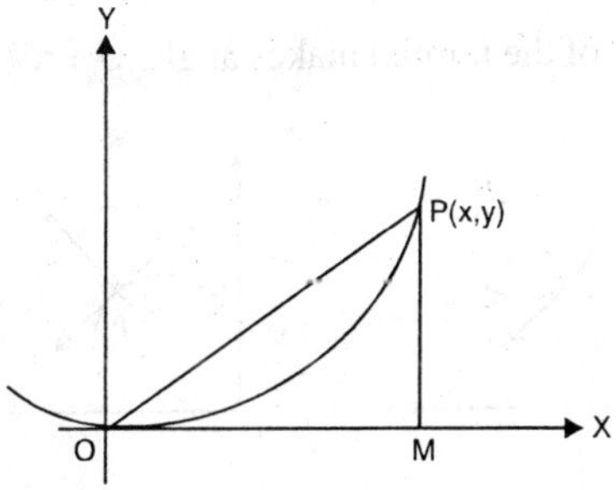

Figure 11.8

(iv) Newton's' Formula for Polar Curves

If the equation of the curve is given in polar form and if the initial line is a tangent to the curve at the origin. Then

$$\rho = \lim_{x \to 0} \frac{x^2}{2y}$$

$$= \lim_{\theta \to 0} \frac{r^2 \cos^2 \theta}{2r \sin \theta}$$

$$= \lim_{\theta \to 0} \frac{r}{2\theta} \cdot \frac{\theta}{\sin \theta} \cdot \cos^2 \theta$$

$$= \lim_{\theta \to 0} \frac{r}{2\theta}$$

11.8. CENTRE OF CURVATURE

The centre of curvature at any point P of a curve is the point which lies on the positive direction of the normal at P and is at a distance, ρ, from it.

The positive direction of the normal is obtained by rotating the positive direction of the tangent through $\pi / 2$ in the anti-clockwise direction.

From the above figures, we see that the centre of curvature at any point of a curve lies on the side towards which the curve is concave.

Determination of the co-ordinates of the centre of curvature for any point

Let $P(x, y)$ be a point of the curve

$$y = f(x)$$

Let the positive direction of the tangent make angle ψ with x-axis so that the positive direction of the normal makes angle $\psi + \pi/2$ with x-axis.

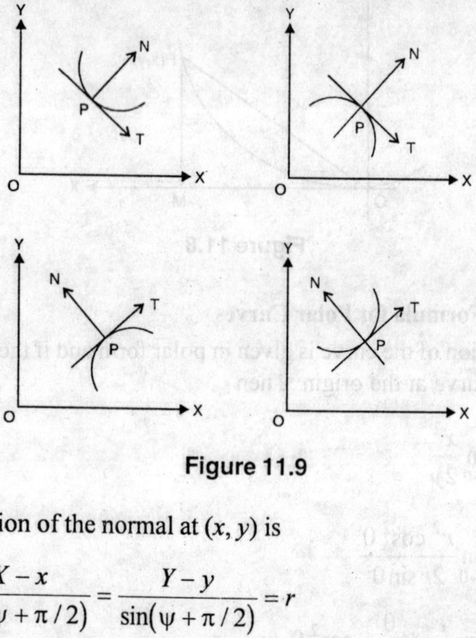

Figure 11.9

The equation of the normal at (x, y) is

$$\frac{X - x}{\cos(\psi + \pi/2)} = \frac{Y - y}{\sin(\psi + \pi/2)} = r$$

$$\Rightarrow \quad \frac{X - x}{-\sin\psi} = \frac{Y - y}{\cos\psi} = r$$

where X, Y are the co-ordinate of any point on the normal and r the variable point (X, Y) from (x,y).

Thus, the co-ordinates (X, Y) of the point on the normal at a distance, r form (x, y) are

$$(x - r\sin\psi, y + r\cos\psi)$$

For the centre of curvature, we have

$$r = \rho$$

Hence, if (x, y) be the centre of curvature, we have

$$X = x - \rho \sin \psi$$

$$Y = y + \rho \cos \psi$$

We know that,

$$\sin \psi = \frac{y_1}{\sqrt{[1 + y_1^2]}}$$

$$\cos \psi = \frac{1}{\sqrt{[1 + y_1^2]}}$$

$$\rho = \frac{(1 + y_1^2)^{3/2}}{y_2}$$

$$\therefore X = x - \frac{y_1(1 + y_1^2)}{y_2}$$

$$Y = y + \frac{1 + y_1^2}{y_2}$$

11.9. PROPERTIES OF THE EVOLUTE

Evolute: *The locus of the centers of curvature of a curve is called its evolute and a curve is said to be an involute of its evolute.*

If $P'(X, Y)$ be the centre of curvature for any point $P(x, y)$ on the given curve, we have

$$X = x - \rho \sin \psi$$

$$Y = y + \rho \cos \psi$$

Differentiating w.r.t. x, we obtain

$$\frac{dX}{dY} = 1 - \rho \cos \psi \frac{d\psi}{dx} - \sin \psi \frac{d\rho}{dx}$$

$$= 1 - \frac{ds}{d\psi} \cdot \frac{dx}{ds} \cdot \frac{d\psi}{dx} - \sin \psi \frac{d\rho}{dx}$$

$$= -\sin\psi \frac{d\rho}{dx} \qquad \ldots(i)$$

$$\frac{dY}{dX} = \frac{dy}{dx} - \rho\sin\psi\frac{d\psi}{dx} + \cos\psi\frac{d\rho}{dx}$$

$$= \frac{dy}{dx} - \frac{ds}{d\psi}\cdot\frac{dy}{ds}\cdot\frac{d\psi}{dx} + \cos\psi\frac{d\rho}{dx}$$

$$= \cos\psi\frac{d\rho}{dx} \qquad \ldots(ii)$$

From (i) and (ii), we obtain

$$\frac{dY}{dX} = -\cot\psi \qquad \ldots(iii)$$

Now $\frac{dY}{dX}$ is the slope of the tangent to the evolute at P' and $-\cot\psi$ is the slope of the normal PP' to the original curve at P.

By (iii) the slopes of two lines, which have a point P' in common, are equal and as such they coincide.

Thus the normals to a curve are the tangents to its evolute.

Again taking the square of (i) and (ii) and adding, we get

$$\left(\frac{dX}{dx}\right)^2 + \left(\frac{dY}{dy}\right)^2 = \left(\frac{d\rho}{dx}\right)^2 \qquad \ldots(iv)$$

Let S, be the length of the arc of the evolute measured from a fixed point on it upto (X, Y) so that

$$\left(\frac{ds}{dx}\right)^2 = \left(\frac{dX}{dx}\right)^2 + \left(\frac{dY}{dx}\right)^2 \qquad \ldots(v)$$

Here, x is a parameter for the evolute

From (iv) and (v)

$$\left(\frac{ds}{dx}\right)^2 = \left(\frac{d\rho}{dx}\right)^2$$

$$\Rightarrow \frac{ds}{dx} = \frac{d\rho}{dx}$$

$$\Rightarrow S = \rho + C$$

where C is a constant.

Let M_1, M_2 be the two points on the evolute corresponding to the points L_1, L_2 on the original curve.

Let ρ_1, ρ_2 be the values of ρ, for L_1, L_2 and

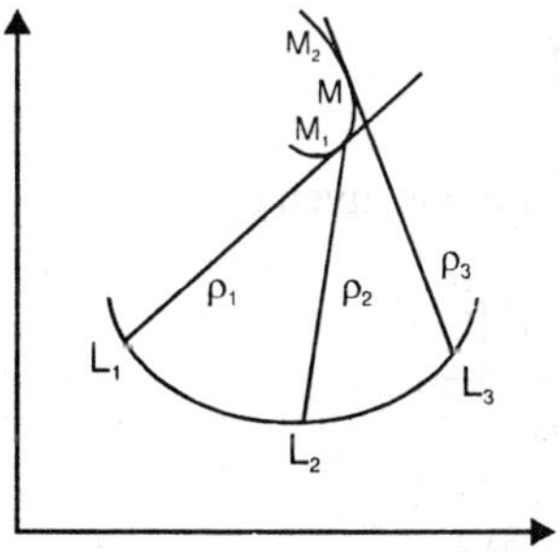

Figure 11.10

S_1, S_2 be the values of S for M_1, M_2.

We have

$$S_1 = \rho_1 + C$$

$$S_2 = \rho_2 + C$$

$$\Rightarrow S_2 - S_1 = \rho_2 - \rho_1$$

$\Rightarrow$ Arc M_1M_2 = difference between the radii of curvatures at L_1, L_2.

Thus, the difference between the radii of curvature at two points of a curve is equal to the length of the arc of the evolute between the two corresponding points.

Examples

Example 1: Show that the curvature at any point of the catenary

$$y = \cosh\frac{x}{c}$$

varies inversely as the square of the ordinate.

Solution: We have,

$$y = \cosh\frac{x}{c}$$

$$\therefore \quad \frac{dy}{dx} = \sinh\frac{x}{c}$$

$$\Rightarrow \quad \frac{d^2y}{dx^2} = \frac{1}{c}\cosh\frac{x}{c}$$

The radius of curvature ρ is given by

$$\rho = \frac{\left[1+(dy/dx)^2\right]^{3/2}}{d^2y/dx^2}$$

$$= \frac{\left[1+\sinh^2\frac{x}{c}\right]^{3/2}}{\frac{1}{c}\cosh\frac{x}{c}}$$

$$= \frac{\left(\cosh^2\frac{x}{c}\right)^{3/2}}{\frac{1}{c}\cosh\frac{x}{c}}$$

$$= c\cosh^2\frac{x}{c}$$

∴ The curvature is equal to

$$\frac{1}{\rho} = \frac{1}{c\cosh^2\frac{x}{c}}$$

$$= \frac{c}{c^2\cosh^2\frac{x}{c}}$$

$$= \frac{c}{y^2}$$

Hence the curvature at any point (x, y) of the given catenary varies inversely as the square of the ordinate y.

Example 2: Prove that for the ellipse

$$\frac{x^2}{a^2} + \frac{y^2}{b^2} = 1,$$

the radius of curvature

$$\rho = \frac{a^2b^2}{p^3}$$

where p is the perpendicular distance from the centre upon the tangent at (x, y).

Solution: The equation of the tangent at any point (x, y) of the ellipse

$$\frac{x^2}{a^2} + \frac{y^2}{b^2} = 1 \text{ is}$$

$$\frac{xX}{a^2} + \frac{yY}{b^2} = 1$$

where X, Y are the current co-ordinates.

The perpendicular distance p from the centre (0,0) to the above tangent is given by

$$p = \frac{1}{\sqrt{x^2/a^4 + y^2/b^4}}$$

$$\Rightarrow p^3 = \frac{1}{\left(x^2/a^4 + y^2/b^4\right)^{3/2}}$$

$$\text{Now, } \rho = \frac{\left(f_x^{\,2} + f_y^{\,2}\right)^{3/2}}{f_{xx}\cdot f_y^{\,2} - 2f_{xy}\cdot f_x\cdot f_y + f_{yy}\cdot f_x^{\,2}} \qquad \text{...(1)}$$

Here, $f_x = \frac{2x}{a^2}$, $f_y = \frac{2y}{b^2}$,

$$f_{xy} = 0\,,\ f_{xx} = \frac{2}{a^2}\,,\ f_{yy} = \frac{2}{b^2}$$

Substituting these values in (1), we have

$$\rho = \frac{\left(4x^2/a^4 + 4y^2/b^4\right)^{3/2}}{\left(2/a^2\right)\left(4y^2/b^4\right) + \left(2/b^2\right)\left(4x^2/a^4\right)}$$

$$= \frac{\left(x^2/a^4 + y^2/b^4\right)^{3/2}}{\left(y^2/b^2 + x^2/a^2\right)\left(1/a^2b^2\right)}$$

$$= \frac{a^2b^2}{p^3}$$

Example 3: Find the radius of curvature at the origin of the curve:

$$x^3 - 2x^2y + 3xy^2 - 4y^3 + 5x^2 - 6xy + 7y^2 - 8y = 0$$

Solution: Here, x-axis is the tangent at the origin.

Dividing by y, we get

$$x.\frac{x^2}{y} - 2x^2 + 3xy - 4y^2 + 5\frac{x^2}{y} - 6x + 7y - 8 = 0$$

Let $x \to 0$ so that

$$\lim_{x\to 0}\left(\frac{x^2}{y}\right) = 2\rho$$

$$\therefore \quad 0.2\rho + 5.2\rho - 8 = 0$$

$$\Rightarrow \quad \rho = 4/5$$

Exmaple 4: Find the radius of curvature for the curve

$$x = a\ln\left\{s + \sqrt{\left(a^2 + s^2\right)}\right\}, \quad y = \sqrt{\left(a^2 + s^2\right)}$$

Solution: Since x and y are given as functions of the arc length s, the radius of curvature ρ is given by

$$\frac{1}{\rho} = \left(\frac{d^2x}{ds^2}\right)^2 + \left(\frac{d^2y}{ds^2}\right)^2$$

Now, $$\frac{dx}{ds} = \frac{a}{s + \sqrt{(a^2 + s^2)}}\left(1 + \frac{s}{\sqrt{(a^2 + s^2)}}\right)$$

$$= \frac{a}{\sqrt{(a^2 + s^2)}}$$

$$\therefore \quad \frac{d^2x}{ds^2} = \frac{-as}{(a^2 + s^2)^{3/2}}$$

Also, $$\frac{dy}{ds} = \frac{s}{\sqrt{(a^2 + s^2)}}$$

$$\therefore \quad \frac{d^2y}{ds^2} = \frac{1}{\sqrt{(a^2 + s^2)}} - \frac{s^2}{(a^2 + s^2)^{3/2}}$$

$$= \frac{a^2}{(a^2 + s^2)^{3/2}}$$

Substituting these values of $\frac{d^2x}{ds^2}$ and $\frac{d^2y}{ds^2}$ in the formula for ρ, we have

$$\frac{1}{\rho} = \frac{a^2s^2}{(a^2 + s^2)^3} + \frac{a^4}{(a^2 + s^2)^3}$$

$$= \frac{a^2(a^2 + s^2)}{(a^2 + s^2)^3}$$

$$= \frac{a^2}{\left(a^2 + s^2\right)^2}$$

$$\Rightarrow \rho = \frac{a^2 + s^2}{a}$$

Example 5: Find the points on the parabola

$$x = at^2, \quad y = 2at$$

where the magnitude of the radius of curvature is equal to the latus rectum.

Solution: We have,

$$x = at^2, \quad y = 2at$$

$$\therefore \frac{dx}{dt} = 2at \quad \frac{dy}{dt} = 2a$$

$$\Rightarrow \frac{d^2x}{dt^2} = 2a \quad \frac{d^2y}{dt^2} = 0$$

Substituting these values in the formula

$$\rho = \frac{\left[(dx/dt)^2 + (dy/dt)^2\right]^{3/2}}{(dx/dt)\left(d^2y/dt^2\right) + (dy/dt)\left(d^2x/dt^2\right)}$$

$$\rho = \frac{\left[4a^2t^2 + 4a^2\right]^{3/2}}{2at.0 + 2a.2a}$$

$$\rho = \frac{(2a)^3(t^2+1)^{3/2}}{4a^2}$$

$$= 2a(t^2+1)^{3/2}$$

Since $\rho =$ latus rectum

$$\therefore 2a(t^2+1)^{3/2} = 4a$$

$$\Rightarrow 4a^2(t^2+1)^3 = 16a^2$$

$$\Rightarrow (t^2+1)^3 = 4$$

$$\therefore t = \pm\sqrt{4^{1/3}-1}$$

Thus, the radius of curve to the parabola at the points $\left(at^2, 2at\right)$ for $t = \pm\sqrt{4^{1/3}-1}$ is equal to the latus rectum.

Example 6: Show that the curvature of the point $\left(\frac{3a}{2}, \frac{3a}{2}\right)$ on the folium $x^3 + y^3 = 3axy$ is $\frac{-8\sqrt{2}}{3a}$

Solution: We have,

$$x^3 + y^3 = 3axy$$

Differentiating, we get

$$3x^2 + 3y^2\frac{dy}{dx} = 3ay + 3ax\frac{dy}{dx} \quad \ldots(i)$$

$$\Rightarrow \frac{dy}{dx} = \frac{ay - x^2}{y^2 - ax}$$

$$\therefore \left[\frac{dy}{dx}\right]_{(3a/2, 3a/2)} = -1$$

Again differentiating (i), we have

$$2x + 2y\left(\frac{dy}{dx}\right)^2 + y^2\frac{d^2y}{dx^2} = a\frac{dy}{dx} + a\frac{dy}{dx} + ax\frac{d^2y}{dx^2}$$

Ssubstituting $\frac{3a}{2}, \frac{3a}{2}, -1$ for $x, y, \frac{dy}{dx}$ respectively, we get

$$\left[\frac{d^2y}{dx^2}\right]_{(3a/2, 3a/2)} = \frac{-32}{3a}$$

Hence the curvature at $(3a/2, 3a/2)$

$$= \frac{d^2y/dx^2}{\left[1+(dy/dx)^2\right]^{3/2}}$$

$$= \frac{-32/3a}{(2)^{3/2}}$$

$$= \frac{-8\sqrt{2}}{3a}$$

Example 7: Prove that for the cardioid

$$r = a(1+\cos\theta), \frac{\rho}{r} \text{ is constant.}$$

Solution: We have,

$$r = a(1+\cos\theta)$$

$$\therefore r_1 = -a\sin\theta$$

and $r_2 = -a\cos\theta$

$$\therefore \rho = \frac{\left(r^2 + r_1^2\right)^{3/2}}{r^2 + 2r_1^2 - rr_2}$$

$$= \frac{\left[a^2(1+\cos\theta)^2 + a^2\sin^2\theta\right]^{3/2}}{a^2(1+\cos\theta)^2 + 2a^2\sin^2\theta - a(1+\cos\theta)(-a\cos\theta)}$$

$$= \frac{a^3[2(1+\cos\theta)]^{3/2}}{3a^2(1+\cos\theta)}$$

$$= \frac{2^{3/2}a}{3}(1+\cos\theta)^{1/2}$$

$$\therefore \rho^2 = \frac{8}{9}a^2(1+\cos\theta)$$

$$\Rightarrow \frac{\rho}{a(1+\cos\theta)} = \frac{8a}{9}$$

$$\Rightarrow \frac{\rho^2}{r} = \frac{8a}{9} = k,$$

where k is constant.

Example 8: For the curve $r^m = a^m \cos m\theta$, prove that

$$\rho = \frac{am}{(m+1)r^{m-1}}$$

Solution: We have,

$$r^m = a^m \cos m\theta$$

$$\Rightarrow \log r^m = \log(a^m \cos m\theta)$$

$$\Rightarrow m \log r = m \log a + \log \cos m\theta$$

Differentiating both sides w.r.t. θ, we get

$$\frac{m}{r}.\frac{dr}{d\theta} = -m\frac{\sin m\theta}{\cos m\theta}$$

$$\Rightarrow r_1 = \frac{dr}{d\theta} = -r \tan m\theta$$

$$\Rightarrow r_2 = \frac{d^2r}{d\theta}$$

$$= -rm\sec^2 m\theta - \tan m\theta.\frac{dr}{d\theta}$$

$$= -rm\sec^2 m\theta + r\tan^2 m\theta$$

Hence, $\rho = \dfrac{\left(r^2 + r^2 \tan^2 m\theta\right)^{3/2}}{r^2 + 2r^2 \tan^2 m\theta + r^2 \sec^2 m\theta - r^2 \tan^2 m\theta}$

$$= \frac{r^3 \sec^3 m\theta}{r^2 \sec^2 m\theta + mr^2 \sec^2 m\theta}$$

$$= \frac{r}{(m+1)(\cos m\theta)}$$

$$= \frac{1}{m+1} \cdot \frac{a^m}{r^{m-1}} \text{ v}$$

Example 9: Show that the radius of curvature for the hyperbola

$$p^2 = a^2 \cos^2 \psi - b^2 \sin^2 \psi$$

is $\dfrac{a^2 b^2}{p^3}$.

Solution: We have,

$$p^2 = a^2 \cos^2 \psi - b^2 \sin^2 \psi$$

Differentiating w.r.t. ψ, we get

$$p\frac{dp}{d\psi} = -(a^2 + b^2) \sin \psi \cos \psi$$

Again differentiating w.r.t. ψ, we have

$$p\frac{d^2 p}{d\psi^2} + \left(\frac{dp}{d\psi}\right)^2 = -(a^2 + b^2)(\cos^2 \psi - \sin^2 \psi)$$

$$\therefore p\frac{d^2 p}{d\psi^2} = -(a^2 + b^2)(\cos^2 \psi - \sin^2 \psi) - \left(\frac{dp}{d\psi}\right)^2$$

$$= -(a^2 + b^2)(\cos^2 \psi - \sin^2 \psi) - \frac{(a^2 + b^2)^2 \sin^2 \psi \cos^2 \psi}{p^2}$$

$$= -(a^2 + b^2)\left[(\cos^2 \psi - \sin^2 \psi) + \frac{(a^2 + b^2)^2 \sin^2 \psi \cos^2 \psi}{p^2}\right]$$

$$= -\frac{(a^2 + b^2)}{p^2}\left[a^2 \cos^4 \psi + b^2 \sin^4 \psi\right]$$

$$= -\frac{1}{p^2}\left[\left(a^2\cos^2\psi - b^2\sin^2\psi\right)^2 + a^2b^2\right]$$

$$= -\frac{1}{p^2}\left(p^4 + a^2b^2\right)$$

$$\therefore \frac{d^2p}{d\psi^2} = -\frac{1}{p^3}\left(p^4 + a^2b^2\right)$$

$$= -p - \frac{a^2b^2}{p^3}$$

Thus, $\rho = p + \frac{d^2p}{d\psi^2} = -\frac{a^2b^2}{p^3}$

$$\Rightarrow |\rho| = \frac{a^2b^2}{p^3}$$

Example 10: For the cycloid

$$x = a(t + \sin t),\ y = a(1 - \cos t)$$

prove that

$$\rho = 4a\cos\frac{t}{2}$$

Solution: We have,

$$x = a(t + \sin t), \qquad y = a(1 - \cos t)$$

$$\Rightarrow \frac{dx}{dt} = a(1 + \cos t) \quad \Rightarrow \frac{dy}{dt} = a\sin t$$

$$\therefore \frac{dy}{dx} = \frac{dy}{dt}\bigg/\frac{dx}{dt}$$

$$= \frac{a\sin t}{a(1 + \cos t)}$$

$$= \frac{2\sin t/2\cos t/2}{2\cos^2 t/2}$$

$$= \tan t/2$$

$$\frac{d^2y}{dx^2} = \frac{1}{2}\sec^2\frac{t}{2}\cdot\frac{dt}{dx}$$

$$= \frac{1}{2}\sec^2\frac{t}{2}\cdot\frac{1}{2a\cos^2\frac{t}{2}}$$

$$= \frac{1}{4a}\cdot\frac{1}{\cos^4\frac{t}{2}}$$

$$\therefore \rho = \frac{\left[1+(dy/dx)^2\right]^{3/2}}{d^2y/dx^2}$$

$$= \frac{\left(1+\tan^2\frac{1}{2}t\right)^{3/2}}{\frac{1}{4a}\cdot\frac{1}{\cos^4\frac{t}{2}}}$$

$$= 4a\cos t/2$$

Example 11: Show that the radii of curvature of the curve

$$y^2(a-x) = x^2(a+x)$$

at the origin are $\pm a\sqrt{2}$.

Solution: Substitute

$$y = px + \frac{qx^2}{2} + \cdots\cdots$$

in the given equation,

Then,

$$\left(px + \frac{qx^2}{2} + \cdots\cdots\right)^2(a-x) = x^2(a+x)$$

Equating co-efficients of x^2 and x^3 on both sides of the above equation, we have

$$ap^2 = a, \quad apq - p^2 = 1$$

$\therefore p = \pm 1$ and $q = \pm 2/a$

For $p = 1$, $q = \dfrac{2}{a}$

$$\rho = \frac{(1+1)^{3/2}}{2/a}$$

$$= a\sqrt{2}$$

For $p = -1$, $q = \dfrac{-2}{a}$,

$$\rho = \frac{(1+1)^{3/2}}{-2/a}$$

$$= -a\sqrt{2}$$

Thus, the radius of curvature at the origin $= \pm a\sqrt{2}$

Example 12: Find the co-ordinates of the centre of curvature at a point (x, y) of the parabola,

$$y^2 = 4ax.$$

Hence obtain its evolute.

Solution: We have,

$$y^2 = 4ax$$

Differentiating, we get

$$2y\frac{dy}{dx} = 4a$$

$$\Rightarrow \frac{dy}{dx} = \frac{2a}{y}$$

$$\Rightarrow \frac{d^2y}{dx^2} = \frac{-2a}{y^2} \cdot \frac{dy}{dx}$$

$$= \frac{-4a}{y^3}$$

If (X, Y) be the centre of curvature, we have

$$X = x - \frac{\frac{2a}{y}\left(1 + \frac{4a^2}{y^2}\right)}{-\frac{4a^2}{y^3}}$$

$$= x + \frac{y^2 + 4a^2}{2a}$$

$$= \frac{2ax + 4ax + 4a^2}{2a}$$

$$\Rightarrow X = 3x + 2a \qquad \ldots\text{(i)}$$

$$Y = y + \frac{1 + \frac{4a^2}{y^2}}{-\frac{4a^2}{y^3}}$$

$$= y - \frac{y(y^2 + 4a^2)}{4a^2}$$

$$= \frac{-y^3}{4a^2}$$

$$= \mp \frac{(4ax)^{3/2}}{4a^2}$$

$$\Rightarrow Y = \mp \frac{2x^{3/2}}{a^{1/2}} \qquad \ldots\text{(ii)}$$

For determination of the evolute, we eliminate x from (i) and (ii). Thus,

$$Y^2 = \frac{4x^3}{a}$$

$$= \frac{4}{a}\left(\frac{X-2a}{3}\right)^3$$

$$\Rightarrow 27aY^2 = 4(X-2a)^3$$

is the required evolute.

Example 13: Find the radius of curvature at the origin for the curve $x^3+y^3-2x^2+6y=0$.

Solution: $y=0$

$\Rightarrow$ x-axis is a tangent to the curve at the origin.

$\therefore$ By Newton's formula,

$$\rho = \lim_{x\to 0}\frac{x^2}{2y}$$

Dividing the given equation by $2y$, we get

$$\frac{x^3}{2y}+\frac{y^2}{2}-\frac{2x^2}{2y}+3=0$$

When $y\to 0, x\to 0$ and $\frac{x^2}{2y}\to\rho$

$$\therefore \rho.0+0-2.\rho+3=0$$

$$\Rightarrow \rho=\frac{3}{2}$$

Example 14: Find the evolute of the asteroid

$$x=a\cos^3\theta,\ y=a\sin^3\theta$$

Solution: We have,

$$\frac{dy}{dx}=-\tan\theta$$

$$\Rightarrow \frac{d^2y}{dx^2} = \frac{1}{3a}\sec^4\theta \operatorname{cosec}\theta$$

$$X = a\cos^3\theta + \frac{\tan\theta(1+\tan^2\theta)}{\sec^4\theta \operatorname{cosec}\theta}.3a$$

$$= a\cos^3\theta + 3a\sin^2\theta.\cos\theta \qquad \ldots\text{(i)}$$

$$Y = a\sin^3\theta + \frac{1+\tan^2\theta}{\sec^4\theta \operatorname{cosec}\theta}.3a$$

$$= a\sin^3\theta + 3a\cos^2\theta.\sin\theta \qquad \ldots\text{(ii)}$$

To eliminate θ, we separately add and subtract (i), (ii), so that we have

$$(X+Y) = a(\cos\theta + \sin\theta)^3$$

$$\Rightarrow (X+Y)^{1/3} = a^{1/3}(\cos\theta + \sin\theta)$$

and $X - Y = a(\cos\theta - \sin\theta)^3$

$$\Rightarrow (X-Y)^{1/3} = a^{1/3}(\cos\theta - \sin\theta)$$

On squaring and adding, we obtain

$$(X+Y)^{2/3} + (X-Y)^{2/3} = 2a^{2/3}$$

is the required evolute.

Example 15: Find the length of the arc of the evolute of the parabola $y^2 = 4ax$ which is intercepted between the parabola.

Solution: The evolute is

$$27ay^2 = 4(x-2a)^3$$

Let *L*, *M* be the points of the evolute *LAM* and the parabola.

$$27a.4ax = 4(x-2a)^3$$

$$\Rightarrow x^3 - 6ax^2 - 15a^2x - 8a^3 = 0$$

Now, $8a, -a, -a$ are the roots of this cubic equation of which

$$x = 8a$$

is the only admissible value.

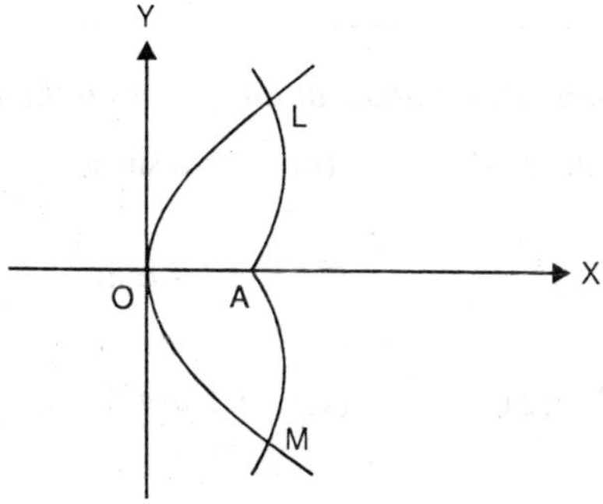

Figure 11.11

$-a$ being negative

$\therefore \left(8a, 4\sqrt{2}a\right)$, $\left(8a, -4\sqrt{2}a\right)$ are the co-ordinates of L, M.

If (X, Y) be the centers of curvature for any point (x, y) on the parabola, we have

$$X = 3x + 2a,$$

$$Y = -y^3 / 4a^2$$

Thus, $A(2a, 0)$ is the centre of curvature for O(0/0) and L $\left(8a, 4\sqrt{2}a\right)$ is the centre of curvature for $P\left(2a, -2\sqrt{2}a\right)$.

The radius of curvature $O = OA = 2a$

The radius of curvature at $P = PL = 6\sqrt{3}a$

arc $AL = PL - OA = 2a(3\sqrt{3} - 1)$

Hence the required length $MAL = 4a\left(3\sqrt{3} - 1\right)$.

Exercise –11

1. *Find the radius of curvature at any point on the following curves:*

   (i) $s = c\tan\psi$ (ii) $s = a\sin\psi$

   (iii) $s = a\sec^3\psi$ (iv) $s = 4a\sin\frac{1}{3}\psi$

   (v) $s = c\ln\sec\psi$ (vi) $s = a\left(e^{m\phi} - 1\right)$

   (vii) $s = 8a\sin^2\frac{\psi}{6}$

   (viii) $s = a\ln(\tan\psi + \sec\psi) + a\tan\psi\sec\psi$

2. Show that the curvature of a straight line at every point thereof is zero.
3. Show that the radius of curvature at any point P on the parabola $y^2 = 4ax$ varies as $(SP)^3$, where S is the focus of the parabola.
4. Find ds/dx for the curves:

   (i) $y = \frac{c\cosh x}{c}$ (ii) $a\log\left[\frac{a^2}{a^2 - x^2}\right]$

5. Prove that $\rho = \frac{r^3}{2k^2}$ for the rectangular hyperbola $xy = k^2$ where r is the distance from the origin of the point considered.
6. Show that the radius of curvature at any point of the astroid $x = a\cos^3\theta$, $y = a\sin^3\theta$ is equal to three times the length of the perpendicular from the origin to the tangent.
7. Show that the radius of curvature at the point $\left(a\cos^3\theta, b\sin^3\theta\right)$ on the curve $x^{2/3} + y^{2/3} = a^{2/3}$ is $3a\sin\theta\cos\theta$.
8. Show that for the curve $x = a\cos\theta(1 + \sin\theta)$, $y = a\sin\theta(1 + \cos\theta)$ the radius of curvature is a, at the point for which the value of the parameter θ is $-\pi/4$.

9. Find the points on the parabola $y^2 = 8x$ at which the radius of curvature is 125/16.

10. Show that the radius of curvature at any point of the curve

$$x = t - c\sinh\frac{t}{c}\cosh\frac{t}{c}\ ,\ y = 2c\cosh\frac{t}{c}\ \text{is}$$

$-2c\cosh^2\left(\frac{t}{c}\right)\sinh\left(\frac{t}{c}\right)$, where t is the parameter.

11. In the cycloid $x = a(\theta + \sin\theta)$, $y = a(1 - \cos\theta)$ prove that ρ is $4a\cos\frac{\theta}{2}$.

12. Show that the radius of curvature at a point of the curve $x = ae^{\theta}(\sin\theta - \cos\theta)$, $y = ae^{\theta}(\sin\theta + \cos\theta)$ is twice the distance of the tangent at the point form the origin.

13. If CP and CD be a pair of conjugate semi-diameters of an ellipse with semi-axes of lengths a and b, prove that the radius of curvature at P is equal to $\frac{CD^3}{ab}$.

14. Prove that the radius of curvature at the point $(-2a, 2a)$ on the curve $x^2y = a(x^2 + y^2)$ is, $-2a$.

15. The tangents at P, Q on the cycloid $x = a(\theta - \sin\theta)$, $y = a(1 - \cos\theta)$ are at right angles, prove that if ρ_1 and ρ_2 be the radii of curvature at these points, then $\rho_1{}^2 + \rho_2{}^2 = 16a^2$.

16. Find the radius of curvature for $\sqrt{(x/a)} - \sqrt{(y/b)} = 1$ at the points where it touches the co-ordinate axes.

17. If ρ_1 and ρ_2 be the radii of curvature at the extremeties of two conjugate diameters of an ellipse, prove that $\left(\rho_1{}^{2/3} + \rho_2{}^{2/3}\right)(ab)^{2/3} = a^2 + b^2$

18. Show that the ratio of the radii of curvature of the curves $xy = a^2$, $x^3 = 3a^2y$ at points which have the same abscissa varies as the square root of the ratio of the ordinates.

19. Show that for the curve in which $s = ae^{x/c}$,

$c\rho = s(s^2 - c^2)^{1/2}$.

20. Show that $3\sqrt{2}/2$ is the least value of $|\rho|$ for $y = \log x$.

21. If ρ be the radius of curvature at any point (x, y) on a plane curve, then

show that $\left(\frac{d^3x}{ds^3}\right)^2 + \left(\frac{d^3y}{ds^3}\right)^2 = \frac{1}{\rho^4}\left[1+\left(\frac{d\rho}{ds}\right)^2\right]$.

22. If ρ_1, ρ_2 are the radii of curvature at the extremities of a focal chord of a parabola whose semi-lactus rectum is l, prove that

$(\rho_1)^{-2/3} + (\rho_2)^{-2/3} = (l)^{-2/3}$.

23. Show that the radius of curvature of the envelope of the line $x\cos\alpha + y\sin\alpha = f(\alpha)$ is $f(\alpha) + f''(\alpha)$

24. Show that for the curve $x = \frac{3}{2}a(\sinh u\cosh u + u)$, $y = a\cosh^2 u$ if the normal at $P(x, y)$ meets the axis of x in G, the radius of curvature at P is equal to $3PG$.

25. Find the radius of curvature at any point of the following curves.

(i) $\frac{x^2}{a^2} - \frac{y^2}{b^2} = 1$

(ii) $x^m + y^m = 1$

26. Find the radius of curvature at the point (r, θ) on each of the following curves:

(i) $r\theta = a$ (ii) $r^n = a^n \sin n\theta$

(iii) $r^2 = a^2\cos 2\theta$ (iv) $\theta = \sqrt{(r^2 - a^2)/a} - \cos^{-1}\frac{a}{r}$

27. Prove that for the cardioides $r = a(1 + \cos\theta)$, ρ^2/r is constant.

28. Show that for the conic

$\frac{1}{r} = 1 + e\cos\theta$

$$\rho = \frac{l\left(1+2e\cos\theta+e^2\right)^{3/2}}{(1+e\cos\theta)^3}$$

29. If ρ_1, ρ_2 be the radii of curvature at the extremities of any chord of the cardioide $r = a(1+\cos\theta)$ which passes through the pole, then

$$\rho_1{}^2+\rho_2{}^2 = \frac{16a^2}{9}$$

30. Show that for the curve

$$a^n p = r^{n+1},$$

ρ varies inversely as the $(n-1)$th power of the radius vector.

31. Show that at the points in which the Archimedean spiral $r = a\theta$ intersects the hyperbolic spiral $r\theta = a$, their curvature are in the ratio 3:1.

32. Show that the evolute of the ellipse

$$x = a\cos\theta,\ y = b\sin\theta$$

$$(ax)^{2/3}+(by)^{2/3} = (a^2-b^2)^{2/3}$$

33. Show that the evolute of the tractrix

$$x = a\left(\cos t+\log\tan\frac{1}{2}t\right),\ y = a\sin t$$

is the catenary $y = a\cosh(x/a)$.

34. Show that the radii of curvature of the curve

$$x = ae^{\theta}(\sin\theta-\cos\theta),\ y = ae^{\theta}(\sin\theta+\cos\theta)$$

and its evolute at corresponding points are equal.

35. Find the radius of curvature at the origin for each of the following curves.

(i) $y = x+x^2+x^3$

(ii) $8x^3-5x^2y+3xy^2-4y^3+5x^2-6xy+7y^2-8y = 0$

(iii) $y^2(y+1) = 4x^2$

36. Show that the circle of curvature, at the point $(am^2, 2am)$ of the parabola $y^2 = 4ax$ for its equation

$$x^2 - y^2 - 6am^2n - 4ax + 4am^3y = 3a^2m^4$$

37. Apply Newton's formula to find the radius of curvature of the curve $r = a\sin\theta$ at the origin.

38. Prove that the evolute of the hyperbola $2xy = a^2$ is

$$(x+y)^{2/3} - (x-y)^{2/3} = 2a^{2/3}.$$

39. Find the radii of curvature at the origin of the two branches of the curve given by the equation

$$x = 1 - t^2,\ y = t - t^3$$

40. Prove that the distance between the pole and the centre of curvature corresponding to any point on the curve $r^n = a^n \cos n\theta$ is

$$\frac{\left[a^{2n} + (n^2 - 1)r^{2n}\right]^{1/2}}{(n+1)r^{n-1}}.$$

41. Show that the whole length of the evolute of the

(i) ellipse $\dfrac{x^2}{a^2} + \dfrac{y^2}{b^2} = 1$ is $4\left(\dfrac{a^2}{b} - \dfrac{b^2}{a}\right)$

(ii) astroid $x = a\cos^3\theta,\ y = x\sin^3\theta$ is $12a$.

12

Concavity and Convexity

12.1. INTRODUCTION

In this chapter, we shall consider special type of points where the curve crosses the tangent.

Let P be a given point on a curve. Draw the tangent to the curve at the point P.

Let l be a given straight line and let θ be the acute angle formed by the tangent at P and the line l.

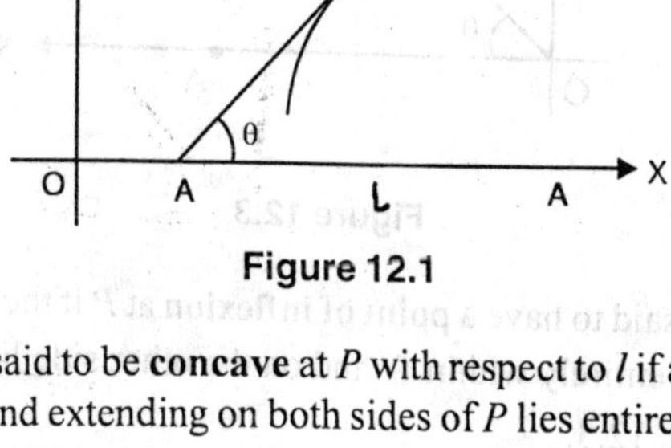

Figure 12.1

The curve is said to be **concave** at P with respect to l if a sufficiently small arc containing P and extending on both sides of P lies entirely within the angle θ as shown in fig. 12.1.

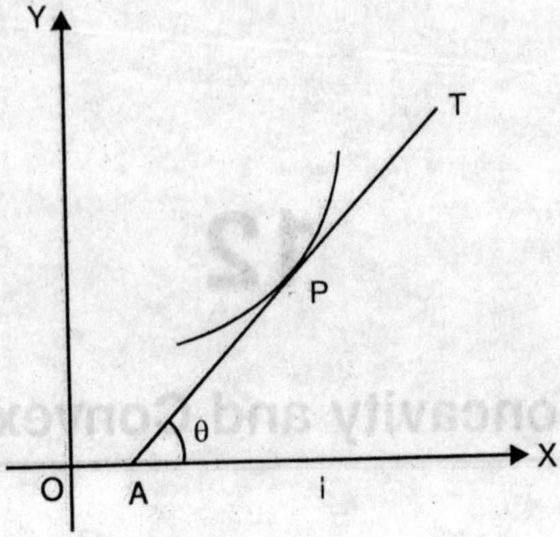

Figure 12.2

The curve is said to be **convex** at P with respect to l if a sufficiently small arc containing P and extending on both sides of P lies entirely without the angle θ as shown in fig. 12.2.

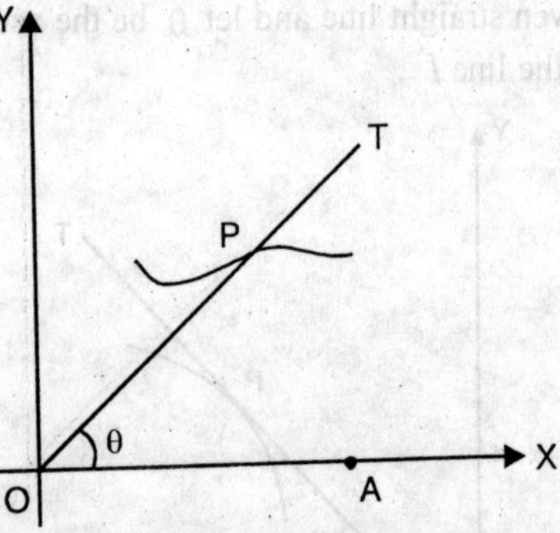

Figure 12.3

The curve is said to have a **point of inflexion** at P if the arc of the curve on one side of P lies entirely within θ and on the other side lies entirely without θ as shown in fig. 12.3.

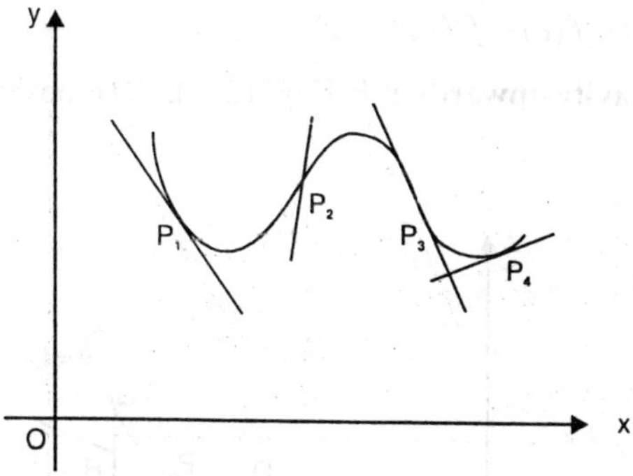

Figure 12.4

In fig. 12.4., the portion P_1P_2 and P_3P_4 of the curve are *concave upwards* and the portion P_2P_3 is *concave downwards*. P_2 and P_3 are *points of inflexion.*

At a point of inflexion, the curve changes form concavity upwards to concavity downwards.

12.2. CRITERIA FOR CONCAVITY UPWARDS, CONCAVITY DOWNWARDS AND INFLEXION AT A GIVEN POINT

Consider a curve $y = f(x)$ and a point $P[c, f(c)]$ there of. Suppose that the tangent at P is not parallel to y-axis.

Take a point $Q[x, f(x)]$ on the curve near the point $P[c, f(c)]$. The point Q will lie to the right or the left of the point P according as $x > c$ or $x < c$.

Draw QM perpendicular to x-axis meeting the tangent at P in R.

The equation of the tangent at P being

$$y - f(c) = f'(c)(x - c)$$

the ordinate MR of the tangent corresponding to the abscissa x is

$$f(c) + f'(c)(x - c)$$

Also, the ordinate MQ of the curve for the abscissa x is $f(x)$.

$\therefore RQ = MQ - MR$

$= f(x) - f(c) - f'(c)(x-c)$

(i) For **concavity upwards** at P, [Fig. 12.5.], RQ is positive when θ lies on either side of P.

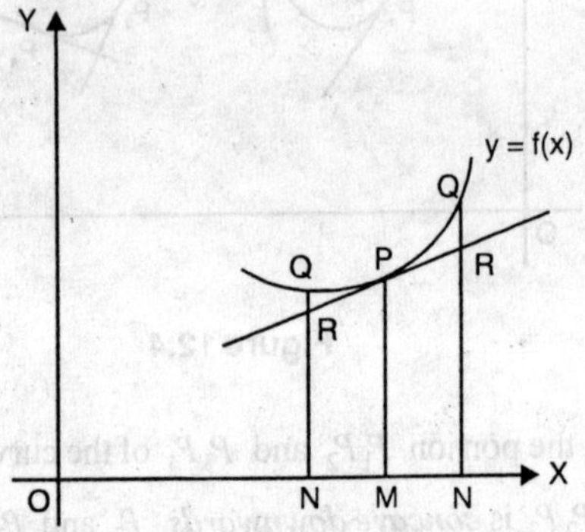

Figure 12.5

(ii) For **concavity downwards** at P, [Fig. 12.6.], RQ is negative when Q lies either side of P.

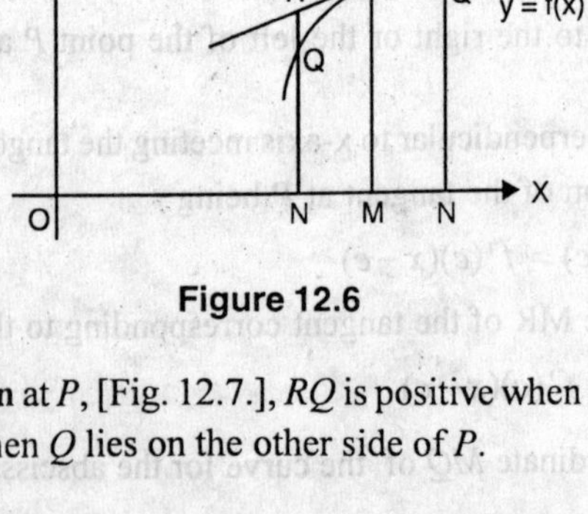

Figure 12.6

(iii) For **inflexion** at P, [Fig. 12.7.], RQ is positive when θ lies on one side of P and negative when Q lies on the other side of P.

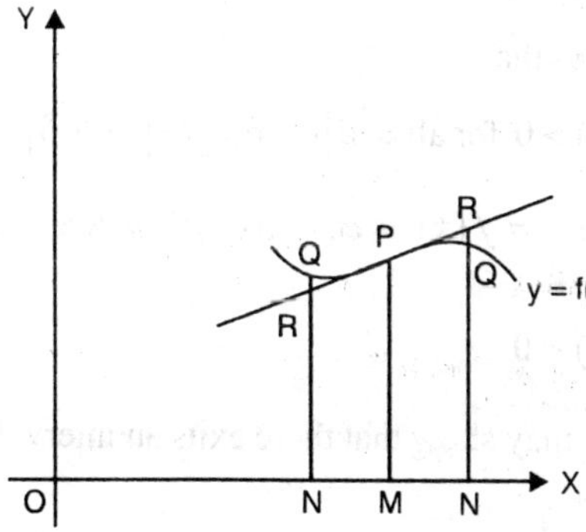

Figure 12.7

Thus, we have to examine the behaviour of the sign of RQ for values of x such that $|x-c|$ is sufficiently small.

$$RQ = \phi(x)$$
$$= f(x) - f(c) - f'(c)(x-c)$$

so that $\phi'(x) = f'(x) - f'(c)$

$$\phi''(x) = f''(x)$$

Now, three cases arises.

Case I. $f''(c) > 0$

Since $\phi''(c) = f''(c)$ is positive, there exists an interval around c for every point x of which $\phi''(x)$ is positive.

Thus, $\phi'(x)$ is strictly increasing in $[c-\delta, c+\delta]$

Also, as $\phi'(c) = 0$, we see that

$$\phi'(x) < 0 \text{ for all } x \in [c-\delta, c[$$

$$\phi'(x) > 0 \text{ for all } x \subset]c, c+\delta]$$

Again it follows that ϕ is strictly decreasing in $[c-\delta, c]$ and strictly decreasing in $[c, c+\delta]$.

Also, $\phi(c) = 0$

Finally, it follows that,

$$RQ = \phi(x) > 0 \text{ for all } x \in [c-\delta, c[\cup]c, c+\delta]$$

Thus, the curve $y = f(x)$ is *concave upwards* at $[c, f(c)]$ if $f''(c) > 0$.

Case II. $f''(c) < 0$

In this case, we may show that there exits an interval $[c-\delta, c+\delta]$ around c such that

$$RQ = \phi(x) < 0 \text{ for all } x \in [c-\delta, c[\cup]c, c+\delta]$$

Thus, the curve, $y = f(x)$ is *concave downwards* at $[c, f(c)]$ if $f''(c) < 0$.

Case iiI. $f''(c) = 0$

A curve has inflexion at *P* if it changes from concavity upwards to concavity downwards, as a point moving along the curve passes through *P*.

Thus, the curve has inflexion at $[c, f(c)]$, if $f''(x)$ changes sign as x passes through c.

Since f'' is assumed to be continuous, it follows that at a point of inflexion $f''(c) = 0$.

Note: To determine the criteria for an extreme value for $x=c$, we have to examine the behaviour of the sign of

$$f(x) - f(c)$$

and to determine the criteria for concavity upwards (downwards), point of inflexion for x = c, we exmaine the behaviour of the sign of

$$f(x) - f(c) - f'(c)(x-c)$$

Thus, the curve $y = f(x)$

(i) is *concave upwards* in $[a, b]$ if

$$f''(x) > 0 \text{ for all } x \in [a,b]$$

(ii) is *concave downward* in $[a,b]$ if

$$f''(x) < 0 \text{ for all } x \in [a,b]$$

(iii) has a *point of inflexion* at $[c, f(c)]$ if

$f''(c) = 0$ and $f''(x)$ changes sign as x passes through c.

12.3. SIGN OF THE RADIUS OF CURVATURE

If we make the convention that s increases with x, then ds/dx is positive, that is, positive sign is taken before the radical

$$\sqrt{\left[1+(dy/dx)^2\right]}$$

Therefore, it is clear from the expression

$$\frac{\left[1+(dy/dx)^2\right]^{3/2}}{d^2y/dx^2}$$

for the radius of curvature ρ that the sign of ρ depends upon the sign of d^2y/dx^2. That is, ρ is positive or negative according as d^2y/dx^2 is positive or negative.

But we have seen that d^2y/dx^2 is positive or negative according as the curve is convex or concave to the x-axis. It follows that the radius of curvature is positive or negative according as the curve is convex or concave to the axes.

Also, at a point of inflexion,

$$\frac{d^2y}{dx^2} = 0$$

Therefore, the curvature of the curve is zero.

Examples

Example 1: Find the points of inflexion on the curve

$$y = (\log x)^3$$

Solution: We have,

$$y = (\log x)^3$$

$$\therefore \frac{dy}{dx} = 3(\log x)^2 . \frac{1}{x}$$

$$\Rightarrow \frac{d^2 y}{dx^2} = 6\log x . \frac{1}{x^2} - \frac{3}{x^2}(\log x)^2$$

$$= \frac{3\log x}{x^2}(2 - \log x)$$

$$= \frac{3\log x}{x^2} . \log \frac{e^x}{x}$$

Thus, $\frac{d^2 y}{dx} = 0$ if $\log x = 0$ or $\log x = 2$

Now, $\log x = 0 \Rightarrow x = e^0 = 1$

and $\log x = 2 \Rightarrow x = e^2$

Thus, we expect points of inflexion on the curve for $x = 1$ and $\Rightarrow x = e^2$.

Now, $d^2 y / dx^2$ changes sign from negative or positive as x passes through 1 and changes sign from positive to negative as x passes through e^2.

Thus, $(1,0)$ and $(e^2, 8)$ are the two points of inflexion of the given curve.

Example 2: Show that the curve $y = x^3$ has a point of inflexion at the origin.

Solution: We have,

$$y = x^3$$

$$\therefore \frac{dy}{dx} = 3x^2$$

$$\Rightarrow \frac{d^2 y}{dx^2} = 6x$$

$\therefore$ At the origin, $\frac{d^2 y}{dx^2} = 0$

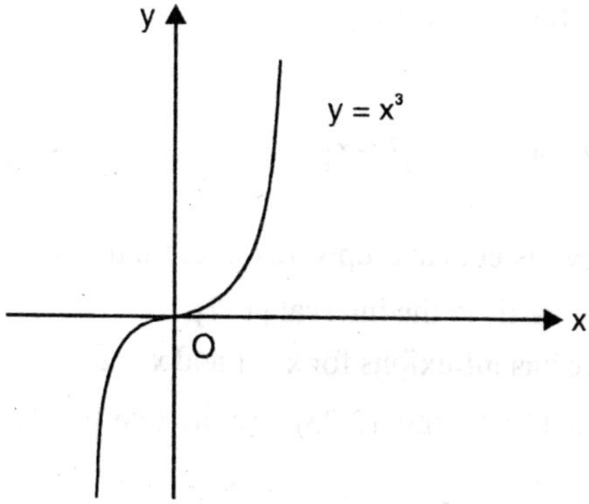

Figure 12.8

But $\frac{d^3y}{dx^3} = 6 \neq 0$

$\therefore$ The curve has a point of inflexion at the origin.

Example 3: Find the ranges of values of x for which the curve

$$y = x^4 - 6x^3 + 12x^2 + 5x + 7$$

is concave upwards or downwards. Also determine the points of inflexion.

Solution: We have,

$$y = x^4 - 6x^3 + 12x^2 + 5x + 7$$

$$\therefore \frac{dy}{dx} = 4x^3 - 18x^2 + 24x + 5$$

$$\frac{d^2y}{dx^2} = 12x^2 - 36x + 24$$

$$= 12(x-1)(x-2)$$

$$\frac{d^2y}{dx^2} > 0 \text{ for all } x \in]-\infty, 1[$$

$$\frac{d^2y}{dx^2} < 0 \text{ for all } x \in]1,2[$$

$$\frac{d^2y}{dx^2} > 0 \text{ for all } x \in]2,+\infty[$$

Thus, the curve is concave upwards in the intervals $[-\infty,1]$ and $[2,+\infty]$ and concave downwards in the interval [1, 2].

Also the curve has inflexions for x = 1 and x = 2.

It follows that (1,19) and (2,33) are the two points of inflexion on the curve.

Example 4: Examine for concavity, convexity and the point of inflexion, the curve

$$y = x^3 - 6x^2 + 11x - 6$$

Solution: We have,

$$y = x^3 - 6x^2 + 11x - 6$$

$$\therefore \frac{dy}{dx} = 3x^2 - 12x + 11$$

$$\Rightarrow \frac{d^2y}{dx^2} = 6x - 12$$

Thus, $\frac{d^2y}{dx^2} > 0$ for all $x \in]2,\infty[$

$\therefore$ The curve is convex to the x-axis at every point in the interval $]2,\infty[$.

Similarly, $\frac{d^2y}{dx^2} < 0$ for all $x \in]-\infty,2[$

$\therefore$ The curve is concave to the x-axis at every point in the interval $]-\infty,2[$.

Also, $\frac{d^2y}{dx^2} = 0$ at $x = 2$

But $\dfrac{d^3y}{dx^3} = 6 \neq 0$

$\therefore$ The curve has a point of inflexion at the point (2, 0).

Example 5: Show that the curve $\left(a^2 + x^2\right)y = a^2x$ has three points of inflexion.

Solution. We have,

$$\left(a^2 + x^2\right)y = a^2x$$

$$\therefore \frac{dy}{dx} = \frac{a^2\left(a^2 + x^2\right) - 2a^2x^2}{\left(a^2 + x^2\right)^2}$$

$$= \frac{a^2\left(a^2 - x^2\right)}{\left(a^2 + x^2\right)^2}$$

$$\frac{d^2y}{dx^2} = a^2.\frac{-2x\left(a^2 + x^2\right)^2 - 4x\left(a^2 + x^2\right)\left(a^2 - x^2\right)}{\left(a^2 + x^2\right)^4}$$

$$= a^2.\frac{-2x(3a^2 - x^2)}{\left(a^2 + x^2\right)^3}$$

$$= 2a^2.\frac{\left(x - \sqrt{3}a\right)x\left(x + \sqrt{3}a\right)}{\left(a^2 + x^2\right)^3}$$

$$\therefore \frac{d^2y}{dx^2} = 0 \text{ for } x = \sqrt{3}a, 0, -\sqrt{3}a$$

It may be seen that

$\frac{d^2y}{dx^2}$ changes sign as x passes through each of these three values of x.

Hence, the curve has inflexion at the corresponding points.

Thus, $\left(\sqrt{3}a, \sqrt{3}a/4\right), (0, 0), \left(-\sqrt{3}a, -\sqrt{3}a/4\right)$

are the three points of inflexion on the curve.

Exercise – 12

1. *Find the points of inflexion on the curves:*

   (i) $y = ax^3 + bx^2 + cx + d$ (ii) $x = 3y^4 - 4y^3 + 5$

   (iii) $y = \frac{x^3 - x}{3x^2 + 1}$ (iv) $y = \frac{x}{x^2 + 2x + 2}$

   (v) $xy = a^2 \log(y/a)$ (vi) $y = x^2 \log\left(x^2/e^3\right)$

   (vii) $x = (y-1)(y-2)(y-3)$ (viii) $54y = (x+5)^2(x^3 - 10)$

   (ix) $y = \frac{a^2(a-x)}{a^2 + x^2}$ (x) $y = \frac{x^3}{a^2 + x^2}$

2. Examine the curve

   $y = \sin x$

   for concavity and convexity in the interval $]0, 2\pi[$

3. Show that $y = e^x$ is everywhere concave upwards and the curve

   $y = \log x$

   is everywhere concave downwards.

4. Show that $y = \log x$ is convex to x-axis in the interval $]0,1[$ and concave to the x-axis in the interval $]1, \infty[$

5. Find the ranges of values of x in which the curves

   (i) $y = 3x^5 - 40x^3 + 3x - 20$

   (ii) $y = \left(x^2 + 4x + 5\right)e^{-x}$

   are concave upwards or downwards. Also find their points of inflexion.

6. Examine the curve $y = x^4 - 6x^3 + 12x^2 + 5x + 7$ for concavity and convexity. Also determine its point of inflexion.

7. Find the intervals in which the curve $y = (\cos x + \sin x)e^x$ is concave upwards or downwards; x varies in the interval $]0, 2\pi[$.

8. Show that the points of inflexion of the curve

$$y^2 = (x-a)^2(x-b)$$

lie on the line $3x + a = 4b$

9. Show that the curve $ay^2 = x(x-a)(x-b)$ has two and only two points of inflexion.

10. Show that the abscissa of the points of inflexion on the curve

$$y^2 = f(x)$$

satisfies the equation

$$[f'(x)]^2 = 2f(x)f'(x)$$

11. Find the points of inflexion on the curves:

(i) $x = 6t^2, \quad y = 4t^3 - 3t$

(ii) $x = a(2\theta - \sin\theta), \quad y = a(2 - \cos\theta)$

(iii) $x = 2a\cot\theta, \quad y = 2a\sin^2\theta$

(iv) $x = a\tan t, \quad y = a\sin t\cos t$

(v) $y = xe^{-x^2}$ (vi) $y = x^2e^{-x}$

(vii) $y = e^{x/2} - 2e^{-x/2}$ (viii) $y = 2e^x + e^{-x}$

(ix) $y = e^{-2x}\cos x$ (x) $y = x(x-3)e^{2x}$

12. Find the points of inflexion on the curves:

(i) $r = \dfrac{a\theta^2}{\theta^2 - 1}$ (ii) $r = \dfrac{ae^{\theta}}{1+\theta}$

13. Show that $r = a\theta^n$ has points of inflexion if and only if n lies between 0 and –1 and they are given by

$$\theta = \pm\sqrt{[-n(n+1)]}$$

14. Show that the curve,

$$re^{\theta} = a(1+\theta)$$

has no point of inflexion.

13

Singular Points

13.1. INTRODUCTION

Consider the following three curves:

(i) $y^2(a-x)=x^3$

- The graph of this curve is as shown in fig. 13.1
- The curve has two branches.
- The point (0, 0) is common to both the branches.
- Both the branches have a common tangent there, namely $y = 0$.
- Such a point on a curve is called a **cusp**.

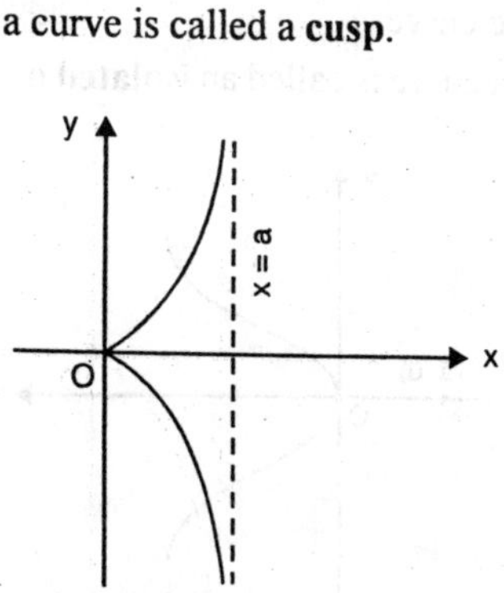

Graph of $y^2(a-x) = x^3$

Figure 13.1

(ii) $(x^2+y^2)x-a(x^2-y^2)=0$

- The graph of this curve is as shown in fig. 13.2.
- The curve has two branches.
- Origin is a point common to the two branches.
- Both the branches have different tangents there.
- Such a point on a curve is called an **isolated** or **conjugate** point.

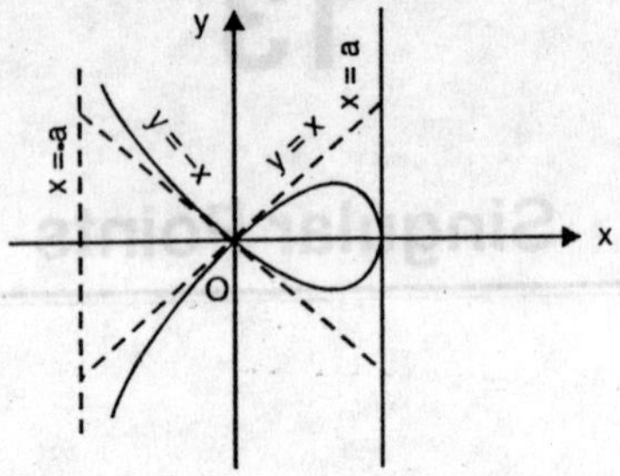

Figure 13.2

(iii) $ay^2=x(x+a)^2,\ a>0$

- The graph of this curve is as shown in fig. 13.3.
- The curve has two branches,
- The tangents to both the branches at $(-a, 0)$ are imaginary.
- There is no point in the immediate neighbourhood of the point $(-a,0)$ which lies on the curve.
- Such a point on a curve is called an **isolated** or **conjugate point**.

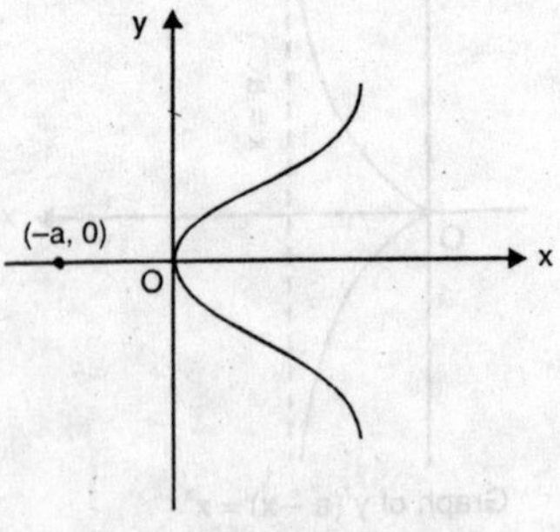

Figure 13.3

13.2. MULTIPLE POINTS

Definition 1: A point (x, y) through which pass k branches of the curve $f(x, y) = 0$ is called a **multiple point** of order *k*.

Multiple points of order two are called double points; multiple points of order three are called triple points.

Definition 2: A double point is called a **node, cusp** or **conjugate point** according as the two tangents at that point are real and distinct, real but coincident or imaginary.

A conjugate point is also called an **isolated point**.

Thus, a multiple point is also, sometimes, called a singular point.

13.3. TANGENTS AT THE ORIGIN

The general equation of rational equation of rational algebraic curve of the nth degree is of the form.

$$f(x, y) = a_0 + (b_1 x + b_2 y) + \left(c_1 x^2 + c_2 xy + c_3 y^2\right) +$$

$$+\left(d_1 x^3 + d_2 x^2 y + d_3 xy^2 + d_4 y^3\right) + \cdots\cdots = 0 \qquad \ldots(1)$$

If the origin $(0,0)$ is a point on (1), then,

$$f(0,0) = 0,$$

and consequently, $a_0 = 0$

The equation (1) becomes

$$f(x, y) = (b_1 x + b_2 y) + \left(c_1 x^2 + c_2 xy + c_3 y^2\right) + \cdots\cdots = 0$$

Let $P(x, y)$ be any point on the curve.

The slope of the chord *OP* is *y*/*x*.

Limiting position of the chord *OP*, when $P \to Q$ is the tangent at 0 so that when $x \to 0$ and $y \to 0$.

$$\lim(y / x) = m$$

the slope of this tangent.

Dividing (1) by *x*, we get

$$\left(b_1+b_2\frac{y}{x}\right)+\left(c_1x+c_2y+c_3y.\frac{y}{x}\right)+\left(d_1x^2+d_2xy+\cdots\right)+\cdots\cdots=0$$

Taking limit as $x \to 0$, we get

$$b_1+b_2m=0$$

so that $m=-b_1/b_2$, if $b_2 \neq 0$

Hence, $\quad y/x=-b_1/b_2$

$$\Rightarrow b_1x+b_2y=0 \qquad \ldots(2)$$

is the tangent at the origin.

If $b_2=0$, $b_1 \neq 0$, then considering the slope of *OP* with reference to *y*-axis, it can be shown that the tangent retains the same form. The equation of the tangent is then again,

$$b_1x+b_2y=0$$

If $b_1=0$, $b_2=0$, the equation of the curve is of the form

$$\left(c_1x^2+c_2xy+c_3y^2\right)+\left(d_1x^3+d_2x^2y+\cdots\right)+\cdots\cdots=0 \qquad \ldots(3)$$

Dividing by x^2 and then taking limit as $x \to 0$, we get

$$c_1+c_2m+c_3m^2=0 \qquad \ldots(4)$$

which is a quadratic equation in *m* and determines as its two roots the slopes of the two tangents so that the origin is a double point in this case.

The equation of either tangent at the origin is

$$y=mx \qquad \ldots(5)$$

where *m* is a root of (4).

Eliminating *m* between (4) and (5), we obtain

$$c_1x^2+c_2xy+c_3y^2=0 \qquad \ldots(6)$$

as the joint equation of the two tangents at the origin.

The equation (6) becomes an identity

if $c_1=c_2=c_3=0$

In this case the second degree terms, also, do not appear in the equation of the curve. Similarly it can be shown that the equation of the tangents can still be written down by equating to zero the terms of the lowest degree which is third in this case.

In general, the equation of the tangent or tangents at the origin is obtained by equating to zero the terms of the lowest degree in the equation of the curve.

Note: The origin is a multiple point if and only if $f(x, y)$ does not conatin any constant term and terms of degree one.

Illustrations:

(i) The equation of the tangents at the origin to the curve

$x^4 + y^4 = a^2 xy$ are $x = 0, y = 0$ so that the origin is a node.

(ii) Thc origin is a cusp on the curve

$(x^2 + y^2)x - 2ay^2 = 0$ and $y = 0$ is the cuspidal tangent.

(iii) The equation of the tangents at the origin to the curve

$$(x^2 + y^2)(x - a) + b^2 x^3 = 0$$

is $x^2 + y^2 = 0$

Since the tangents are imaginary

$$y = \pm ix$$

therefore the origin is a conjugate point on the curve.

(iv) The origin is a triple point on the curve

$$2y^5 + 5x^5 - 3x(x^2 - y^2) = 0$$

and $x = 0, x = y, x = -y$

are the three tangents there at.

13.4. MULTIPLE POINTS ON THE CURVE $f(x, y) = 0$

The slope of the tangent to the curve

$$f(x, y) = 0$$

is given by the equation

$$fx + fy\frac{dy}{dx} = 0 \qquad \ldots(1)$$

At a multiple point, the curve has at least two branches, and therefore at least two tangents.

Therefore at a multiple point (1) must be satisfied by two values of dy/dx.

The equation (1) is of first degree in dy/dx.

So, it can be satisfied by more than one value of dy/dx, if and only if,

$$f_x = 0, \quad f_y = 0$$

Thus, we have the following results:

A point (x,y) on a curve $f(x,y)=0$ is a multiple point if $f_x(x,y)=0$, $f_y(x,y)=0$.

To find multiple points (x,y), we have therefore to find the values of (x,y) which simultaneously satisfy the three equations

$$fx(x,y)=0, \; fy(x,y)=0, \; f(x,y)=0.$$

13.5. SLOPES OF THE TANGENTS AT A DOUBLE POINT

Differentiating the relation

$$f_x + f_y \frac{dy}{dx} = 0$$

with respect to x, we have

$$\frac{\partial}{\partial x}\left[f_x + f_y \frac{dy}{dx}\right] + \frac{\partial}{\partial y}\left[f_x + f_y \frac{dy}{dx}\right]\frac{dy}{dx} = 0$$

$$\Rightarrow f_{xx} + \left(f_{xy}\frac{dy}{dx} + f_y \frac{\partial}{\partial x}\left(\frac{dy}{dx}\right)\right) + f_{xy}\frac{dy}{dx}$$

$$+\left(f_{yy}\frac{dy}{dx} + f_y \frac{\partial}{\partial y}\left(\frac{dy}{dx}\right)\right)\frac{dy}{dx} = 0 \quad \ldots(2)$$

$$\Rightarrow f_{xx} + 2f_{xy}\frac{dy}{dx} + f_{yy}\left(\frac{dy}{dx}\right)^2 = 0 \quad \ldots(3)$$

since $f_x = 0, f_y = 0, f_{xy} = f_{xy}$

In case f_{xx}, f_{xy}, f_{yy}, are not all zero and

$$f_x = 0 = f_y,$$

the point (x, y) will be a double point and will be a node, cusp or conjugate according to as the values of dy/dx are real and distinct, equal or imaginary.

i.e., according as

$$(f_{xy})^2 - f_{xx} \cdot f_{yy} > 0, = 0, < 0$$

If $f_{xx} = f_{xy} = f_{yy} = 0$, the point (x, y) will be multiple point of order highest than the second.

13.6. TYPES OF CUSPS

Two branches of a curve have a common tangent at a cusp.

A cusp is said to be single or double according as both the branches of the curve lie on only one side of the normal or at least one of them extends to both the sides of the normal.

A cusp is also said to be of the first species or second species according as the two branches of the curve are on opposite, sides of the tangent or on the same side of the tangent.

There are five different ways in which the two branches stand in relation to the common tangent and the common normal as illustrated by the following figures.

(1) A curve is said to have a **single cusp of the first species** at a point if both the branches of the curve lie on one side of the normal at the point but are on opposite sides of the tangent at that point, as shown in Fig. 13.4.

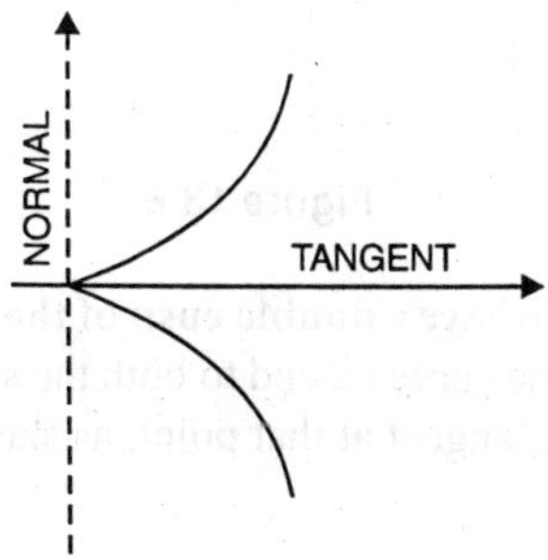

Figure 13.4

(2) A curve is said to have a **single cusp of the second species** at a point if both the branches of the curve lie on the same side of the normal at that point, as shown in Fig. 13.5.

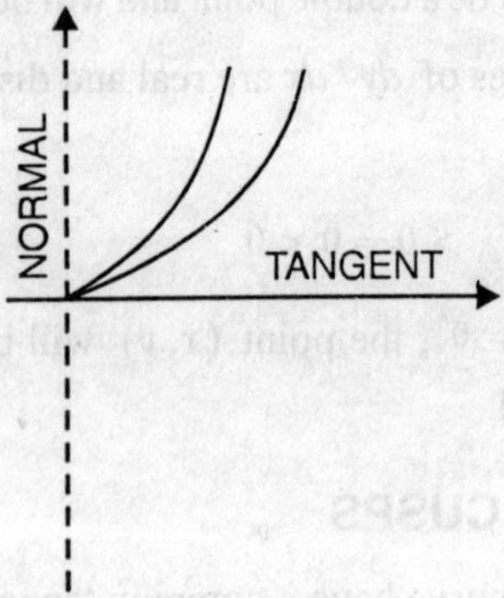

Figure 13.5

(3) A curve is said to have a **double cusp of the first species** at a point if both the branches of the curve extend to both the sides of the normal at the point and lie on opposite sides of the tangent at that point, as shown in Fig. 13.6.

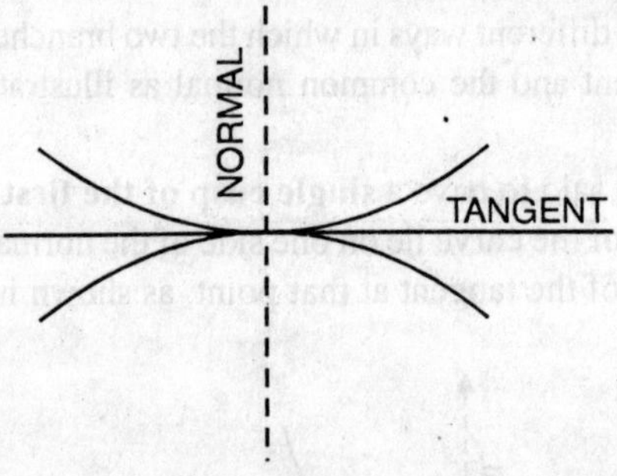

Figure 13.6

(4) A curve is said to have a **double cusp of the second species** at a point if both the branches of the curve extend to both the sides of the normal but are on the same sides of the tangent at that point, as shown in Fig. 13.7.

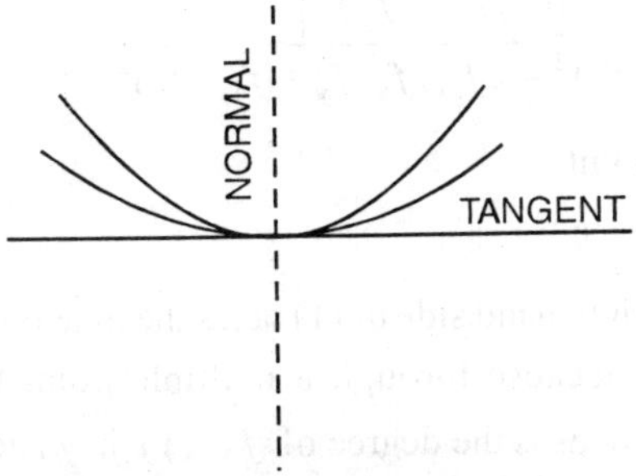

Figure 13.7

(5) A curve is said to have a **point of oscu-inflexion** if there is a cusp of the first kind on one side of the normal and a cusp of the second kind on the other side of the normal. In such a case the point is also a point of inflexion, as the name suggests, as shown in Fig. 13.8.

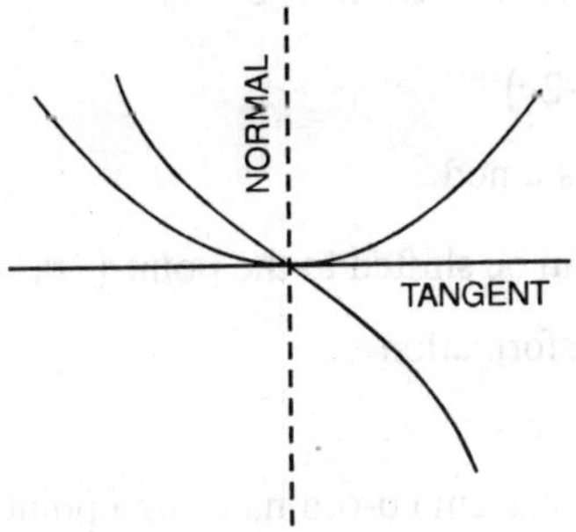

Figure 13.8

Thus, it will be seen that the cusp is single or double according as the two branches lie on the same or different sides of the common normal, also it is of the first or second species according as the branches lie on the different or the same side of the common tangent.

13.7. RADII OF CURVATURE AT MULTIPLE POINTS

The radius of curvature at a point (x, y) of the curve

$$f(x, y) = 0$$

is given by the formula

$$\rho = \pm \frac{\left[(f_x)^2 + (f_y)^2\right]^{3/2}}{f_{xx}(f_y)^2 - 2f_{xx}\cdot f_x\cdot f_y + f_{yy}(f_x)^2} \qquad \ldots(1)$$

At a multiple point,

$$f_x = 0, f_y = 0$$

Therefore the right hand side of (1) takes the indeterminate form 0/0.

This happens because through a multiple point there pass as many branches of the curve as is the degree of $f(x,y)$ in y and consequently there are as many branches of the curve as is the degree of $f(x, y)$ in y and consequently there are as many radii of curvature.

Examples

Example 1: Find the equation of the tangents to the curve

$$y^4 + 4cy^3 + 4c^2y^2 + 3c^2x^2 - c^4 = 0$$

at the point $(-c,-2c)$

and show that it is a node.

Solution: Let the origin be shifted to the point $(-c,-2c)$.

By using the transformation

$$x = x - c \quad \text{and} \quad y = y - c,$$

where x, y are the current co-ordinates of a point with respect to the new set of axes.

Now, the equation of the curve transforms to

$$(y-2c)^4 + 4c(y-2c)^3 + 2c(x-c)^3 + 4c^2(y-2c)$$
$$+3c^2(x-c)^2 - c^2 = 0$$

$$\Rightarrow y^4 - 4cy^3 + 2cx^3 + 4c^2y^2 - 3c^2x^2 = 0$$

Equating to zero the lowest degree terms we get

$$4c^2y^2 - 3c^2x^2 = 0$$

so that the tangents are

$$y = \pm\left(\sqrt{3}/2\right)x.$$

Substituting, $x = x+c$ and $y = y+2c$, we find

the equations of the tangents at $(-c,-2c)$ to the given curve are

$$y+2c = \pm\left(\sqrt{3}/2\right)(x+c)$$

Since the two tangents are distinct, the point $(-c,-2c)$ is a node.

Example 2: Find the equation of the tangent at $(-1,-2)$ to the curve

$$x^3+2x^2+2xy-y^2+5x-2y=0.$$

and show that this point is a cusp.

Solution: Let the origin be shifted to the point $(-1,-2)$.

By using the transformation

$x = x-1, \quad y = y-2,$

where x, y are the current co-ordinates of a point with respect to the new set of axes.

Now, the equation of the curve transforms to

$$(x-1)^3+2(x-1)^2+2(x-1)(y-2)-(y-2)^2+5(x-1)-2(y-2)=0$$

$$\Rightarrow x^3-x^2+2xy-y^2=0$$

Equating to zero the lowest degree terms, we get

$$-x^2+2xy-y^2=0$$

$$\Rightarrow (y-x)^2=0$$

which are two coincident lines.

Therefore, the point is a cusp and the cuspidal tangent.

The tangent at the cusp with reference to the new axes is

$$y-x=0$$

Substituting, $x = x+1$ and $y = y+2$, we find that the equations of the tangent at $(-1,-2)$ to the given curve are

$(y+2)-(x+1)=0$

$\Rightarrow y = x\text{-}1$

Example **3:** Find the multiple points on the curve

$$x^4-2ay^3-3a^2y^2-2a^2x^2+a^4=0$$

Also, find the tangent at the multiple point.

Solution: Let $f(x,y)=x^4-2ay^3-3a^2y^2-2a^2x^2+a^4$

$\therefore f_x=4x^3-4a^2x$

$f_y=-6ay^2-6a^2y$

$f_x=0$ gives $x=0,\ \ a,\ -a$

$f_y=0$ gives $y=0,\ \ -a$

Therefore, $f_x=0$, $f_y=0$ give the six points

$(0,0)$, $(0,-a)$, $(a,0)$, $(a,-a)$, $(-a,0)$, $(-a,-a)$

of these six points, only three satisfy the equation

$f(x,y)=0$

These are the points

$(a,0)$, $(-a,0)$, $(0,-a)$

Hence, these are the only three multiple points on the curve.

To find the equations of the tangents to the curve at these points. We proceed as follows.

First Method

We have,

$f_{xx}=12x^2-4a^2$,

$f_{xy}=0$,

$f_{yy}=-12ay-6a^2$

At the point $(a,0)$

$f_{xx}=8a^2$

$$f_{xy} = 0$$

$$f_{yy} = -6a^2$$

The equation

$$f_{yy}\left(\frac{dy}{dx}\right)^2 + 2f_{xy}\frac{dy}{dx} + f_{xx} = 0$$

the values of dy/dx at $(a, 0)$ are given by

$$-6a^2(dy/dx)^2 + 8a^2 = 0$$

$$\Rightarrow \frac{dy}{dx} = \pm\frac{2}{\sqrt{3}}$$

The two values are real.

So the point is a node.

The tangents at $(a, 0)$ are

$$y = \pm(2/\sqrt{3})(x - a)$$

Thc equation of the tangents at $(-a, 0)$ and $(0, -a)$ are

$$y = \pm\sqrt{\frac{4}{3}}(x + a),$$

$$y + a = \pm\sqrt{\frac{2}{3}}x$$

Second Method:

Differentiating the given equation w.r.t, x, we get

$$4x^3 - 6ay^2y_1 - 6a^2yy_1 - 4a^2x = 0$$

which identically vanishes for the multiple points.

Differentiating again, we get

$$12x^2 - 12ayy_1^2 - 6ay^2y_2 - 6a^2y_1^2 - 6a^2yy_2 - 4a^2 = 0$$

From this, we see that

(i) for $(a,0)$, $y_1^2 = \frac{3}{4} \Rightarrow y_1 = \pm\sqrt{\frac{3}{4}}$

(ii) for $(-a, 0)$, $y_1^2 = \frac{3}{4}$, $\Rightarrow y_1 = \pm\sqrt{\frac{3}{4}}$

(iii) for $(0, -a)$, $y_1^2 = \frac{2}{3}$, $\Rightarrow y_1 = \pm\sqrt{\frac{2}{3}}$

Knowing the slopes of the tangents, we can now put down their equations.

Third method:

For finding the tangents at $(a, 0)$, we shift the origin to this point.

$$x = X + a, \quad y = Y + 0$$

The transformed equation is

$$(X+a)^4 - 2aY^3 - 3a^2Y^2 - 2a^2(X+a)^2 + a^4 = 0$$

$$\Rightarrow X^4 + 4X^3a - 2aY^3 + 4a^2X^2 - 3a^2Y^2 = 0$$

The tangents at the new origin are

$$4a^2X^2 - 3a^2Y^2 = 0$$

$$\Rightarrow y = \pm\sqrt{(4/3)}X$$

The tangents at the multiple point $(a, 0)$ are

$$y = \pm\sqrt{(4/3)}(x-a)$$

Similarly, it may be shown that

$$y = \pm\sqrt{(4/3)}(x+a)$$

and $$y + a = \pm\sqrt{(2/3)}x$$

are the tangents at the multiple points on the curve are nodes.

Example 4: Show that the curve

$$y^2 = 2x^2y + x^3y + x^3$$

has a single cusp of the first species at the origin.

Solution: Equating to zero the lowest degree terms, we see that the origin is a cusp and

$y = 0$ is the cuspidal tangent.

The given equation can be written in the form

$$y^2 - yx^2(2+x) - x^3 = 0$$

and solving, we have

$$y = \frac{x^2(2+x) \pm \sqrt{\left[x^4(2+x)^2 + 4x^3\right]}}{2}$$

For positive values of x, we have

$$x^4(2+x)^2 + 4x^3 > x^4(2+x)^2$$

$$\Rightarrow \sqrt{\left[x^4(2+x)^2 + 4x^3\right]} > x^2(2+x)$$

So that two values of x correspond two values of y with opposite signs. Thus the two branches lie on opposite sides of x-axis when x is positive. Again, we have

$$x^4(2+x)^2 + 4x^3 = x^3\left(4 + 4x + 4x^2 + x^3\right)$$

For values of x which are sufficiently small in numerical value,

$$4 + 4x + 4x^2 + x^3$$

is positive for the same $\rightarrow 4$ when $x \rightarrow 0$

Thus, for negative values of x which are sufficiently small in numerical value,

$$x^3\left(4 + 4x + 4x^2 + x^3\right)$$

is negative so that the values of y are imaginary.

Thus x cannot take up negative values.

Hence, the curve has a single cusp of first species at the origin.

Example 5: Find the position and nature of the cusp on the curve

$$y^2 + 3x^4 = 4x^2y + x^5$$

Solution: Writing the equation of the curve, as

$$f(x,y) = y^2 + 3x^4 - 4x^2y - x^5 = 0,$$

We have,

$$f_x = 12x^3 - 8xy - 5x^4$$

$$f_y = 2y - 4x^2$$

Solving the equations $f(x,y) = 0$, $f_x = 0$, $f_y = 0$ together, we get that $(0,0)$ is the only multiple point.

Equating to zero the lowest degree terms, we find that $y = 0$ is a cuspidal tangent.

Thus the origin is a cusp.

Solving $f(x,y) = 0$, as a quadrate in y, we have

$$y = 2x^2 \pm x^2\sqrt{(1+x)} \qquad \ldots(1)$$

For each small positive value of x, there are two values of y which are both positive.

Thus, in the neighbourhood of the origin, the curve lies on both sides of the y-axis.

Hence the origin is a double cusp of the second species.

Example 6: Find the radii of curvature at the origin for the curve

$$x^3 + y^3 = 3axy$$

Solution. The tangents at the origin are given by

$$xy = 0,$$

so that the origin is a node.

The radius of curvature ρ_1 of the branch having $y = 0$, as the tangent at the origin is given by

$$\rho_1 = \lim_{x \to 0}\left(x^2 / 2y\right)$$

Replacing y by $x^2/2\rho$ in the equation of the curve, we have

$$x^3 + \left(x^3/2\rho\right)^3 = 3ax\left(x^2/2\rho\right)$$

Dividing throughout by x^3 and taking limit as $x \to 0$, we have

$$1 = \frac{3}{2}a/\rho_1$$

so that

$$\rho = \frac{3}{2}a$$

Again, the radius of curvature ρ_2 of the branch having $x = 0$ as the tangent at the origin is given by

$$\rho_2 = \lim_{y \to 0}\left(y^2 / 2x\right)$$

Replacing x by $y^2 / 2\rho$ in the equation of the curve, we have

$$\left(y^2 / 2\rho\right)^3 + y^3 = 3a\left(y^2 / 2\rho\right)y$$

Dividing throughout by y^3 and taking limits as $y \to 0$, we have

$$1 = \frac{3}{2}a / \rho_2$$

so that

$$\rho_2 = \frac{3}{2}a$$

Thus the radius of curvature at the origin for each branch is $\frac{3}{2}a$.

Example 7: Find the radii of curvature at the origin of the branches of the curve

$$y^4 + 2axy^2 = ax^3 + x^4$$

Solution: Here,

$$2xy^2 = x^3$$

$\Rightarrow x = 0$, $y = \pm\left(1/\sqrt{2}\right)$ are the three tangents at the origin so that it is a triple point.

For finding ρ, for the branch which touches $x = 0$, we find $\lim\left(y^2 / 2x\right)$.

To do this, we write

$$\left(y^2 / 2x\right) = \rho_1$$

$$\Rightarrow \quad x = y^2/2\rho_1$$

and substitute this value of x in the given equation.

$\lim \rho_1 = \rho_2$ is the radius of curvature of the corresponding branch at the origin.

$$y^4 + \frac{2ay^4}{2\rho_1} = a\frac{y^6}{8\rho_1 3} + \frac{y^8}{16\rho_1 4}$$

$$\Rightarrow 1 + \frac{a}{\rho_1} = a\frac{y^2}{8\rho_1 3} + \frac{y^4}{16\rho_1 4}$$

Let $y \to 0$ so that we have

$$1 + \frac{a}{\rho_1} = 0$$

Thus, ρ for this branch $= -a$

To find, ρ, for the other branches we proceed as follows.

Suppose that the equation of either branch is given by

$$y = f(0) + xf'(0) + \frac{x^2}{2!} + \frac{x^2}{2!} f''(0) + \cdots\cdots$$

We have,

$$f(0) = 0$$

Also, we write,

$$f'(0) = p,$$

$$f''(0) = q.$$

Thus, we have,

$$y = px + \frac{1}{2}qx^2 + \cdots\cdots$$

Making substitution in the given equation, we get

$$\left(px + \frac{1}{2}qx^2 + \cdots\cdots\right)^4 + 2ax\left(px + \frac{1}{2}qx^2 + \cdots\cdots\right)^2 = ax^3 + x^4$$

Equating co-efficients of x^3 and x^4, we get

$$2ap^2 = a,$$

$$p^4 + 2apq = 1$$

which give

$$p = \frac{1}{\sqrt{2}}, \quad q = \frac{3\sqrt{2}}{8a}$$

$$p = -\frac{1}{\sqrt{2}}, \quad q = -\frac{3\sqrt{2}}{8a}$$

$$\therefore \quad \rho = \frac{\left(1+p^2\right)^{3/2}}{q}$$

$$= \pm 2\sqrt{3}a$$

for the two branches.

Example 8: Find the radii of curvature of the curve

$$2x^5 - cx^2y^2 + 2cx^3y - 4c^2xy^2 + c^2y^3 = 0$$

Solution: The tangents at the origin are given by

$$-4c^2xy^2 + c^2y^3 = 0$$

$$\Rightarrow \quad y^2(4-4x) = 0$$

so that the origin is a triple point.

$y = 0$ is cuspidal tangent and $y = 4x$ is a tangent to the third branch of the curve.

To find the radii of curvature, let us assume that the equation of each of the three branches of the curve can be written in the form

$$y = px + \frac{qx^2}{2!} + \cdots\cdots \qquad \ldots(1)$$

Substituting the value of y from (i) into the equation of the curve, we have

$$2x^5 - cx^2\left(px + q\frac{x^2}{2!} + \cdots\cdots\right)^2 + 2cx^3\left(px + q\frac{x^2}{2!} + \cdots\cdots\right)$$

$$-4c^2x\left(px+q\frac{x^2}{2!}+\cdots\cdots\right)+c^2\left(px+q\frac{x^2}{2!}+\cdots\cdots\right)^3=0 \qquad \ldots(2)$$

Equating the coefficient of x^3 and x^4 on both sides, we have

$$-4c^2p^2+c^2p^3=0 \qquad \ldots(3)$$

$$-cp^2+2cp-4c^2pq+\frac{3}{2}c^2p^2q=0 \qquad \ldots(4)$$

From (3), we have

$p=0$ or 4

When $p=0$, (4) vanishes identically.

Comparing the coefficients of x^5 on both sides of (2), we have

$$2-cpq+cq-c^2q^2+\frac{3}{2}pq^2c^2=0 \qquad \ldots(5)$$

Corresponding to $p=0$, we have form (5),

$$2+cq-c^2q^2=0$$

$\Rightarrow \quad q=2/c$ or $-1/c$.

The radii of curvature are given by

$$\frac{\left(1+p^2\right)^{3/2}}{q}$$

and are therefore $c/2$ and $-c$.

When $p=4$, (4) gives

$$-16c+8c-16c^2q+24c^2q=0$$

so that $q=1/c$

The corresponding radius of curvature is

$$\frac{\left(1+p^2\right)^{3/2}}{q}=(17)^{3/2}c$$

Thus the radii of curvature of the three branches at the origin are $\frac{1}{2}c$, $-c$ and $(17)^{3/2}$.

Example 9: Show that the pole is a triple point on the curve

$$r = a(2\cos\theta + \cos 3\theta)$$

and the radii of curvature of the three branches are

$$\sqrt{3}a/2, a/2, \sqrt{3}a/2$$

Solution. The radius vector, r, vanishes for the values of θ, given by

$$2\cos\theta + \cos 3\theta = 0$$

$$\Rightarrow \quad 2\cos\theta + 4\cos^3\theta - 3\cos\theta = 0$$

$$\Rightarrow \quad \cos\theta\left(4\cos^2\theta - 1\right) = 0$$

$$\Rightarrow \quad \cos\theta = 0, \cos\theta = \frac{1}{2}, \cos\theta = -\frac{1}{2}$$

Thus, $r = 0$ when $\theta = \pi/3,\ \pi/2,\ 2\pi/3$ so that the pole is a triple point.
We have,

$$r_1 = a(-2\sin\theta - 3\sin 3\theta),$$

$$r_2 = a(-2\cos\theta - 9\cos 3\theta)$$

For $\theta = \pi/3$,

$$r_1 = -\sqrt{3}a$$

$$r_2 = 8a$$

For $\theta = \pi/2$,

$$r_1 = a,\quad r_2 = 0$$

For $\theta = 2\pi/3$,

$$r_1 = -\sqrt{3}a,\quad r_2 = -8a$$

Also, $r = 0$ for each of these branches.

Putting these values in the formula

$$\rho = \frac{\left(r^2 + r_1^{\,2}\right)^{3/2}}{r^2 + 2r_1^{\,2} - rr_2}$$

We get the required result.

Exercise – 13

Find the equation of the tangents at the origin to the following curves:

1. $a^2y^2 = x^2(a^2 - x^2)$
2. $a^2(x^2 - y^2) = x^2y^2$
3. $y^2(a^2 - x^2) = x^2(b - x)^2$
4. $x^2(x^2 - 4a^2) = y^2(x^2 - a^2)$
5. $(x^2 + y^2)^3 = a^2(x^2 - y^2)^2$
6. $(x^2 + y^2)^2 = 4a^2xy$
7. $x^5 + 2a^2x^2y - a^3y^2 = 0$
8. $(x^2 + y^2)(2a - x) = b^2x$
9. $y^5 + ax^2y^2 - ay^3x - 2a^2x^2y + a^2x^3 = 0$
10. $(x-2)^2 = y(y-1)^2$ at $(2,1)$
11. $x^4 - 4ax^3 - 2ay^3 + 4a^2x^2 + 3a^2y^2 - a^4 = 0$ at $(a, 0)$ and $(2a, a)$.
12. Find the equation of the tangent at $(-1,-2)$ to the curve

    $$x^3 + 2x^2 + 2xy - y^2 + 5x - 2y = 0$$

    and show that this point is a cusp.
13. Show that the origin is a node; a cusp or a conjugate point on the curve

    $$y^2 = ax^2 + ax^3$$

    according as, a is positive, zero or negative.
14. Find the equations of the tangents to the curve

    $$x^3 + xy^2 + 2ax^2 + 2ay^2 + 4axy + 5a^2x + 8a^2y + 8a^3 = 0$$

    at the point $(-a,-2a)$ and show that it is a node.

Find the position and nature of the multiple points on the curves:

15. $x^3 + y^3 = 3axy$
16. $x^4 + y^4 = 4a^2xy$
17. $x^3 + y^3 = 6x^2$
18. $x^4 - y^4 = 4a^2xy$
19. $y^2 = (x-1)(x-2)^2$
20. $ay^2 = (x-a)^2(x-b)^2$

21. $(y+1)^2 = (x-1)^2(x-4)$ 22. $y^2(x-a) = x^2(x+a)$

23. $x^4 - 4ax^3 + 2ay^3 + 4a^2x^2 - 3a^2y^2 - a^4 = 0$

24. $x^4 + 4ax^3 + 4a^2x^2 - b^2y^2 - 2b^3y - a^4 - b^4 = 0$

25. $x^2y^2 = (a+y)^2(b^2 - y^2)$

26. $(x+y)^3 - \sqrt{2}(y-x+2)^2 = 0$

27. $x^3 + y^3 + a^3 = 3axy$

28. $(2y+x+1)^2 - 4(1-x)^5 = 0$

29. $x^4 + y(y+4a)^3 + 2x^2(y-5a)^2 = 5a^2x^2$

30. $(y^2 - a^2)^3 + x^4(2x+3a)^2 = 0$

Find the position and nature of multiple points of the following curves. Also find the equations of the tangents at each multiple point.

31. $x^4 - 4ax^3 - 2ay^3 + 4a^2x^2 + 3a^2y^2 - a^4 = 0$

32. $x^4 + y^4 - 2x^2 - 2y^2 + 1 = 0$

33. $x^4 - 8x^3 + 12x^2y + 16x^2 + 48xy + 4y^2 - 64y = 0$

34. $x^3 + y^3 + 3(x^2 + y^2) = 3xy + 1$

35. $(y-2)^2 = x(x-1)^2$

36. Show that each of the curves

$$(x\cos\alpha - y\sin\alpha - b)^3 = c(x\sin\alpha + y\cos\alpha)^2$$

for all different values of α, has a cusp, show that all the cusps lie on a circle.

37. Show that the origin is a triple point on the curve

$$x^4 - a^2yx^2 + by^3 = 0$$

one branch touching the x-axis and the others inclined to it at angles $\pm\tan^{-1}(a/b)$

38. Show that the origin is a node, a cups or an isolated point on the curve

$$y^2 = ax^3 + bx^3$$

according as a is positive, zero or negative.

39. Show that the point $(a,0)$ is a node, cusp or conjugate point on the curve

$$y^2 = (x-a)^2(x-c)$$

according as $a > c$, $a = c$ or $a < c$.

Find the position and nature of cusp on the following curves:

40. $x^3 + y^3 = 2ax^2$

41. $y^3 = x^2(x+a)$

42. $\left(y-x^2\right)^2 = x^3$

43. $\left(y-x^2\right)^2 = x^5$

44. $x^2(x-y)+y^2 = 0$

45. $x^2(x+y)-y^2 = 0$

46. $(y-x)^2 + x^6$

47. $y^2(3a-x) = (x-a)^3$

48. $x^6 - ayx^4 - a^3x^2y + a^4y^2 = 0$

49. $x^4 - 2x^2y - xy^2 - 2x^2 - 2xy + y^2 - x + 2y + 1 = 0$

50. $x^5 - ax^3y - a^2x^2y + a^2y^2 = 0$

51. Examine the curve

$$x^5 + 16x^2y - 64y^2 = 0$$

for singularities.

52. Show that the curve

$$y^3 = (x-a)^2(2x-a)$$

has a single cusp at $(a,0)$.

Find the radii of curvature at the origin of the following curves:

53. $y^3 = (x+y)^2$

54. $x^3 + y^3 = 3axy$

55. $a\left(x^2 - y^2\right) = 2x^3 + y^3$

56. $y^2(a-x) = x^2(a+x)$

57. $x^2 - 3xy - 4y^2 + y^3 + y^4x + x^5 = 0$

58. $2x^4 - 5ax^2y - 2axy^2 + 2a^2y^2 = 0$

59. $x^5 + ax^2y^2 - ax^3y - 2a^2xy^2 + a^2y^2 = 0$

60. $ax^3 + b^2xy + cy^3 = x^4 + y^4$

61. Show that the radius of curvature at the origin for both the branches of the curve

$$y^2(a-x) = x^2(a+x)$$

is $\sqrt{2}a$.

62. Find the radii of curvature, at the origin, of the two branches of the curve given by the equations

$$x = 1 - t^2, \; y = t - t^3$$

63. Show that $(a,0)$ in polar co-ordinates, is a triple point on the curve

$$r = a\left(1 + 2\sin\frac{\theta}{2}\right),$$

and find the radii of curvature at the point.

14

Asymptotes

14.1. DEFINITION

A straight line is said to be a rectilinear asymptote or simply an asymptote of an infinite branch of a curve, if as the point P tends to infinity along the branch, the perpendicular distance of P from the straight line tends to zero.

Illustrations

1. The lines $y = \pm\pi/2$ are asymptotes to the curve $y = \tan^{-1} x$.

2. The line $x = a$ is an asymptote of the cissoid $y^2(a-x) = x^3$.

3. The lines $x = 0$ and $y = 0$ are asymptotes to the curve $y = 1/x$.

14.2. DETERMINATION OF ASYMPTOTES

The equation of a line which is not parallel to x-axis is of the form

$$y = mx + c$$

Let $P(x,y)$ be a point on an infinite branch of the curve $f(x,y) = 0$.

Let $p = PM$ be the perpendicular distance of any point $P(x,y)$ on the infinite branch of a given curve,

Then, $P = \dfrac{|y-mx-c|}{\sqrt{1+m^2}}$

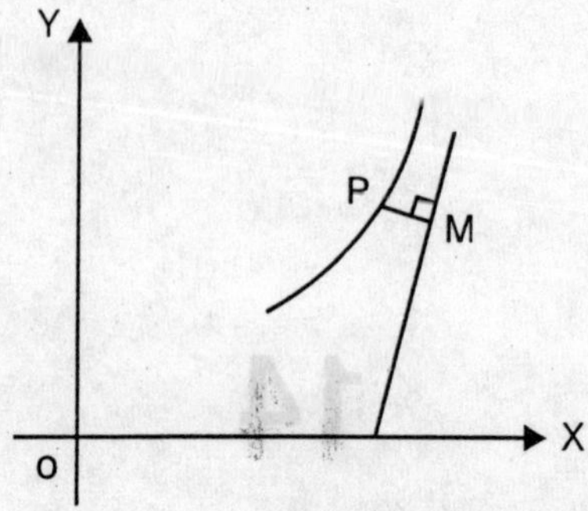

Figure 14.1

If $y = mx + c$ is an asymptote, then $PM \to 0$ as P tends to infinity along the curve.

Now, as P tends to infinity, $x \to \infty$

Therefore, we have,

$$\lim_{x\to\infty} (y - mx - c) = 0$$

$$\Rightarrow \quad \lim_{x\to\infty} (y - mx) = c \qquad \ldots(1)$$

$$\text{Also,} \quad \frac{y}{x} - m = (y - mx)\left(\frac{1}{x}\right)$$

$$\text{Therefore,} \quad \lim_{x\to\infty} (y/x - m) = \lim_{x\to\infty}\left[(y - mx)\left(\frac{1}{x}\right)\right]$$

$$= c - 0 = 0$$

$$\Rightarrow \quad \lim_{x\to\infty} (y/x) = m \qquad \ldots(2)$$

From (1) and (2), we have the following method to determine oblique asymptotes:

i) Find $\lim_{x\to\infty} (y/x)$ in $f(x, y) = 0$ and denote this limit by m.

ii) Find $\lim_{x\to\infty} (y - mx)$ and denote this limit by c.

Then $y = mx + c$ is an asymptote of $f(x, y) = 0$

Note:

(I) The values of y for different branches of the curve $f(x,y)=0$ will be different for a given value of x. therefore, we may get several different values of m and correspondingly several different values of

$$\lim_{x\to\infty}(y-mx)$$

Thus a curve may have more than one asymptote.

(II) This method will determine all the asymptote except those which are parallel to y-axis. To determine such asymptotes, we start with the equation $x=my+d$ which can represent every straight line not parallel to x-axis and so that when $y\to\infty$,

$$m=\lim(x/y)$$

and $$d=\lim(x-my)$$

The asymptote not parallel to any axis can be obtained either way.

14.3. DETERMINATION OF THE ASYMPTOTES PARALLEL TO THE CO-ORDINATE AXES

14.3.1. Asymptote parallel to Y-axis.

Let

$$x=k$$

be an asymptote of the curve.

Here, y, alone tends to infinity since a point $P(x,y)$ recedes to infinity along the curve.

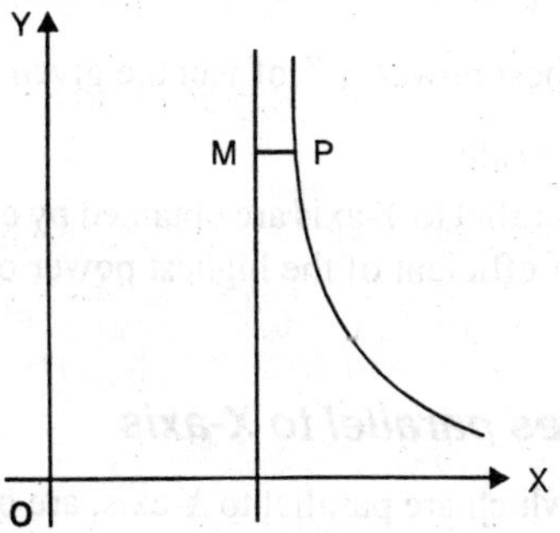

Figure 14.2

The distance PM of any point $P(x,y)$ on the curve is equal to $x-k$.

$$\lim(x-k)=0 \text{ when } y\to\infty$$

or,
$$\lim x = k \text{ when } y\to\infty$$

which gives k.

Arranging the equation of the curve in descending powers of y, so that it takes the form

$$y^m\phi(x)+y^{m-1}\phi_1(x)+y^{m-2}\phi_2(x)+\cdots\cdots=0 \qquad \ldots(i)$$

where, $\phi(x)$, $\phi_1(x)$, $\phi_2(x)$ etc. are polynomial in x.

Dividing the equation (i) by y^m, we get

$$\phi(x)+(1/y).\phi_1(x)+\left(1/y^2\right).\phi_2(x)+\cdots\cdots=0 \qquad \ldots(ii)$$

Let $y\to\infty$

Let us write

$$\lim x = k$$

The equation (ii) gives

$$\phi(k)=0$$

so that k, is a root of the equation $\phi(x)=0$.

Let k_1, k_2, be the roots of $\phi(x)=0$. Then the asymptotes parallel to Y-axis are

$$x=k_1,\ x=k_2 \text{ etc}$$

We know that $(x-k_1)$, $(x-k_2)$ etc., are the factors of $\phi(x)$ which is the co-efficient of the highest power y^m of y in the given equation.

Hence we have the rule:

The asymptotes parallel to Y-axis are obtained by equating to zero the real linear factors in the co-efficient of the highest power of, y, in the equation of the curve.

14.3.2. Asymptotes parallel to X-axis

The asymptotes, which are parallel to X-axis, are obtained by equating to zero the real linear factors in the co-efficient of the highest power of x, in the equation of the curve.

14.4. GENERAL METHOD FOR FINDING OBLIQUE ASYMPTOTES

Let the equation of a rational algebraic curve of the nth degree be of the form.

$$U_n + U_{n-1} + U_{n-2} + \cdots\cdots + U_2 + U_1 + U_0 = 0 \qquad \ldots(1)$$

Where U_r is a homogenous polynomial of degree r in x, y.

Let us write

$$Ur = x^r \phi_r(y/x)$$

where $\phi_r(y/x)$ is a polynomial of degree at the most r in x, y (1) can be written in the form

$$x^n \phi_n(y/x) + x^{n-1}\phi_{n-1}(y/x) + x^{n-2}\phi_{n-2}(y/x) + \cdots$$
$$+ x\phi_1(y/x) + \phi_0(y/x) = 0 \qquad \ldots(2)$$

Dividing (2) by x^n, we get

$$\phi_n(y/x) + (1/x)\phi_{n-1}(y/x) + (1/x^2)\phi_{n-2}(y/x) + \cdots\cdots$$
$$+ (1/x^n)\phi_0(y/x) = 0 \qquad \ldots(3)$$

On taking limits of both sides of (3) as $x \to \infty$, $y/x \to m$, we have the equation

$$\phi_n(m) = 0 \qquad \ldots(4)$$

which determines the slopes of the asymptotes.

Let α be one of the roots of the equation $\phi_n(m) = 0$.

Writing $y - \alpha x = \beta$ and substituting this value of y/x in (3), we have

$$x^n \phi_n(\alpha + \beta/x) + x^{n-1}\phi_{n-1}(\alpha + \beta/x) + x^{n-2}\phi_{n-2}(\alpha + \beta/x) + \cdots\cdots$$
$$+ x\phi_1(\alpha + \beta/x) + \phi_0(\alpha + \beta/x) = 0$$

Expanding each term of the expression on the left hand side by Taylor's series, we have

$$x^n\left[\phi_n(\alpha) + (\beta/x)\phi_n'(\alpha) + \frac{1}{2}(\beta/x)^2\phi_n''(\alpha) + \cdots\right] + x^{n-1}[\phi_{n-1}(\alpha) +$$

$$(\beta/x)\phi_{n-1}'(\alpha)+\cdots]+x^{n-2}[\phi_{n-2}(\alpha)+(\beta/x)\phi_{n-2}'(\alpha)+\cdots]+\cdots=0$$

Arranging the terms according to descending powers of x, we get

$$x^n\phi_n(\alpha)+x^{n-1}[\beta\phi_n'(\alpha)+\phi_{n-1}(\alpha)]$$

$$\left[\frac{1}{2}\beta^2\phi_n''(\alpha)+\beta\phi_{n-1}'(\alpha)+\phi_{n-2}(\alpha)\right]+\cdots=0 \qquad \ldots(5)$$

Putting $\phi_n(\alpha)=0$ in (5), and then dividing by x^{n-1}, we get

$$[\beta\phi'_n(\alpha)+\phi_{n-1}(\alpha)]+\left[\frac{\beta^2}{2}\phi_n''(\alpha)+\beta\phi'_{n-1}(\alpha)+\phi_{n-2}(\alpha)\right]\frac{1}{x}+\cdots=0$$

Taking limits as $x\to\infty$, we have

$$(\lim\beta)\phi_n'(\alpha)+\phi_{n-1}(\alpha)=0$$

or, $\lim(y-\alpha x)=-\phi_{n-1}(\alpha)/\phi n'(a)$, if $\phi_n'(\alpha)\neq 0$

Writing $\lim\beta=c_1$,

Therefore, $c_1\phi_n'(\alpha)+\phi_{n-1}(\alpha)=0$

$$\Rightarrow \quad c_1=-\frac{\phi_{n-1}(\alpha)}{\phi'_n(\alpha)}$$

$$\therefore \quad y=\alpha x-\frac{\phi_{n-1}(\alpha)}{\phi'_n(\alpha)}$$

is the asymptote corresponding to the slope α, if $\phi_n'(\alpha)\neq 0$

Similarly,

$$y=\alpha_2 x-\frac{\phi_{n-1}(\alpha_2)}{\phi'_n(\alpha_2)};\quad y=\alpha_3 x-\frac{\phi_{n-1}(\alpha_3)}{\phi'_n(\alpha_3)},\text{ etc.}$$

are the asymptote of the curve corresponding to the slopes of α_2, α_3 etc., which are the roots of

$\phi_n(\alpha)=0$ if $\phi'_n(\alpha_2),\phi'_n(\alpha_3)$ etc., are not zero.

Note:

(1) If $\phi_n(\alpha) = \phi'_n(\alpha) = 0$, but $\phi_{n-1} \neq 0$, then there is no asymptote parallel to $y = \alpha x$.

(2) A rational algebraic curve of degree n has at all the most n asymptotes.

Examples

Example 1: Examine the Folium

$$x^3 + y^3 - 3axy = 0$$

for asymptotes.

Solution: The given equation is of the third degree.

$$x^3 + y^3 - 3axy = 0 \quad \text{...(i)}$$

Dividing x^3 on both sides, we get

$$1 + \left(\frac{y}{x}\right)^3 - 3a\frac{y}{x}.\frac{1}{x} = 0$$

Let $x \to \infty$. We then find

$$1 + m^3 = 0$$

$$\Rightarrow \quad (m+1)(m^2 - m + 1) = 0$$

$$\therefore \quad m = -1$$

The roots of $m^2 - m + 1 = 0$ are not real.

Put $y + x = p$ so that, p is a variable which $\to c$ when $x \to \infty$.

Putting $p - x$ for y in the equation (i), we have

$$x^3 + (p-x)^3 - 3ax(p-x) = 0$$

$$\Rightarrow \quad 3(p+a)x^2 - 3(p^2 + ap)x + p^3 = 0$$

which is of the second degree in x.

Dividing by x^2, we get

$$3(p+a) - 3(p^2 + ap).\frac{1}{x} + p^3.\frac{1}{x^2} = 0$$

Let $x \to \infty$. We then have

$$3(c+a)=0$$

$$\Rightarrow c=-a$$

Hence, $y=-x-a$

$$\Rightarrow x+y+a=0$$

is the only asymptote of the given curve.

If we start with $x = my + d$, we get no new asymptotes. Thus,

$$x+y+a=0$$

is the only asymptote of the curve.

Example 2: Find the asymptotes of the curve $x^2y^2 = 9x^2 + 4y^2$ parallel to the axes.

Solution: We have,

$$x^2y^2 = 9x^2 + 4y^2$$

$$\Rightarrow x^2y^2 - 9x^2 - 4y^2 = 0$$

$$\Rightarrow x^2\left(y^2-9\right)-4y^2=0$$

The co-efficient of the highest power of x is $\left(y^2-9\right)$

Therefore the asymptotes parallel to the x-axis are given by

$$y^2-9=0$$

$$\Rightarrow y=\pm 3$$

Similarly, the asymptotes parallel to the y-axis are given by $x=\pm 2$

Thus the asymptotes parallel to the axes are

$$y=\pm 3$$

$$x=\pm 2$$

Example 3: Find the asymptotes of the curve

$$3x^3+2x^2y-7xy^2+2y^3-14xy+7y^2+4x+5y=0$$

Solution: We have,

$$\phi_3(m) = 3 + 2m - 7m^2 + 2m^3$$

$$\phi_2(m) = -14m + 7m^2$$

$$\phi_3'(m) = 2 - 14m + 6m^2$$

Now, $\phi_3(m) = 0$ gives

$$3 + 2m - 7m^2 + 2m^3 = 0$$

$$\Rightarrow \quad (1-m)(3+5m-2m^2) = 0$$

$$\Rightarrow \quad (1-m)(1+2m)(3-m) = 0$$

so that $m = -\frac{1}{2}, 1, 3$.

The 'c' giving equation is $c\phi_n'(m) + \phi_{n-1}(m) = 0$, which, in the present case, becomes

$$c(2 - 14m + 6m^2) + (-14m + 7m^2) = 0$$

so that $c = (14m - 7m^2)/(2 - 14m + 6m^2)$

Thus, When $m = -\frac{1}{2}$, $\quad c = -\frac{5}{6}$

When $m = 1$, $\quad c = -\frac{7}{6}$

When $m = 3$, $\quad c = -\frac{3}{2}$

Therefore, the asymptotes are

$$y = -\frac{1}{2}x - \frac{5}{6},$$

$$y = x - \frac{7}{6}, \text{ and}$$

$$y = 3x - \frac{3}{2}.$$

Example 4: Find the asymptotes of the cubic curve

$$2x^3 - x^2 y + 2xy^2 + y^3 - 4x^2 + 8xy - 4x + 1 = 0$$

Solution:

Putting $x = 1,\ y = 1$ in the third degree and second degree terms separately, we get

$$\phi_3(m) = 2 - m - 2m^2 + m^3$$

$$\phi_2(m) = -4 + 8m$$

The slopes of the asymptotes are given by

$$\phi_3(m) = m^3 - 2m^2 - m + 2 = 0$$

$$\Rightarrow \quad (m+1)(m-1)(m-2) = 0$$

$$\therefore \quad m = -1, 1, 2$$

Again, c is given by

$$c\phi'_3(m) + \phi_2(m) = 0$$

$$c\left(1 - 4m + 3m^2\right) + (-4 + 8m) = 0$$

Putting $m = -1, 1, 2$, we get

$c = 2, 2, -4$ respectively.

Therefore, the asymptotes are

$$y = -x + 2,$$

$$y = x + 2,$$

$$y = 2x - 4.$$

Example 5: Find the asymptotes of the curve

$$(x+2y)(x-2y)(x-y) - 2x(x-4y) + 2x = 0$$

Solution: The curve is of the third degree.

The terms of the third degree are

$$(x+2y)(x-2y)(x-y)$$

Since the highest degree terms can be factorized as the product of three distinct linear factors, therefore the asymptotes are parallel to

$$x+2y=0,$$

$$x-2y=0, \text{ and}$$

$$x-y=0.$$

To find the asymptotes parallel to $x+2y=0,$, we write the equation of the curve in the form

$$x+2y=\frac{2x(x-4y)-2x}{(x-2y)(x-y)}$$

$$=\frac{2(1-4y/x)-2/x}{(1-2y/x)(1-y/x)} \quad \begin{pmatrix}\text{Dividing the numerator and} \\ \text{denominator by } x^2\end{pmatrix}$$

Taking limits of the right hand side as $x\to\infty$, $y/x\to-\frac{1}{2}$, we find that

$x+2y=2$ is an asymptote.

Similarly, the equation of the asymptote parallel to $x-2y=0$ is

$$x-y=\lim_{\substack{x\to\infty \\ y/x\to\frac{1}{2}}}\frac{2x(x-4y)-2x}{(x+2y)(x-2y)}$$

$$=-2$$

The equation of the asymptote parallel to $x-y=0$ is

$$x-y=\lim_{\substack{x\to\infty \\ y/x\to 1}}\frac{2x(x-4y)-2x}{(x-2y)(x+2y)}$$

Hence the asymptotes are

$$x+2y=2,$$

$$x-2y=-2, \text{ and}$$

$$x-y=2$$

Example 6: Find the asymptotes of

$$x^3-x^2y-xy^2+y^3+2x^2-4y^2+2xy+x+y+1=0$$

Solution: Here,

$$\phi_3(m) = 1 - m - m^2 + m^3$$

$$= (1-m) - m^2(1-m)$$

$$= (1-m^2)(1-m)$$

$$\phi_2(m) = 2 + 2m - 4m^2$$

The slopes of the asymptotes, given by

$$\phi_3(m) = 0$$

are $1, 1, -1$

To determine, c, we have,

$$c\phi_3(m) + \phi_2(m) = 0$$

$$\Rightarrow c(-1 - 2m + 3m^2) + (2 + 2m - 4m^2) = 0 \qquad \ldots\text{(i)}$$

For $m = -1$, this gives $c = 1$

Therefore, $y = -x + 1$ is the corresponding asymptote.

For $m = 1$, the equation (i) becomes

$$0.c + 0 = 0$$

which is identically true.

In this exceptional case,

c is determined from the equation

$$(c^2/2).\phi_3''(m) + c\phi'_2(m) + \phi_1(m) = 0$$

$$\Rightarrow \quad (c^2/2)(-2 + 6m) + c(2 - 8m) + 1(1 + m) = 0$$

For $m = 1$, this becomes

$$2c^2 - 6c + 2 = 0$$

$$\Rightarrow \quad c = (3 \pm \sqrt{5})/2$$

Hence, $y = x + (3 \pm \sqrt{5})/2$ are the two parallel asymptotes corresponding to the slope 1.

Example 7: Find the asymptotes of the curve

$$x^3+x^2y+xy^2+y^3+2x^2+3xy-4y^2+7x+2y=0$$

Solution: $\phi_3(m)=1+m+m^2+m^3$

$$\phi_2(m)=2+3m-4m^2$$

$$\phi_3'(m)=1+2m+3m^2$$

The slopes of the asymptotes are given by

$$1+m+m^2+m^3=0$$

$$\Rightarrow (1+m)\left(1+m^2\right)=0$$

$$\Rightarrow m=-1,\ \pm i$$

Since, only one of the values of m is real, therefore there is one real asymptote.

The 'i' giving equation is

$$c\phi_3'(m)+\phi_2(m)=0$$

$$\Rightarrow \quad c=\frac{-\phi_2(m)}{\phi_3'(m)}$$

$$=-\frac{\left(2+3m-4m^2\right)}{1+2m+3m^2}$$

$$=\frac{5}{2} \text{ when } m=-1$$

Therefore the only asymptote is

$$y=-x+\frac{5}{2}$$

$$\Rightarrow x+y-\frac{5}{2}=0$$

Exercise – 14.1

Find the asymptotes parallel to co-ordinate axes of the following curves:

1. $x^2y = 1 - y$
2. $x^2y = 1 + y$
3. $y^2x - a^2(x-a) = 0$
4. $xy = x^2 + 1$
5. $x^2y - 3x^2 - 5xy + 6y + 2 = 0$
6. $x^3 + xy^2 = ay^2$
7. $a^2/x^2 + b^2/y^2 = 1$
8. $(x-2)(x-3)(x+y-2) + x^2 - x + 1 = 0$
9. $y = \dfrac{x}{x^2 - 1}$
10. $y = \dfrac{2 + x - x^2}{x^2 - 1}$

Find the asymptotes of the following curves:

11. $x^2 - y^2 - 3x + y + 4 = 0$
12. $x(y-x)^2 = x(y-x) + 2$
13. $x^2(x-y)^2 + a^2(x^2 - y^2) = a^2xy$
14. $x^3 - y^3 = 3xy$
15. $(x-y)^2(x^2+y^2) - 10(x-y)x^2 + 12y^2 + 2x + y = 0$
16. $x^2y + xy^2 + y^2 + 3x = 0$
17. $(x-y+1)(x-y-2)(x+y) = 8x - 1$
18. $x^3 + y^3 = 3ax^2$
19. $y^3 - x^2y + 2xy^2 - y + 1 = 0$
20. $y(x-y)^2 = x + y$
21. $(x+y)(x-y)(x-2y) + 4y(2x-y) + 4y = 0$

22. $x^2y^2(x^2-y^2)^2=(x^2+y^2)^2$

23. $y(y-1)^2-x^2=1$

24. $2y^3+3y^2x-3yx^2-2x^3-3x^2+3y^2+x-3=0$

25. $xy^2=(x+y)^2$

26. $(x+y)(x-y)(2x-y)-4x(x-2y)+4x=0$

27. $y^3+x^2y+2xy^2-2x^3+7xy+3y^2+2x^2-2x+2y+1=0$

28. $2x(y-3)^2=3y(x-1)^2$

29. $(y-a)^2(x^2-a^2)=x^4+a^4$

30. $x^4-y^4+xy=0$

31. $(x+y)^2(x^2+xy+y^2)=a^2x^2+a^2(y-x)$

32. $(x-y)^2(x-2y)(x-3y)-2a(x^3-y^3)-2a^2(x-2y)(x+y)=0$

33. $4(x^4+y^4)-17x^2y^2-4x(4y^2-x^2)+2(x^2-2)=0$

34. $y^2=x^4/(a^2-x^2)$

35. *Show that the following curves have no asymptotes.*

   (i) $x^2(y^2+x^2)=a^2(x^2-y^2)$

   (ii) $x^4+y^4=a^2(x^2-y^2)$

   (iii) $y^2=x(x+1)^2$

   (iv) $a^4y^2=x^5(2a-x)$

36. Find the equation of the tangent to the curve

   $x^3+y^3=3ax^2$

   which is parallel to its asymptote.

14.5. THE CASE OF PARALLEL ASYMPTOTES

If α is a repeated root of the equation

$$\phi_n(m) = 0,$$

so that $\phi_n(\alpha) = 0$, $\phi_n'(\alpha) = 0$, then two different cases arise:

Case I: $\phi_{n-1}(\alpha) \neq 0$,

There will be no asymptote parallel to $y = \alpha x$.

Case II: $\phi_{n-1}(\alpha) = 0$

In this case, the equation

$$\lim \beta\phi'_n(\alpha) + \phi_{n-1}(\alpha) = 0$$

vanishes identically.

Now, we can write from (5)

$$x^{n-2}\left[\frac{1}{2}\beta^2\phi_n''(\alpha) + \beta\phi_{n-1}'(\alpha) + \phi_{n-2}(\alpha)\right] + \cdots\cdots = 0$$

Dividing throughout by x^{n-2} and taking limits as $x \to \infty$, we have

$$\frac{1}{2}c^2\phi_n''(\alpha) + c\phi_{n-1}'(\alpha) + \phi_{n-2}(\alpha) = 0 \qquad \ldots(6)$$

Dividing throughout by x^{n-2} and taking limits as $x \to \infty$, we have

$$\frac{1}{2}c^2\phi_n''(\alpha) + c\phi_{n-1}'(\alpha) + \phi_{n-2}(\alpha) = 0 \qquad \ldots(7)$$

where $\quad c = \lim\limits_{x \to \infty} \beta$

If $\phi_n''(a) \neq 0$, then (7) gives two values of c, say c_1 and c_2 and there are two asymptotes parallel to $y = \alpha x$, namely

$$y = \alpha x + c_1,$$

$$y = \alpha x + c_2$$

If $\phi_n''(\alpha) = \phi_{n-1}'(\alpha) = \phi_{n-2}(\alpha) = 0$,

$\Rightarrow$ the co-efficient of x^{n-2} in (5) vanishes identically independent of β, then we have to consider (5) again, which now reads

$$x^{n-3}\left[\frac{1}{3!}\beta^3\phi_n'''(\alpha)+\frac{1}{2!}\beta^2\phi_{n-1}''(\alpha)+\beta\phi_{n-2}'(\alpha)+\phi_{n-3}(\alpha)\right]$$

$$+x^{n-4}\left[\frac{1}{4!}\beta^4\phi_n^{iv}(\alpha)+\cdots\cdots\right]+\cdots\cdots=0$$

Dividing throughout by x^{n-3} and taking limits as $x\to\infty$, we have

$$\frac{1}{6}c^3\phi_n'''(\alpha)+\frac{1}{2}c^2\phi_{n-1}''(\alpha)+c\phi'_{n-2}(\alpha)+\phi_{n-3}(\alpha)=0 \qquad \ldots(8)$$

which is a cubic in $c\left(=\lim\limits_{x\to\infty} p\right)$ giving three values of c provided

$$\phi_n'''(\alpha)\neq 0$$

Since $\phi_n''(\alpha)=0$

$$\phi_n'''(\alpha)\neq 0,$$

therefore α is a triple root of

$$\phi_n(m)=0$$

and we have three asymptotes parallel to $y=\alpha x$, namely

$$y=\alpha x+c_1,$$

$$y=\alpha x+c_2,$$

$$y=\alpha x+c_3,$$

where c_1, c_2, c_3 are the roots of (9).

If $\phi_n'''(\alpha)=0$, we can proceed further in the same manner as above.

14.6. SPECIAL METHODS FOR FINDING ASYMPTOTES

If the equation of a curve be of the form

$$(ax+by+c)P_{n-1}+F_{n-1}=0$$

where Pr and Fr denote rational algebraic expression, which contain terms of rth degree or lower, then the asymptote parallel to $ax+by=0$ is given by

$$ax+by+c=-\lim_{\substack{x,y\to\infty\\ y/x\to -a/b}}\left(F_{n-1}/P_{n-1}\right)$$

14.6.1. Asymptotes by Inspection

Theorem: If the equation of a curve of the nth degree can be put in the form

$$F_n + F_{n-2} = 0\,;$$

where F_{n-2} is of degree $(n-2)$ at the most, then every linear, factor of F_n, when equated to zero will give an asymptote, provided that no straight line obtained by equating to zero any other linear factor of F_n is parallel to it or coincident with it.

Proof: Let $ax+by+c=0$ be a non-repeated factor of F_n.

We write,

$$F_n = (ax+by+c)F_{n-1}$$

where F_{n-1} is of degree (n–1).

The asymptote parallel to $ax+by+c=0$ is

$$ax+by+c+\lim\frac{F_{n-2}}{F_{n-1}}=0$$

when $x\to\infty$ and $y/x\to -a/b$

For the determination of the limit $\left(F_{n-2}/F_{n-1}\right)$, we divide the numerator as well as the denominator by x^{n-1}

$$\lim_{\substack{x,y\to\infty \\ y/x\to -a/b}} \left(-F_{n-2}/F_{n-1}\right) = \lim\left[\frac{1}{x}-\frac{\left(F_{n-2}/x^{n-2}\right)}{F_{n-1}/x^{n-1}}\right]=0$$

Since $\lim\frac{1}{x}=0$, $\lim F_{n-2}/x^{n-2}$ exists and is finite and $\lim F_{n-1}/x^{n-1}$ exists and is finite and non-zero. Therefore, $ax+by+c=0$ is an asymptote.

14.7. INTERSECTION OF A CURVE AND ITS ASYMPTOTES

Theorem: Any asymptote of a curve of the nth degree cuts the curve in $(n-2)$ points.

Proof: Let $y=mx+c$ be an asymptote of a rational algebraic curve of degree n whose equation is

$$x^n\phi_n(y/x)+x^{n-1}\phi_{n-1}(y/x)+x^{n-2}\phi_{n-2}(y/x)+\cdots\cdots=0$$

To find the points of intersection, we have to solve the two equation simultaneously.

The abscissa of the points if intersection, are the roots of the equation

$$x^n\phi_n(m+c/x)+x^{n-1}\phi_{n-1}(m+c/x)+x^{n-2}\phi_{n-2}(m+c/x)+\cdots\cdots=0\ \ldots\text{(i)}$$

Expanding each term by Taylor's theorem and arranging in descending powers of x, we have

$$\phi_n(m)x^n+\left[c\phi_n'(m)+\phi_{n-1}(m)\right]x^{n-1}+\left[\frac{1}{2}c^2\phi_n''(m)+c\phi_{n-1}'(m)\right.$$

$$\left.+\phi_{n-2}(m)\right]x^{n-2}+\cdots\cdots=0 \quad \ldots\text{(ii)}$$

Since $y=mx+c$ is an asymptote, therefore

$$\phi_n(m)=0$$

$$c\phi_n'(m)+\phi_{n-1}(m)=0$$

Therefore (ii) reduces to

$$\left[\frac{1}{2}c^2\phi_n''(m)+c\phi_{n-1}'(m)+\phi_{n-2}(m)\right]x^{n-2}+\cdots\cdots=0$$

which of the (n–2)th degree and therefore gives (n–2) values of x and consequently, the asymptote cuts the curve in (n–2) points.

Corollary 1: The n, asymptote of a curve of the nth degree cut in $n(n-2)$ points.

Corollary 2: If the equation of a curve of the nth degree can be put in the form

$$F_n+F_{n-2}=0$$

where F_{n-2} is of degree (n–2) at the most and F_n consists of n, non-repeated linear factors, then the $n(n-2)$ points of intersection of the curve and its asymptotes lies on the curve.

$$F_{n-2}=0$$

This follows from the fact that the joint equation of the asymptotes is $F_n=0$. So that the points of intersection satisfy the equations $F_n=0$ and $F_n+F_{n-2}=0$ and consequently they satisfy the equation $F_{n-2}=0$

Corollary 3: For a cubic, $n = 3$ and therefore the asymptotes cut the curve in $3(3-2)-3$ points which lie on a curve of degree $3-2=1$ i.e., the three points of intersection of a cubic curve and its asymptotes lie on a straight line.

Corollary 4: For a quadratic, $n=4$ and, therefore the asymptotes cut the curve in $4(4-2)=8$ points which lie on a curve of degree $4-2=2$ i.e., the eight points of intersection of a quadratic curve and its asymptotes lie on a conic.

14.8. ASYMPTOTES BY EXPANSION

Theorem: If the equation of a curve can be put in the form

$$y = mx + c + A/x + B/x^2 + C/x^3 + \cdots\cdots$$

then $y = mx + c$ is an asymptote to the curve.

Proof: For the determination of the slope of a possible asymptote, we have to determine $\lim y/x$ as x, y tend to infinity.

$$y = mx + c + A/x + B/x^2 + \cdots\cdots, \qquad \ldots\text{(i)}$$

Dividing by x and letting $x \to \infty$, we have

$$y/x = m + C/x + A/x^2 + \cdots\cdots\cdots$$

Therefore m is the slope of a possible asymptote.

From (1), we have

$$\lim_{\substack{x\to\infty \\ y/x\to m}} (y - mx) = \lim_{\substack{x\to\infty \\ y/x\to m}} \left[c + A/x + B/x^2 + \cdots\cdots\right]$$

From (1) and (2), we deduce that

$$y = mx + c$$

is an asymptote.

14.9. POSTION OF A CURVE WITH RESPECT TO AN ASYMPTOTES

Theorem: The curve $y = mx + c + A/x + B/x^2 + \cdots\cdots$ lies above or below the asymptote $y = mx + c$ in the right half of the plane according as $A > 0$or $A < 0$. In the left half of the plane, it lies above or below the asymptote $y = mx + c$ according as $A < 0$ or $A > 0$.

If $A = 0$, the curve lies above or below the asymptote according as $B > 0$ or $B < 0$.

Proof: Let y_1 and y_2 denote the ordinates of the curve and the asymptote respectively, for a given value of x.

Then,

$$y_1 = mx + c + A/x + B/x^2 + \cdots$$

$$y_2 = mx + c$$

So that, $y_1 - y_2 = A/x + B/x^2$ ...(1)

For large values of x, the term on the right hand side is A/x and determincs the sign of $y_1 - y_2$

$$y_1 - y_2 > 0 \text{ or } < 0$$

according as $A/x > 0$ or < 0.

In the right half of the plane, $x > 0$,

So that $y_1 - y_2 > 0$ or < 0

according as $A > 0$ or $A < 0$.

$\Rightarrow$ the curve is above or below the asymptote according as

$A > 0$ or $A < 0$

In the left half-plane, $x < 0$

so that $y_1 - y_2 > 0$ or < 0

according as $A < 0$ or $A > 0$

If $A = 0$, the predominant term on the right hand side of (1) is B/x^2 so that for all x other than zero.

$$y_1 - y_2 > 0 \text{ or } < 0$$

according as $B > 0$ or $B < 0$.

14.10. ASYMPTOTES IN POLAR CO-ORDINATES

The polar equation of any line is

$$p = r\cos(\theta - \alpha),$$

where, p is the length of the perpendicular from the pole to the line and α is the angle which this perpendicular makes with the initial line.

Let OY be the perpendicular on the given line; Y being its foot.

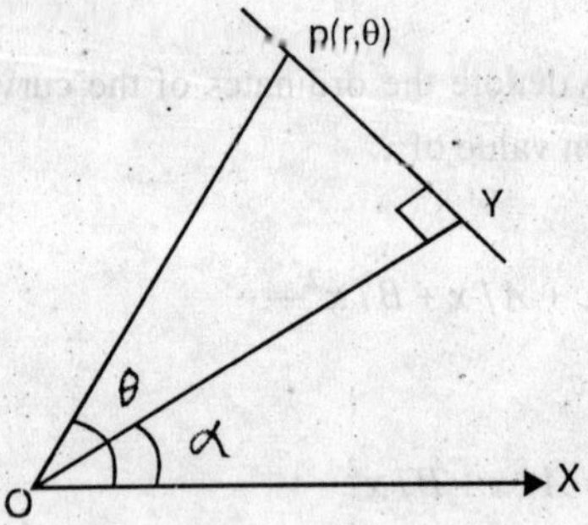

Figure 14.3

Given that,

$$OY = p, \quad \angle XOY = \alpha$$

If $P(r,\theta)$ be any point on the line, we have

$$\angle YOP = \theta - \alpha$$

Now, $$\frac{OY}{OP} = \cos\angle YOP$$

$$\therefore \quad \frac{p}{r} = \cos(\theta - \alpha)$$

$$\Rightarrow \quad p = r\cos(\theta - \alpha)$$

Determination of the asymptote of the curve $r = f(\theta)$

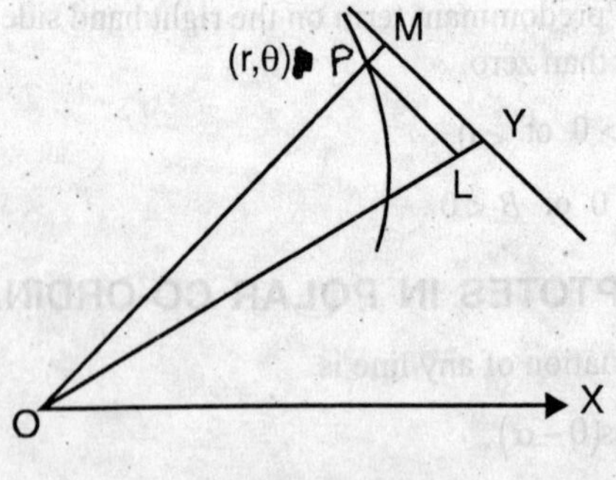

Figure 14.4

Let $P(r,\theta)$ be any point on the curve

$$r = f(\theta)$$

Draw OY perpendicular to the line

$$p = r\cos(\theta - \alpha)$$

Draw $PL \perp OY$ and $PM \perp$ to the given line.

Now,

$$\begin{aligned} PM &= LY \\ &= OY - OL \\ &= p - OP\cos(\theta - \alpha) \\ &= p \quad r\cos(0 - u) \end{aligned}$$

Now, $r \to \alpha$ as the point recedes to infinity the curve.

Let $\theta \to \theta_1$, when $r \to \infty$, then by the theory of indeterminate forms, we have

$$\begin{aligned} p &= \lim_{\theta\to\theta_1}\left[[-\sin(\theta-\alpha)]\Big/\frac{du}{d\theta}\right] \\ &= -\sin(\theta_1-\alpha)\Big/\lim_{\theta\to\theta_1}(du/d\theta) \\ &= -\sin(\pi/2)\Big/\lim_{\theta\to\theta_1}(du/d\theta) \\ &= \lim_{\theta\to\theta_1}\left(-\frac{d\theta}{du}\right) \end{aligned}$$

Therefore, the equation of the asymptote is

$$\begin{aligned} \lim_{\theta\to\theta_1}\left(-\frac{d\theta}{du}\right) &= r\cos(\theta-\alpha) \\ &= r\cos[\theta-(\theta_1-\pi/2)] \\ &= r\sin(\theta_1-\theta) \end{aligned}$$

$$\Rightarrow \quad \lim_{\theta\to\theta_1}\left(-\frac{d\theta}{du}\right) = r\sin(\theta_1-\theta)$$

Working rule for obtaining asymptote to polar curve

Rule: Substitute $r = \frac{1}{u}$ in the equation of the curve, and find these values of θ for which u tends to zero.

If θ_1 be such a value, find out $-(d\theta/du)$ and its limit as $u \to 0$ and $\theta \to \theta_1$. If this limit be p, the desired equation is

$$p = r\sin(\theta_1 - \theta)$$

Examples

Example 1: Find the asymptotes of

$$2x^3 - 5x^2y + 4xy^2 - y^3 + 6x^2 - 7xy + y^2 - x + 5y - 3 = 0$$

Solution: Here,

$$\phi_3(m) = 2 - 5m + 4m^2 - m^3$$

$$\phi_3'(m) = -5 + 8m - 3m^2$$

$$\phi_2(m) = 6 - 7m + m^2$$

$$\phi_3''(m) = 8 - 6m$$

$$\phi_1(m) = -1 + 5m$$

$$\phi_2'(m) = -7 + 2m$$

$\phi_3(m) = 0$ gives

$$2 - 5m + 4m^2 - m^3 = 0$$

$$\Rightarrow (2 - m)(1 - m)^2 = 0$$

so that $m = 1, 1, 2$.

'c' giving equation is

$$c\phi_n''(m) + \phi_{n-1}(m) = 0$$

which in this case becomes

$$c\phi_3(m) + \phi_2(m) = 0$$

$$\Rightarrow \quad c\left(-5+8m-3m^2\right)+\left(6-7m+m^2\right)=0$$

$$\Rightarrow \quad c=-4 \text{ when } m=2$$

So that the corresponding asymptote is

$$y=2x-4$$

When $m=1$, thc 'c' giving equation becomes

$$\frac{1}{2}c^2\phi_3''(m)+c\phi_2'(m)+\phi_1(m)=0$$

$$\Rightarrow \quad \frac{1}{2}c^2(8-6m)+c(-7+2m)+(-1+5m)=0$$

When $m=1$ this becomes,

$$c^2-5c+4=0$$

so that $c=1,4$.

Therefore, the corresponding asymptotes are

$$y=x+1 \text{ and}$$

$$y=x+4$$

Example 2: Find the asymptotes of

$$x^3+4x^2y+4xy^2+5x^2+15xy+10y^2-2y+1=0$$

Solution: Equating to zero the co-efficient of the highest power of y^2 of y we see that ,

$$4x+10=0$$

$$\Rightarrow \quad 2x+5=0$$

is one asymptote.

Factorizing the highest degree terms, we get

$$x(2y+x)^2+5x^2+15xy+10y^2-2y+1=0$$

Here $2y+x$ is a repeated linear factor of highest degree terms i.e., third degree.

Therefore, there will be no asymptote parallel to $2y+x=0$ if $(2y+x)$ is not a factor of the 2nd degree terms also.

But this is not the case.

In fact, the equation is

$$x(2y+x)^2+5(x+y)(x+2y)-2y+1=0$$

Therefore, the curve has two asymptotes parallel to

$$2y+x=0$$

Let $\lim\left(y+\frac{1}{2}x\right)=c$ when $x\to\infty$ and $y/x\to-\frac{1}{2}$ so that

$\lim(2y+x)=2c$,

Dividing by x, the equation becomes

$$(2y+x)^2+5(2y+x)(1+y/x)-2y/x+1/x=0$$

In the limit,

$$4c^2+5.2c\left(1-\frac{1}{2}\right)-2\left(-\frac{1}{2}\right)+0=0 \quad \ldots(i)$$

$$\Rightarrow\ 4c^2+5c+1=0$$

$$\therefore\ c=-\frac{1}{4},-1.$$

Hence, $y=-\frac{1}{2}x-\frac{1}{4}$ and

$$y=-\frac{1}{2}x-1$$

$$\Rightarrow\ 4y+2x+1=0 \text{ and}$$

$$2y+x+2=0$$

are the two more asymptotes.

Now, the asymptotes are

$$(2y+x)^2+5(2y+x).\lim(1+y/x)+\lim(-2y/x+1/x)=0$$

$$\Rightarrow\ (2y+x)^2+5(2y+x).\frac{1}{2}+1=0$$

$$\Rightarrow\ 2(2y+x)^2+5(2y+x)+2=0$$

which gives

$$2y+x+2=0 \text{ and}$$

$$4y+2x+1=0$$

Example 3: Find the asymptotes of the quadratic curve

$$x^4-2x^3y+2xy^3-y^4+x^3-4x^2y+5xy^2-2y^3+x^2-3xy+2y^2+1=0$$

Solution: We have

$$\phi_4(m)=1-2m+2m^3-m^4 \qquad \phi_4'(m)=-2+6m^2-4m^3$$

$$\phi_3(m)=1-4m+5m^2-2m^3 \qquad \phi_4''(m)=12m-12m^2$$

$$\phi_2(m)=1-3m+2m^2 \qquad \phi_4'''(m)=12-24m$$

$$\phi_1(m)=0 \qquad \phi_3'(m)=-4+10m-6m^2$$

$$\phi_3''(m)=10-12m$$

$$\phi_2'(m)=-3+4m$$

$$\phi_4(m)=0$$

$$\Rightarrow (1-m)^3(1-m)=0$$

$$\Rightarrow m=-1,1,1,1$$

'c' giving equation is

$$c\phi_4'(m)+\phi_3(m)=0$$

$$\Rightarrow c\left(-2+6m^2-4m^3\right)+1-4m+5m^2-2m^3=0$$

For $m=-1$, $c=-3/2$

The corresponding asymptote is

$$y=-x-3/2$$

When $m=1$, we consider the equation

$$\frac{1}{2}c^2\phi_4''(m)+c\phi_3'(m)+\phi_2(m)+0$$

$$\Rightarrow \frac{1}{2}c^2\left(12m-12m^2\right)+c\left(-4+10m-6m^2\right)+\left(1-3m+2m^2\right)=0$$

which also vanishes identically when $m = 1$.

We, therefore consider the equation

$$\frac{1}{6}c^3\phi_4'''(m) + \frac{1}{2}c^2\phi_3''(m) + c\phi_2'(m) + \phi_1(m) = 0$$

$$\Rightarrow \frac{1}{6}c^3(12m - 24m) + \frac{1}{2}c^2(10 - 12m) + c(-3 + 4m) = 0$$

When $m = 1$, this becomes

$$-2c^3 - c^2 + c = 0$$

$$\Rightarrow \quad c = 0, \frac{1}{2}, -1.$$

Therefore, there are three asymptotes parallel to $y = x$, namely

$$y = x, \quad y = x + \frac{1}{2}, \quad y = x - 1$$

Example 4: Find the asymptotes of

$$(x - y)^2(x^2 + y^2) - 10(x - y)x^2 + 12y^2 + 2x + y = 0$$

Solution: The asymptotes parallel to the two imaginary lines $x^2 + y^2 = 0$ are imaginary.

To obtain the two asymptotes parallel to the lines

$$x - y = 0$$

We rewrite the equation, on dividing it by $(x^2 + y^2)$ as

$$(x - y)^2 - 10(x - y)\frac{1}{1 + (y/x)^2} + \frac{12(y/x)^2 + 2/x + y/x.1/x}{1 + (y/x)^2} = 0$$

We take the limits when $x \to \infty$ and $y/x \to 1$.

Therefore, the asymptotes are

$$(x - y)^2 - 10(x - y)\lim\frac{1}{1 + (y/x)^2} + \lim\frac{12(y/x)^2 + 2/x + y/x.1/x}{1 + (y/x)^2} = 0$$

$$\Rightarrow \quad (x-y)^2 - 5(x-y) + 6 = 0$$

$$\Rightarrow \quad x-y-2=0, \quad x-y-3=0$$

Example 5: Find the asymptotes of the curve

$$(x+y+1)\left(x^2-xy+y^2\right)-6xy+2x^2+y^2+3x-7y+4=0$$

Solution: The asymptote parallel to $x+y+1=0$ is given by

$$x+y+1 = -\lim_{\substack{x,y\to\infty \\ y/x\to -1}} \frac{-6xy+2x^2+y^2+3x-7y+4}{x^2-xy+y^2}$$

$$= -\lim_{\substack{x,y\to\infty \\ y/x\to -1}} \frac{-6(y/x)+2+(y/x)^2+3/x-7\left(y/x^2\right)+4\left(1/x^2\right)}{1-(y/x)+(y/x)^2}$$

$$=3$$

$$\Rightarrow x+y-2=0$$

Since, $\phi_3(m) = (2+m)\left(1-m+m^2\right)$

therefore, $\phi_3(m)=0$ has no real roots other than -1 and therefore $x+y-2=0$ is the only asymptote.

Example 6: Find the asymptotes of the curve

$$(x+y)^2\left(x^2+y^2\right)-8(x+y)y^2+4x^2-3xy-y^2+4x+3=0$$

Solution: The two asymptotes parallel to $x+y=0$ are given by

$$(x+y)^2-(x+y)\lim_{\substack{x\to\infty \\ y/x\to -1}} \frac{8y^2}{x^2+y^2}$$

$$+\lim_{x,\,y\to\infty} \frac{4x^2-3xy-y^2+4x+3}{x^2+y^2}=0$$

$$\Rightarrow (x+y)^2 - 4(x+y) + 3 = 0$$

so that the asymptotes parallel to $x+y=0$ are

$$x+y-1=0,$$

$$x+y-3=0$$

Since $\phi_4(m) = (1+m)^2(1+m^2)$ has only two roots real, therefore there are no other asymptotes.

Example 7: Find the asymptotes of the curve

$$(x^2-y^2)^2 - x^2 - 3y^2 + 2x = 0$$

Solution: The equation of the curve is of the form

$$F_4 + F_2 = 0$$

Since the factors of F_4 are repeated, therefore corresponding to each repeated factor, there are two parallel asymptotes.

Writing the equation of the curve as

$$(x-y)^2 = (x^2+3y^2-2x)/(x+y)^2$$

We find that the two asymptotes parallel to $x-y=0$ are

$$x-y = \pm\left[\lim_{\substack{x,y\to\infty \\ y/x\to -1}} (x^2+3y^2-2x)/(x+y)^2\right]^{1/2}$$

$$= \pm 1$$

Similarly the two asymptotes parallel to $x+y=0$ are

$$x+y = \pm 1$$

Example 8: Find the asymptotes of the curve

$$x^2y - xy^2 + xy + y^2 + x - y = 0$$

and show that they cut the curve again in the three points which lie on the line $x+y=0$.

Solution: The asymptotes of the given curve, are

$$y = 0, \quad x = 1, \; x - y + 2 = 0$$

The joint equation of the asymptotes is

$$y(x-1)(x-y+2) = 0$$

$$\Rightarrow \; x^2y - xy^2 + xy + y^2 - 2y = 0$$

The equation of the curve can be written as

$$x^2y - xy^2 + xy + y^2 - 2y + (x+y) = 0$$

Here, $F_3 = x^2y - xy^2 + xy + y^2 - 2y$

$$F_1 = x + y$$

Hence, the points of intersection lie on the line

$$F_1 = x + y = 0$$

Example 9: Show that the four asymptotes of the quadratic curve

$$(x^2 - y^2)(x^2 - 4y^2) + 2x^3 - 3x^2y - 5xy^2 + 6y^3 + y^2 - 3xy + 1 = 0$$

cut the curve in points which lie on a circle of unit radius.

Solution: The asymptotes of the curve

$$(x^2 - y^2)(x^2 - 4y^2) + 2x^3 - 3x^2y - 5xy^2 + 6y^3 + y^2 - 3xy + 1 = 0 \text{ are}$$

$$x - y = 0$$

$$x + y + 1 = 0$$

$$x - 2y = 0$$

$$x + 2y + 1 = 0$$

The joint equation of the asymptotes is

$$F_4 = (x-y)(x+y+1)(x-2y)(x+2y+1)$$

$$= (x^2 - y^2)(x^2 - 4y^2) + 2x^3 - 3x^2y - 5xy^2 + 6y^3$$

$$+ x^2 - 3xy + 2y^2 = 0$$

The equation of the curve can be re-arranged and written as

$$F_4 - (x^2 + y^2 - 1) = 0$$

The eight points of intersection of

$$F_4 = 0 \text{ and}$$

$$F_4 - (x^2 + y^2 - 1) = 0$$

lie on the unit circle $x^2 + y^2 - 1 = 0$.

Example 10: Find the equation of the cubic which has the same asymptotes as the curve

$$x^3 - 6x^2y + 11xy^2 - 6y^3 + x + y + 1 = 0$$

and which passes through the points, $(0,0,)$, $(1,0)$, and $(0,1)$

Solution: We write,

$$F_3 = x^3 - 6x^2y + 11xy^2 - 6y^3$$

$$= (x - y)(x - 2y)(x - 3y)$$

$$F_1 = x + y + 1$$

The equation of the curve can be written in the form

$$F_3 + F_1 = 0$$

where F_3 has non-repeated linear factors.

Thus, $F_3 = 0$ is the joint equation of the asymptotes of the cubic.

The general equation of the cubic is of the form

$$F_3 + ax + by + c = 0$$

$$\Rightarrow \quad x^3 - 6x^2y + 11xy^2 - 6y^3 + ax + by + c = 0$$

where $ax + by + c$ is the general linear expression.

In order that it may pass through the points $(0,0,)$, $(1,0)$, and $(0,1)$ we must have

$$c = 0$$

$$1 + a = 0 \quad \Rightarrow a = -1$$

$$-6+b=0 \Rightarrow b=6$$

Thus the required cubic is

$$x^3-6x^2y+11xy^2-6y^3-x+6y=0$$

Example 11: Find the asymptotes of the curve

$$y^2=x(x-a)(x-2a)/(x+3a)$$

and determine on which side of the asymptotes the curve lies.

Solution: We have,

$$y=\pm\sqrt{\left[\frac{x(x-a)(x-2a)}{x+3a}\right]}$$

$$\pm x\left(1-\frac{a}{x}\right)^{1/2}\left(1-\frac{2a}{x}\right)^{1/2}\left(1-\frac{3a}{x}\right)^{-1/2}$$

$$\pm x\left(1-\frac{a}{2x}-\frac{a^2}{8x^2}\cdots\right)\left(1-\frac{a}{x}-\frac{a^2}{2x^2}\cdots\right)\left(1-\frac{3a}{2x}-\frac{27a^2}{8x^2}\cdots\right)$$

$$\pm x\left(1-\frac{3a}{x}-\frac{11a^2}{2x^2}\cdots\right)$$

thus, we have two values of y.

$$y=x-3a+\frac{11}{2}a^2x\ldots.$$

$$y=-x+3a-\frac{11}{2}a^2x\ldots.$$

therefore,

$$y=x-3a,$$

$$y=-x+3a$$

are the two asymptotes.

The difference between the ordinate of the curve and that of the asymptote $y=x-3a$ being

$$\frac{11}{2}u^2x \ldots\ldots\ldots$$

We see that the curve lies above the asymptote when x is positive and below it when x is negative.

Similarly, it may be seen that the curve lies below the second asymptote when x is positive and above it when x is negative.

It is easy to see that

$$x - 3a$$

is also an asymptote of the curve.

For determination of the position of the curve relative to this asymptote, we suppose that

$$x = -3a + A/y + B/y^2 + C/y^3 + \cdots$$

Substituting this value of x in the equation of the curve, we have

$$y^2\left[\frac{A}{y} + \frac{B}{y^2} + \frac{C}{y^3} + \ldots\ldots\right]$$

$$= \left[-3a + \frac{A}{y} + \cdots\right]\left[-4a + \frac{A}{y} + \cdots\right]\left[-5a + \frac{A}{y} + \cdots\right]$$

Equating the co-efficients of like powers of y, we have

$$A = 0$$

$$B = -60a^3, \text{ etc.}$$

$$\therefore \quad X = -3a \frac{-60a^3}{y^2} + \cdots\cdots$$

The difference between the abscissa of the curve and the asymptote $x - 3a$, for the same value of y, being

$$-60a^3/y^2 + \cdots\cdots$$

Which is negative whether y be positive or negative.

Example 12: Find the oblique asymptotes of the curve $x^3 - xy^2 + y^2 = 0$ and find the position of the curve relative to them.

Solution: The equation of the curve can be written as

$$y^2 = x^3/(x-1)$$

$$\Rightarrow \quad y = \pm x\left(1-\frac{1}{x}\right)^{-1/2}$$

$$= \pm x\left[1+\frac{1}{2}(1/x)+\frac{3}{8}(1/x)^2+\frac{5}{16}(1/x)^3+\cdots\right]$$

$$= \pm x\left[x+\frac{1}{2}+\frac{3}{8}(1/x)+\frac{5}{16}(1/x)^2\cdots\right]$$

showing that the curve has two branches whose equations can be written as

$$y = x+\frac{1}{2}+\frac{3}{8}(1/x)+\frac{5}{16}(1/x)^2+ \qquad \ldots(1)$$

and $y = -x-\frac{1}{2}-\frac{3}{8}(1/x)-\frac{5}{16}(1/x)^2- \qquad \ldots(2)$

From (1), we find that

$$y = x+\frac{1}{2}$$

is an asymptote to this branch (say A)

From (2), we find that

$$y = -x-\frac{1}{2}$$

is an asymptote to this branch (say B)

For branch A,

$$y_1 - y_2 = \frac{3}{8}(1/x)+\frac{5}{16}(1/x)^2+\cdots\cdots$$

so that the curve is above the asymptote when $x > 0$ and below the asymptote when $x < 0$.

For branch B,

$$y_1 - y_2 = -\frac{3}{8}(1/x)-\frac{5}{16}(1/x)^2\cdots\cdots$$

so that the curve is below the asymptote when $x > 0$ and above the asymptote when $x < 0$.

Example 13: Find the asymptote of the curve

$$r = \frac{a}{\frac{1}{2} - \cos\theta}$$

Solution: Here,

$$u = \frac{1}{r}$$

$$= \frac{1}{a}\left(\frac{1}{2} - \cos\theta\right)$$

when $u \to 0, \left(\frac{1}{2} - \cos\theta\right) \to 0$

so that $\cos\theta \to \frac{1}{2}$

$$\therefore \quad \theta_1 = \pm \pi/3.$$

now, $\frac{dv}{d\theta} = \frac{1}{a}\sin\theta$

$$\Rightarrow \quad -\frac{d\theta}{du} = -\frac{a}{\sin\theta}$$

$$\therefore \quad -\frac{d\theta}{dv} \to -\frac{2a}{\sqrt{3}} \quad \theta \to \frac{\pi}{3},$$

and $\quad -\frac{d\theta}{du} \to -\frac{2a}{\sqrt{3}} \quad \theta \to -\frac{\pi}{3}.$

$$\frac{-2a}{\sqrt{3}} = r\sin\left(\frac{\pi}{3} - \theta\right) \quad 4a = r\left(\sqrt{3}\sin\theta - 3\cos\theta\right),$$

and $\quad \frac{-2a}{\sqrt{3}} = r\sin\left(-\frac{\pi}{3} - \theta\right) \quad 4a = r\left(\sqrt{3}\sin\theta + 3\cos\theta\right)$

are the two asymptotes.

Example 14: Find the asymptotes of the curve $r = a\theta^2/(\theta - 1)$

Solution: Putting $r = 1/u$ in the equation of the curve, the equation becomes

$$u = (\theta - 1)/(a\theta^2)$$

now, $u = 0$ if $\theta = 1$

therefore $\theta_1 = 1$

also, $\dfrac{du}{d\theta} = \dfrac{\theta^2 - 2\theta(\theta - 1)}{a\theta^4}$

$$\therefore \quad p = \lim_{\theta \to 1}\left(\frac{-d\theta}{du}\right)$$

$$= -a$$

Therefore the equation of the asymptote is

$$r\sin(\theta_1 - \theta) = p$$

$$r\sin(1 - \theta) = -a$$

$$r\sin(\theta - 1) = a$$

Exercise – 14.2

Find the asymptotes of the following curves:

1. $x(x+y=0)^2 - x + 1 = 0$.

2. $xy(x+y) = a\left(x^2 - y^2\right)$.

3. $y^3 - xy^2 - x^2y + x^3 + x^2 - y^2 - 1 = 0$.

4. $(x-1)(x-2)(x+y) + x^2 + x + 1 = 0$.

5. $(x-y)^2\left(x^2+y^2\right) - 10(x-y)x^2 + 12y^2 + 2x + y = 0$.

6. $y^3 - x^3 + y^2 + x^2 + y - x + 1 = 0$.

7. $(2x-3y)^2(x+y+1) = x + 9y + 8$.

8. $x\left(y^2 - 3by + 2b^2\right) = y^3 - 3bx^2 + b^3$.

9. $x^3 + 4x^2y + 5xy^2 + 2y^3 + 2x^2 + 4xy + 2y^2 - x - 9y + 1 = 0$.

10. $x^2(3y+x)^2 + (3y+x)\left(x^2+y^2\right) + 9y^2 + 6xy + 9y - 6x + 9 = 0$.

11. $y^3 + 4xy^2 + 4x^2y + 5y^2 + 15xy + 10x^2 - 2x + 1 = 0$.

12. $\left(y^2 + xy - 2x^2\right)^2 + \left(y^2 + xy - 2x^2\right)(2y + x)$

    $-7y^2 - 19xy - 28x^2 + x + 2y + 3 = 0$.

13. $\left(x^2 - y^2\right)^2 - 2\left(x^2 + y^2\right) + x - 1 = 0$.

14. $x(y-3)^3 = 4y(x-1)^3$.

15. $x^4 + 2x^3y - 2xy^3 - y^4 - x^2 + y^2 + 3x + 3y = 0$.

16. $(a+x)^2\left(b^2 + x^2\right) = x^2y^2$.

17. $x^4 - 2xy^3 + 2x^3y - x^4 - 3x^3 + 3x^2y + 3xy^2 - 3y^3 - 2x^2 + 2y^2 - 1 = 0$

18. $y^3 - 5xy^2 + 8x^2y - 4x^3 - 3y^2 + 9xy - 6x^2 + 2y - 2x + 1 = 0$.

19. $y\left[(y-1)^2 - x^2\right] = 1$.

20. Show that the asymptotes of the cubic

    $x^3 - xy^2 - 2xy + 2x - y = 0$.

    cut the curve again in points which lie on the line.

    $3x - y = 0$.

21. Find the asymptotes of the curve $x^2y - xy^2 + xy + y^2 + x - y = 0$ and show that they cut the curve again in three points , which lie on the curve $x + y = 0$.

22. Show that $y = x + a$ is the only asymptote meets the curve in P, Q; show that the midpoint of PQ lies on the hyperbola

    $x(x - y) + ay = 0$.

23. Find the equation of the cubic which has the same asymptotes as the curve $x^3 - 6x^2y + 11xy^2 - 6y^3 + x + y + 1 = 0$ and which passes through the points $(0,0), (1,0)$ and $(0,1)$.

24. Find the equation of the straight line on which lie three points of intersection of the cubic

    $x^3 + 2x^2y - xy^2 - 2y^3 + 4y^2 + 2xy + y - 1 = 0$.

    and its asymptotes.

25. Find the equation of the quadratic curve which has $x = 0, y = 0, y = x, y = -x$ as asymptotes, which passes through the point (a, b) and cuts its asymptotes again in eight points lying on the circle

    $x^2 + y^2 = a^2$.

26. Show that the four asymptotes of the curve

    $xy(x^2 - y^2) - 25x^2 - 9y^2 + 144 = 0$.

    cut it again in eight points lying on an ellipse whose eccentricity is $4/5$.

27. Find the asymptotes of the curve

$$(2x-3y+1)^2(x+y)-8x+2y-9=0.$$

and show that they intersect the curve again in three points, which lie on a straight line. Obtain the equation of the line.

28. Show that the asymptotes of the quadratic curve

$$x^2y^2-a^2\left(x^2+y^2\right)-a^3(x+y)+a^4=0.$$

form a square, two of whose vertices lie on the curve.

29. Find the asymptotes of the curve

$$4x^4-13x^2y^2+9y^4+32x^2y-42y^3$$

$$-20x^2+74y^2-56y-4x+16=0.$$

and show that they pass through the intersection of the curve with the parabola

$$y^2=4x.$$

30. Find the asymptotes of the curve

$$4\left(x^4+y^4\right)-17x^2y^2-4x\left(4y^2-x^2\right)+2\left(x^2-2\right)=0.$$

and show that they pass through the points of intersection of the curve with the ellipse $x^2+4y^2=4$.

31. *Find the asymptotes and their position with regard to the following curves*

(i) $x^3+y^3=3ax^2$.

(ii) $x^3+y^3=3axy,\ a>0$.

(iii) $x^2(x-y)+y^2=0$.

(iv) $x^3-y^3+a^2(x-2y)=0$.

(v) $y^2(x+4a)=x\left(x^2-a^2\right)$.

(vi) $y^2x-x^3+3a\left(x^2+y^2\right)-2a^2x=0$.

Find the asymptotes of each of the following curves

32. $r\theta=a$.

33. $r\ln\theta=a$.

34. $r=a\log\theta$.

35. $r\log\theta=a$.

36. $r\sin\theta = 2\cos 2\theta$.

37. $r\sin 2\theta = a$.

38. $r = a + b\cot n\theta$.

39. $r = a\sec\theta + b\tan\theta$.

40. $r\sin\theta = ae^{\theta}$.

41. $r\cos\theta = a\sin^2\theta$.

42. $r^2 = a^2\left(\sec^2\theta + \operatorname{cosec}^2\theta\right)$.

43. $2r^2 = \tan 2\theta$.

44. $r(\pi + \theta) = ae^{\theta}$.

45. $r\left(\theta^2 - n^2\right) = 2a\theta$.

46. $r^n \sin n\theta = a^n$.

47. $r\sin 4\theta = a\sin 3\theta$.

48. $r = \dfrac{a\theta}{\theta - 1}$

49. $r = \dfrac{3a\sin\theta.\cos\theta}{\sin^3\theta + \cos^3\theta}$.

50. $r = \dfrac{3a\sin\theta.\cos\theta}{\cos^3\theta + \sin^3\theta}$.

51. Find the equation of the asymptutes of the curve given by the equation

$$r^n f_n(\theta) + r^{n-1} f_{n-1}(\theta) + \cdots\cdots + f_0(\theta) = 0.$$

52. Show that all the asymptotes of the curve

$r\tan n\theta = a$.

touch the circle

$r = a/n$.

53. Find the asymptotes of the curve

$r\cos 2\theta = a\sin 3\theta$.

15

Curve Tracing

15.1. INTRODUCTION

We have seen that how to determine the tangents and normals, asymptotes, maxima and minima, points of inflexion, singular points etc. of a curve in the previous chapters. The general problem of curve tracing in its elementary aspects, will be taken up in this chapter.

15.2. PROCEDURE FOR TRACING CARTESIAN EQUATIONS

1. ***Symmetrics:***

(i) A curve $f(x, y) = 0$ is said to be symmetrical with respect ot the x-axis if $(x, -y)$ is also a point on the graph whenever (x, y) is a point on it.

If a curve is symmetrical with respect to the x-axis, it is enough to first sketch the portion of the graph above the x-axis and then sketch its reflection in the x-axis.

Example: The curve $y^2 = x$ is symmetrical with respect to the x-axis.

(ii) *If a curve is symmetrical with respect ot the y-axis, it is enough to first sketch the portion of the graph to the right of the y-axis, and then sketch reflection in the y-axis.*

Example: The curve $y - x^2 = 0$ is symmetrical with respect to the y-axis.

(iii) *If a curve is symmetrical with respect to both the axes, then it is enough to first sketch the portion of the graph in the first quadrant, and then take its reflection in the y-axis to get the portion of the graph in the second quadrant. To complete the graph, we take the reflection in the x-axis of the portions of the graph in the first and the second quadrants.*

Example: The curve $x^2y^2 = a^2x^2 + b^2y^2$ is symmetrical with respect to both the axes.

(iv) *If the equation of a curve remains unaltered when x is replaced by –x and y is replaced by –y throughout, then the portion of the graph in the third quadrant can be obtained by rotating the portion of the graph in the first quadrant through an angle of 180° about the origin.*

Example: The curve $y = x^3$ is symmetrical with respect to the origin.

(v) *If a curve is symmetrical with respect to the line $y = x$, it is enough to sketch the portion of the graph on one side of the line $y = x$ and then complete the graph by taking the reflection of this portion in the line $y = x$.*

Example: The curve $x^3 + y^3 = 3axy$ is symmetrical with respect to the line $y = x$.

2. *Quadrants:*

Sometimes it is possible to see that graph lies in certain quadrants only and that no part of the graph lies in certain other quadrants. For example, no portion of the graph of the curve $y^2(4 - x) = x^3$ lie in the second and the third quadrants.

3. *Point of Intersection with the axes of co-ordinates:*

To find the points on the x-axis, solve $f(x, y) = 0$ and $y = 0$ together.

To find the points on the y-axis, solve $f(x, y) = 0$ and $x = 0$ together. The points on the graph lying on the axes of co-ordinates act as gates through which the graph enters from one quadrant to another.

4. *Discontinuities:*

Find the values of x which give discontinuitievs. Find the behavior of the curve near the points of discontinuity.

5. *Shape of the curve bear the origin:*

If the curve passes through origin, the shape of the curve near the origin can be obtained by retaining only the most predominant terms and dropping out the other terms.

6. ***Behavior for large x:***

The appıuximatr shape of the curve for large values of x can be obtained by retaining the term involving the hlglicst power of x.

7. ***Asymptotes:***

Find the asymptotes of the curve. Find the position of the curve with respect to each asymptote.

8. ***Points of Inflexion:***

Examine the curve for points of inflexion by calculating d^2y/dx^2. Recall that at a point of inflexion (x_0, y_0), d^2y/dx^2 vanishes and changes sign as (x, y) passes through (x_0, y_0).

9. ***Multiple Points:***

Examine the curve for multiple points, if any. For each multiple point find whether it is a node, cusp or conjugate point. For each cusp, find whether it is of the first or second species, whether it is a single or a double cusp or whether it is a point of oscu-inflexion. Also, determine the shape of the curve near the cusp.

10. ***Maxima and Minima:***

Examine the curve for cirtical points. For each critical point, find whether the curve has a maximum or a minimum there.

15.3. EQUATIONS OF THE FORM $y = f(x)$

It may be pointed out that a curve $y = f(x)$ where f is a polynomial function has no asymptotes.

Example 1: Trace the curve $y = x^2$.
Solution: We have, $y = x^2$

The following points about the curve can be noted.

(i) The graph is symmetric with respect to the y-axis.

(ii) Since y is real for all $x \in R$, therefore the domain is R.

(iii) No portion of the graph lies in the third and the fourth quadrants.

(iv) The curve meets the axes of co-ordinates only at the origin.

(v) The tangent at the origin is the line $y = 0$.

(vi) $y \to +\infty$ as $x \to +\infty$ or $x \to -\infty$

(vii) dy/dx vanishes at $x = 0$, and changes sign from negative to positive as x passes through zero. Therefore y has a minimum at $x = 0$.

(viii) There are no points of inflexion.

(ix) There are no asymptotes.

A rough sketch fo the graph can be easily drawn in Fig. 15.1.

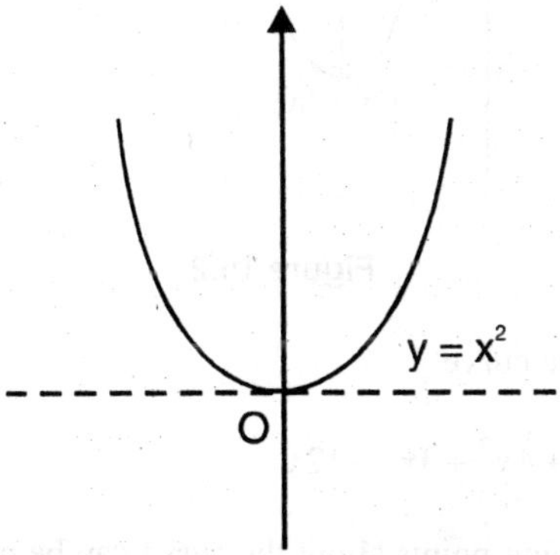

Figure 15.1

Example 2: Trace the curve $y = x^2(x - 3a),\ a > 0$.

Solution: The following points about the given curve can be noted.

(i) The curve does not possess any symmetries.

(ii) No portion of the graph lies in the second quadrant.

(iii) The curve meets the x-axis at points (0, 0), ($3a$, 0).
It meets the y-axis only at the origin.

(iv) The tangent at the origin is the line $y = 0$.

(v) $\dfrac{dy}{dx} = 3x(x - 2a)$

since $dy/dx > 0$ when $x > 0$, therefore y is increasing in $]-\infty, 0]$
since $dy/dx < 0$ in $]0, 2a[$, therefore y is decreasing in $[0, 2a]$
Again since $dy/dx > 0$ if $x > 2a$, therefore y is increasing in $]2a, \infty[$

(vi) The curve has a maximum at (0, 0) and a minimum at $(2a, -4a^3)$

(vii) $y \to -\infty$ as $x \to -\infty$ and
$y \to +\infty$ as $x \to +\infty$

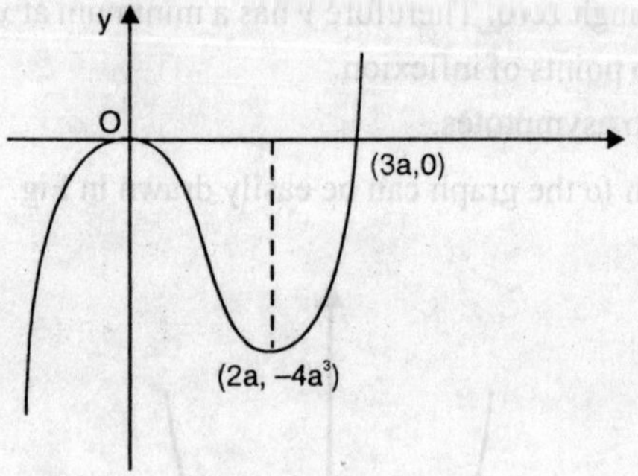

Figure 15.2

Example 3: Trace the curve

$$y = -\frac{3}{2}x^4 + 4x^3 + 3x^2 - 12x$$

Solution: The following points about the curve can be noted.

(i) The curve is not symmetrical about any line.

(ii) Origin lies on it and it meets co-ordinate axes in origin only.

(iii) $$\frac{dy}{dx} = -6x^3 + 12y^2 + 6y - 12$$

$$= 6(x-2)(x-1)(x+1)$$

$$\frac{d^2y}{dx^2} = -18x^2 + 24x + 6$$

$$= 6\left(-3x^2 + 4x + 1\right)$$

The following is the table of variations:

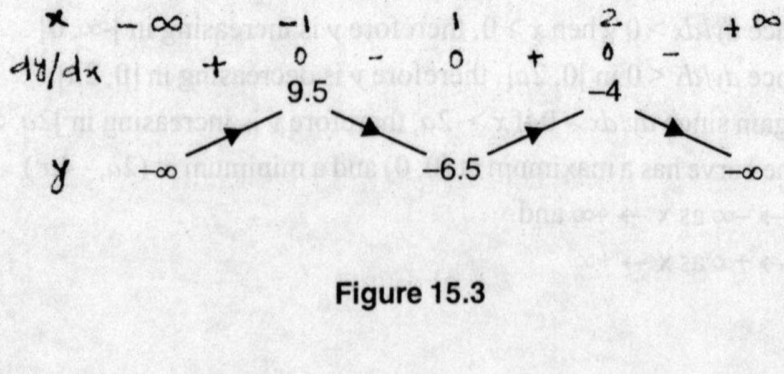

Figure 15.3

Thus, y is strictly increasing in $]-\infty, -1]$ and $[1, 2]$ and decreasing in $[-1, 1]$ and $[2, +\infty[$.

Also y is maximum for $x = -1$ and 2 and minimum for $x = 1$.,

(iv) $d^2y/dx^2 = 0$ if and only of $x = \frac{1}{3}\left(2 \pm \sqrt{7}\right)$

since $\sqrt{7}$ =2.6 ...., we may see that

$\frac{1}{3}\left(2-\sqrt{7}\right) = -.2$ .... and

$\frac{1}{3}\left(2+\sqrt{7}\right) = 1.5$ ....

Also, d^2y/dx^2 changes sign as x passes through $\frac{1}{3}\left(2-\sqrt{7}\right)$ and $\frac{1}{3}\left(2+\sqrt{7}\right)$.

We have,

$$d^2y/dx^2 < 0 \ \forall x \in \left]-\infty, \tfrac{1}{3}\left(2-\sqrt{7}\right)\right[$$

$$d^2y/dx^2 < 0 \ \forall x \in \left[\tfrac{1}{3}\left(2+\sqrt{7}\right), +\infty\right[$$

$$d^2y/dx^2 > 0 \ \forall x \in \left]\tfrac{1}{3}\left(2-\sqrt{7}\right), \tfrac{1}{3}\left(2+\sqrt{7}\right)\right[$$

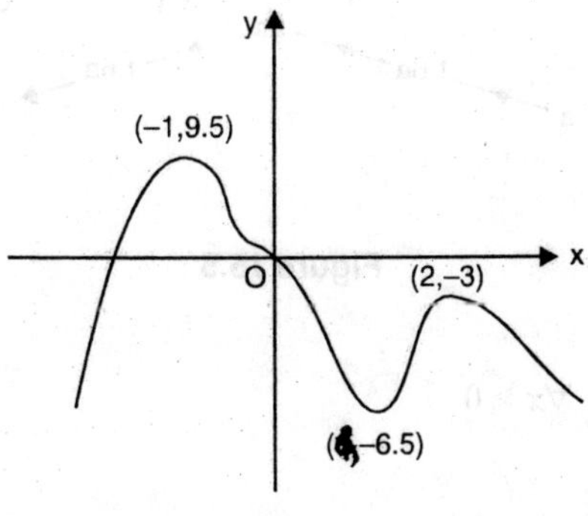

Figure 15.4

Thus, the curve is concave downwards in $\left]-\infty, \frac{1}{3}\left(2-\sqrt{7}\right)\right[$, upwards in $\left]\frac{1}{3}\left(2-\sqrt{7}\right), \frac{1}{3}\left(2+\sqrt{7}\right)\right[$ and downwards in $\left[\frac{1}{3}\left(2+\sqrt{7}\right), +\infty\right[$.

Also the curve has two points of inflexion corresponding to

$$x = \frac{1}{3}\left(2-\sqrt{7}\right) \text{ and } x = \frac{1}{3}\left(2+\sqrt{7}\right).$$

Example 4: Trace the curve

$$y = \frac{8a^3}{x^2+4a^2}$$

Solution: The following points about the curve can be noted.

(i) The curve is symmetrical about y-axis.

(ii) The curve does not pass through the origin.

(iii) The curve meets y-axis at $(0, 2a)$ and $y = 2a$ is the tangent there at.

(iv) $y = 0$ is the only asymptote to the curve.

(v) y is positive for all values of x. Therefore the curve lies in first and second quadrant only.

(vi) $$\frac{dy}{dx} = \frac{-16xa^3}{\left(x^2+4a^2\right)^2}$$

The following is the table of variations of y.

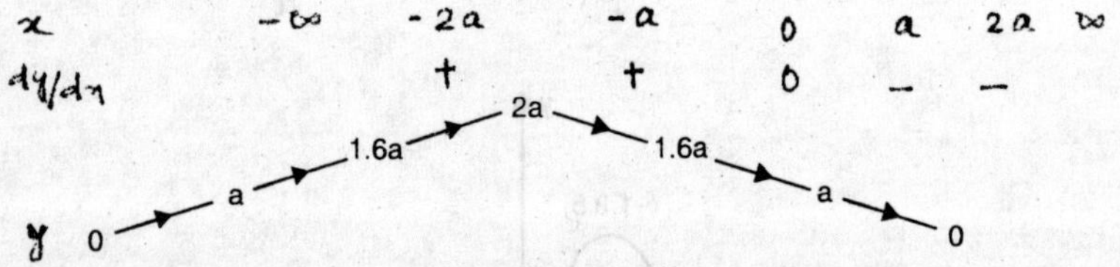

Figure 15.5

$$\therefore \frac{dy}{dx} < 0 \quad \forall x > 0$$

and $$\frac{dy}{dx} > 0 \quad \forall x < 0$$

Thus y is strictly increasing in $]-\infty, 0]$ and strictly decreasing in $[0, \infty[$. Also y is maximum for $x = 0$ and it value there is $2a$.

Thus, we have the curve as given in the figure.

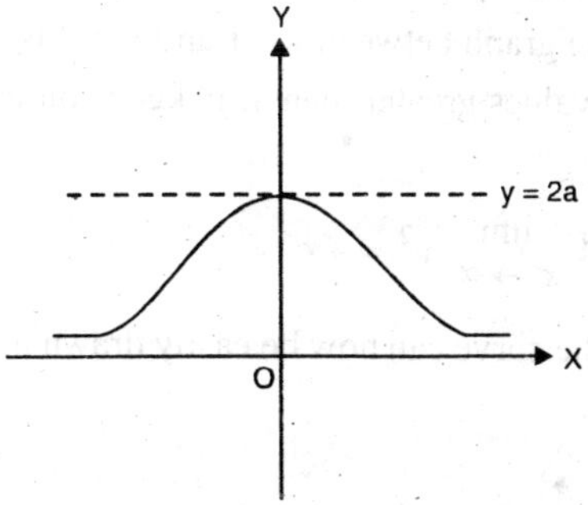

Figure 15.6

15.4. EQUATION OF THE FORM $y^2 = f(x)$

Example 1: Trace the curve

$$y^2 = (x-1)(x-2)(x-4)$$

Solution: The following points about the curve can be noted.

(i) The curve is symmetrical about the x-axis.

(ii) The curve meets the x-axis at the points (1, 0), (2, 0) and (4, 0).

(iii) The curve does not meet the y-axis.

(iv) No part of the graph lies in the second and the third quadrant.

(v) The equations of the tangents at the points (1, 0), (2, 0) and (4, 0) are $x = 1, x = 2$ and $x = 4$ respectively.

(vi) There are no multiple points.

(vii) There are no asymptotes.

(viii) Since y vanishes at $x = 1$ and $x = 2$. therefore y^2 must have a maximum for some value of x lying between 1 and 2. In fact $\frac{d}{dx}\left(y^2\right) = 3x^2 - 14x + 14$, so that

$$\frac{d}{dx}\left(y^2\right) = 0 \text{ when } x = \tfrac{1}{3}\left(7 \pm \sqrt{7}\right)$$

Let $x_1 = \frac{1}{3}\left(7-\sqrt{7}\right)$, $x_2 = \frac{1}{3}\left(7+\sqrt{7}\right)$

Then $1 < x_1 < 2$ and $x_2 > 2$

Also y^2 is a maximum when $x = x_1$.

The portion of the graph between $x = 1$ and $x = 2$ is therefore an oval.

(ix) When x takes values greater than 4, y^2 keeps on increasing. For larger values of x,

$$y^2 \approx x^3 \text{ and } \lim_{x \to \infty} y^2 = +\infty.$$

The graph of the curve can now be easily drawn in fig. 15.7.

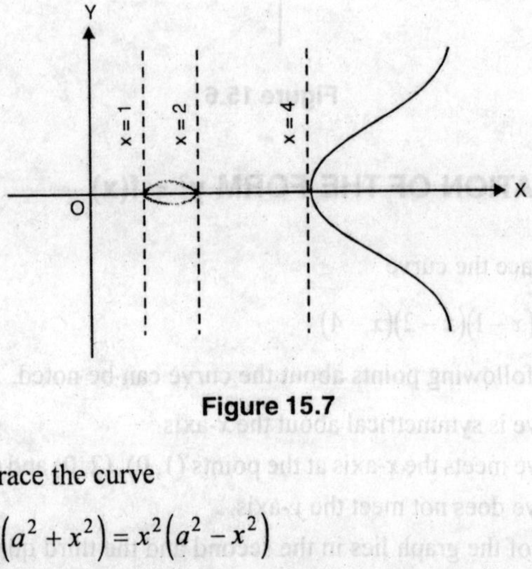

Figure 15.7

Example 2: Trace the curve

$$y^2\left(a^2+x^2\right) = x^2\left(a^2-x^2\right)$$

Solution: The following points about the given curve can be noted.

(i) It is symmetrical about both the axes.

(ii) It passes through the origin and $y = \pm x$ are the two tangents there at. Thus the origin is a node.

(iii) The curve meets x-axis at $(a, 0)$, $(0, 0)$ and $(-a, 0)$. and meets y-axis at $(0, 0)$ only. The tangents at $(a, 0)$ and $(-a, 0)$ are $x = a$ and $x = -a$ respectively.

(iv) The curve has no asymptotes.

(v) $y = \pm x\sqrt{\dfrac{a^2-x^2}{a^2+x^2}}$

y is defined for those values for which

$a^2 - x^2 \geq 0$

$\Leftrightarrow -a \leq x \leq a$

so that the curve lies between the two lines $x = a$ and $x = -a$.

(vi) $$\frac{dy}{dx} = \frac{a^4 - 2a^2x^2 - x^4}{\left(a^2 + x^2\right)^{3/2}\left(a^2 - x^2\right)^{1/2}}$$

$$\frac{dy}{dx} \to \infty \text{ as } x \to -a \text{ when } x \to a$$

Also, $\frac{dy}{dx} = 0$ when $a^4 - 2a^2x^2 - x^4 = 0$

Again,

$$a^4 - 2a^2x^2 - x^4 = -\left[x^4 + 2a^2x^2 - a^4\right]$$

$$= -\left[x^2 - \left(-1 + \sqrt{2}\right)a^2\right]\left[x^2 - \left(-1 - \sqrt{2}\right)a^2\right]$$

$$\left[x^2 + \left(1 + \sqrt{2}\right)a^2\right]$$

Thus, $\frac{dy}{dx} = 0$ for $x = \sqrt{\left(-1 + \sqrt{2}\right)a}$ and $x = -\sqrt{\left(-1 + \sqrt{2}\right)a}$

$$\frac{dy}{dx} < 0 \text{ for } x \in \left]-a, -\sqrt{\left(-1 + \sqrt{2}\right)a}\right[\cup \left]\sqrt{\left(-1 + \sqrt{2}\right)a}, a\right[$$

and $$\frac{dy}{dx} > 0 \text{ when } x \in \left]-\sqrt{\left(-1 + \sqrt{2}\right)a}, \sqrt{\left(-1 + \sqrt{2}\right)a}\right[$$

Thus y is strictly decreasing in $\left[-a, -\sqrt{\left(1 + \sqrt{2}\right)a}\right]$ and $\left[\sqrt{\left(-1 + \sqrt{2}\right)a}, a\right]$ and strictly increasing in $\left[-\sqrt{\left(-1 + \sqrt{2}\right)a}, \sqrt{\left(-1 + \sqrt{2}\right)a}\right]$

We may note that

$$\sqrt{\left(-1+\sqrt{2}\right)} = b$$

Thus, the curve,

$$y^2 = \frac{x^2\left(a^2 - x^2\right)}{a + x^2}$$

is shown in Fig. 15.8.

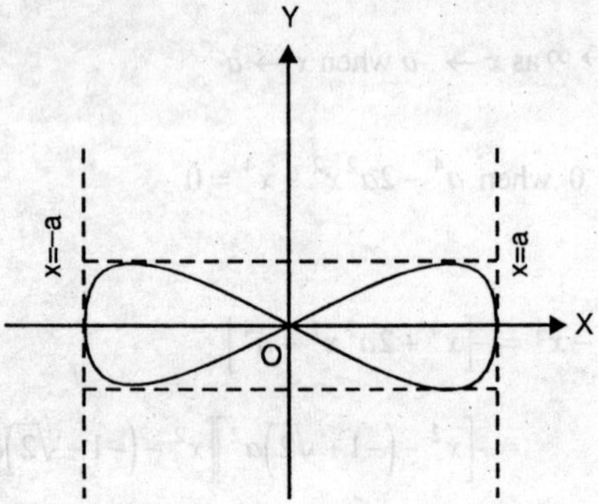

Figure 15.8

Example 4: Trace the curve

$$\left(x^2 - a^2\right)\left(y^2 - b^2\right) = a^2b^2$$

Solution: The following points about the given curve can be noted.

(i) The curve is symmetrical about both the axes.

(ii) Origin is a point on the curve. Since the tangents at the origin are increasing, therefore the origin is a conjugate point.

(iii) The curve does not meet the axes at any other point.

(iv) Rewriting the equation in the form

$$y = \pm\sqrt{\frac{b^2x^2}{x^2 - a^2}}$$

We find that $|x| > a$

Similarly, it can be seen that $|y| > b$

Therefore, no portion of the graph lies in the region bounded by the lines $x = \pm a$, $y = \pm b$

(v) The lines $x = \pm a$, $y = \pm b$ are the asymptotes of the curve. The portion of the curve lying in the first quadrant is above the asymptote $y = b$ and on the right of the asymptote $x = a$.

(vi) In the 1st quadrant, y decreases as x increases.
Also $y \to +\infty$ as $x \to a$ and $y \to b$ as $x \to +\infty$

(vii) The portion of the graph lying in the first quadrant can now be drawn. By using symmetrics, the graph can be completed.
The graph is as shown in fig. 15.9.

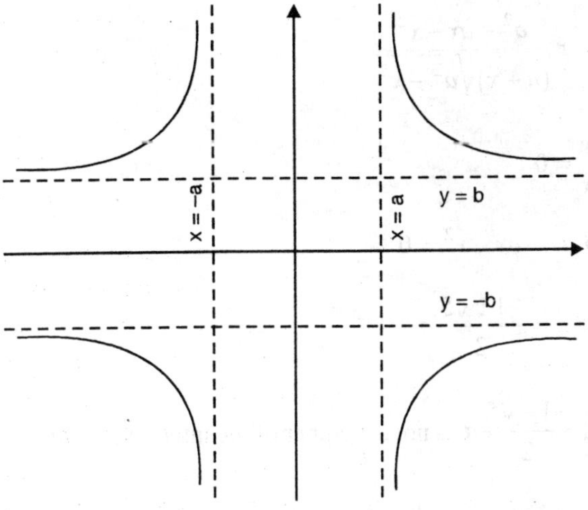

Figure 15.9

Example 5: Trace the curve

$$\left(x^2 + y^2\right)x - a\left(x^2 - y^2\right) = 0, \quad (a > 0)$$

Solution: The following points about the given curve can be noted.

(i) The curve is symmetrical about x-axis.

(ii) Origin lies in the curve. $y = \pm x$ are the two tangents there at so that origin is a node.

(iii) (0, 0) and $(a, 0)$ are its only points of intersection with co-ordinate axes, $x = a$ is the tangent at $(a, 0)$.

(iv) $x = -a$ is an asymptote of the curve.

(v) $y^2 = x^2\left(\dfrac{a-x}{a+x}\right)$

$$\Rightarrow y = \pm x\sqrt{\frac{a-x}{a+x}},\ x \neq -a$$

y is defined only when $x \in]-a, a]$ so that the curve lies between the lines $x = -a$ and $x = a$.

(vi) $\dfrac{dy}{dx} = \pm\dfrac{a^2 - ax - x^2}{(a+x)\sqrt{a^2 - x^2}}$

$$\frac{dy}{dx} = 0$$

$$\Rightarrow a^2 - ax - x^2 = 0$$

$$\Rightarrow x = \frac{-1 \pm \sqrt{5}}{2}a$$

Now, $\dfrac{-1-\sqrt{5}}{2}a$ is not a member of the interval $]-a, a]$

Also $\dfrac{dy}{dx} > 0\ \forall x \in \left]-a, \dfrac{-1+\sqrt{5}}{2}a\right[$

and $\dfrac{dy}{dx} < 0\ \forall x \in \left]\dfrac{-1+\sqrt{5}}{2}a, a\right[$

Thus, y is strictly increasing in $\left]-a, \dfrac{-1+\sqrt{5}}{2}a\right]$ and strictly decreasing in $\left[\frac{-1+\sqrt{5}}{2}a,\ a\right]$.

The point corresponding to $x = \frac{-1+\sqrt{5}}{2}a$ gives a maximum value of y.

Combining the above facts, the curve is as given in the Fig. 15.10.

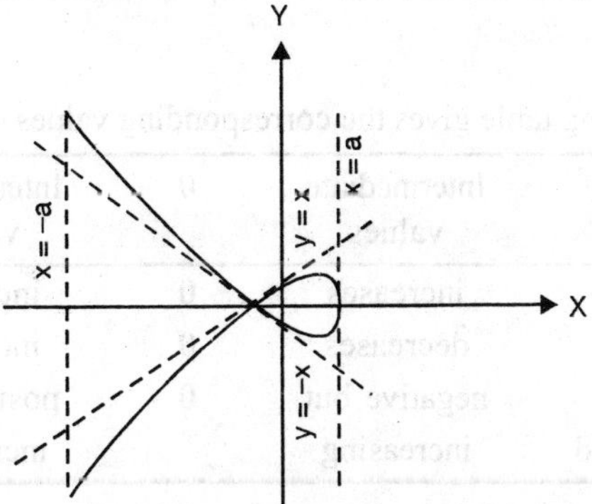

Figure 15.10

15.5. PARAMETRIC EQUATIONS

Example 1: Trace the cycloid.

(i) $x = a(\theta + \sin\theta)$, $y = a(1-\cos\theta)$, as θ varies from $-\pi$ to π.

(ii) $x = a(\theta - \sin\theta)$, $y = a(1-\cos\theta)$, as θ varies from 0 to 2π.

Solution: (i) We have

$$\frac{dx}{d\theta} = a(1+\cos\theta)$$

$$\frac{dy}{d\theta} = a\sin\theta$$

$$\therefore \quad \frac{dy}{dx} = \tan\frac{\theta}{2}$$

Since $\dfrac{dx}{d\theta}$ is positive for all values of θ, therefore as θ increases x increases.

Also, $\dfrac{dy}{d\theta}$ is negative if $-\pi < \theta < 0$ and positive if $0 < \theta < \pi$.

Therefore y decreases as θ increases in $[-\pi, 0]$ and y increases as θ increases in $[0, \pi]$.

The following table gives the corresponding values of θ, x, y and dy/dx.

θ	$-\pi$	Intermediate values	0	Intermediate values	π
x	$-a\pi$	increases	0	increases	$a\pi$
y	$2a$	decreases	0	increases	$2a$
dy/dx	not defined	negative but increasing	0	positve and increasing	not defined

When $\theta = -\pi$, $x = -a\pi$, $y = 2a$ so that $(-a\pi, 2a)$ is a point on the cycloid.

Also, $\dfrac{dy}{dx}$ is not defined but $\dfrac{dy}{dx}$ is zero, so that the tangent is parallel to the y-axis at this point. When θ increases from $-\pi$ to 0, x increases and y decreases, so that the point (x, y) moves to the right and downwards till it reaches $(0, 0)$, where $\dfrac{dy}{dx}=0$ so that the x-axis is the tangent there.

When θ increases from 0 to π, x and y both increases, so that (x, y) moves to the right and upwards till it reaches $(a\pi, 2a)$.

When $\theta = \pi$, $x = a\pi$, $y = 2a$, but $\dfrac{dy}{dx}$ is not defined. Since $\dfrac{dy}{dx}$ is zero, therfore the tangent at $(a\pi, 2a)$ is parallel to the y-axis.

Thus, (x, y) describes the curve C'OC as θ varies form $-\pi$ to π. It can be easily seen that (x, y) describe congurent portion to the left of the portion C'OC as θ varies in the intervals ...., $[-5\pi, -3\pi]$, $[-3\pi, -\pi]$, and to the right of the portion C'OC as θ varies in the intervals $[\pi, 3\pi]$, $[3\pi, 5\pi]$......

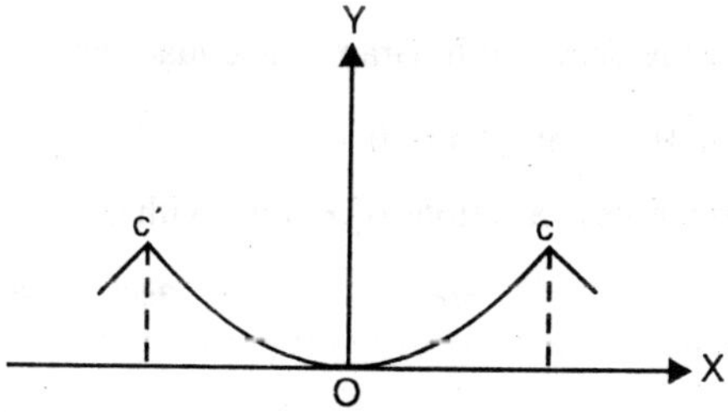

Figure 15.11

(ii) Let (x', y') be the current co-ordinates of any point on the given curve and θ', the parameter, $0 \le q' \le 2\pi$ so that

$$x' = a(\theta' - \sin\theta'), \quad y' = a(1 - \cos\theta')$$

Writing

$$x' = a(\theta + \sin\theta), \quad y' = a(1 + \cos\theta)$$

we find that

$$x' - a\pi = a[(\theta' - \pi) + \sin(\theta' - \pi)], \quad y'' = a[1 + \cos(\theta'' - \pi)]$$

Therefore, $x' - a\pi = a(\theta + \sin\theta)$,

$$y' = a(1 + \cos\theta),$$

where we have written $\theta' - \pi = \theta$

Now as θ' varies from 0 to 2π, $\theta' - \pi$ will vary from $-\pi$ to π;

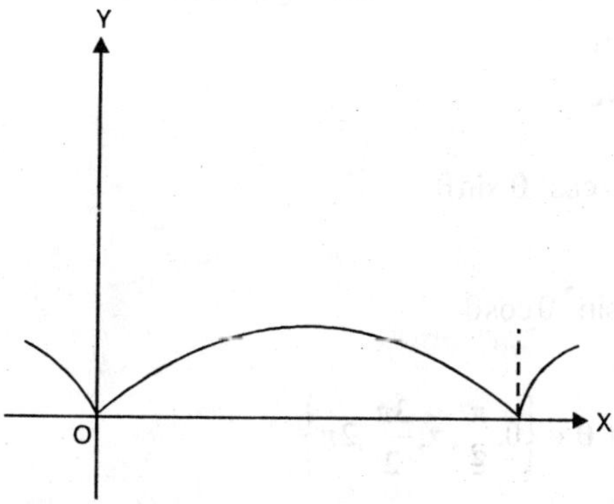

Figure 15.12

Example 2: Trace the curve with parametric equations

$$x = a\cos^3\theta, \qquad y = b\sin^3\theta$$

The following table gives variations of x and y with θ,

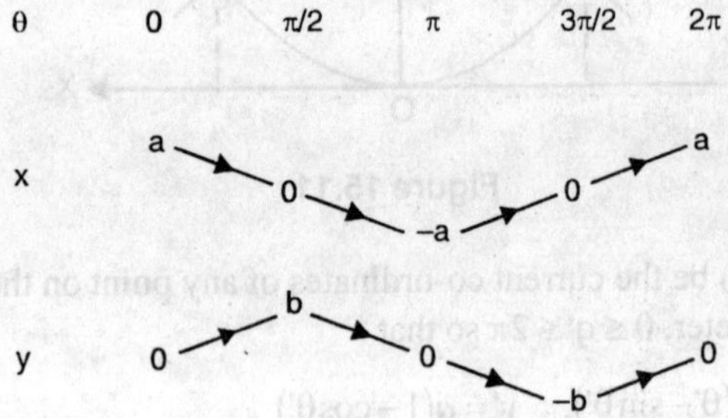

Figure 15.13

so that we have the following results

θ varies in $\left[0, \frac{\pi}{2}\right] \Rightarrow (x, y)$ starting from $(a, 0)$ moves to the left and upwards to (o, b).

θ varies in $\left[\frac{\pi}{2}, \pi\right] \Rightarrow (x, y)$ starting from $(0, b)$ moves to the left and downwards to $(-a, 0)$.

θ varies in $\left[\pi, \frac{3\pi}{2}\right] \Rightarrow (x, y)$ starting from $(-a, 0)$ moves to the right and downwards to $(0, -b)$.

θ varies in $\left[\frac{3\pi}{2}, 2\pi\right] \Rightarrow (x, y)$ starting from $(0, -b)$ moves to the right and downwards to $(a, 0)$.

Again, we have

$$\frac{dx}{d\theta} = -3a\cos^2\theta . \sin\theta$$

$$\frac{dy}{d\theta} = 3b\sin^2\theta\cos\theta$$

$$\frac{dx}{d\theta} = 0 \text{ if } \theta \in \left\{0, \frac{\pi}{2}, \pi, \frac{3\pi}{2}, 2\pi\right\}$$

It follows that

$$\frac{dy}{dx} = \frac{dy}{d\theta} \Big/ \frac{dx}{d\theta}$$

$$= \frac{-b}{a} \frac{\sin^2 \theta \cos\theta}{\cos^2 \theta \sin \theta}$$

$$= \frac{-b}{a} \tan \theta$$

for the values of θ fore which

$$\frac{dy}{d\theta} \neq 0.$$

Also

(i) $\theta \to 0 \quad \Rightarrow \quad \frac{dy}{dx} \to 0$

(ii) $\theta \to \pi/2 \quad \Rightarrow \quad \frac{dy}{dy} \to \infty$

(iii) $\theta \to \pi \quad \Rightarrow \quad \frac{dy}{dy} \to 0$

(iv) $\theta \to 3\pi/2 \quad \Rightarrow \quad \frac{dy}{dy} \to \infty$

Thus the tangent at each of the points $(a,0)$,$(-a,0)$ is the x-axis and the tangent at each of the points $(0,b)$,$(0,-b)$ is the y-axis.

The curve is as given in the following diagram.

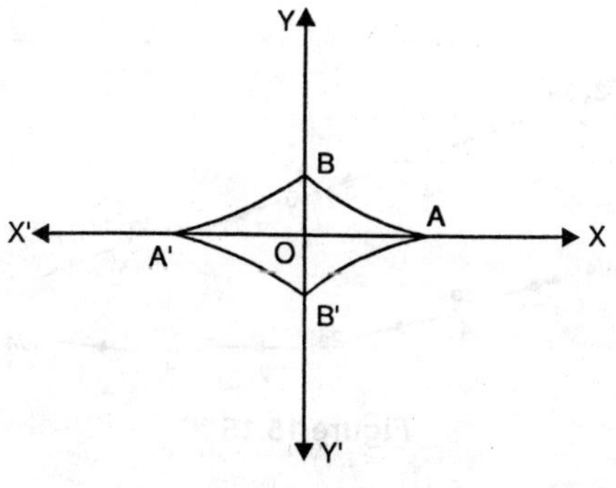

Figure 15.14

Example 3: Trace the curve

$$x = a\sin 2\theta(1+\cos 2\theta),\ \ y = a\cos 2\theta(1-\cos 2\theta)$$

Solution: Since, *x*, *y* are perodic functions of θ, with π as their period the values of *x*, *y* will repeat themselves as θ vareies in the intereval

$$]\pi, 2\pi[,\ \ ]2\pi, 3\pi[\ \ \text{etc.}$$

We thus confine our attention to $\theta \in [0, \pi[$ only.

We have,

$$\frac{dx}{d\theta} = 4a\cos 3\theta . \cos\theta$$

$$\frac{dy}{d\theta} = 4a\cos 3\theta . \sin\theta$$

$$\frac{dx}{d\theta} = 0 \text{ for } \theta = \frac{\pi}{6}, \frac{\pi}{2}, \frac{5\pi}{6}$$

$$\therefore \quad \frac{dy}{dx} = \tan\theta \text{ if } \theta \notin \left\{\frac{\pi}{6}, \frac{\pi}{2}, \frac{5\pi}{6}\right\}$$

Thus, the tangent at any point corresponding to the value θ of the parameter makes an angle θ with *x*-axis.

Table of variation of *x* and *y* with θ

θ	0		$\pi/6$		$\pi/3$		$\pi/2$		$2\pi/3$		$5\pi/6$		π
$\frac{dx}{d\theta}$		+		–		–		–		–		+	
$\frac{dy}{d\theta}$		+		–		–		+		+		–	

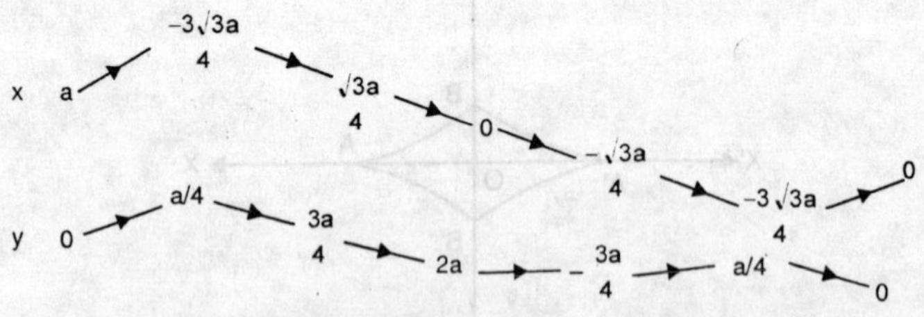

Figure 15.15

The curve is as given in Fig. 15.16

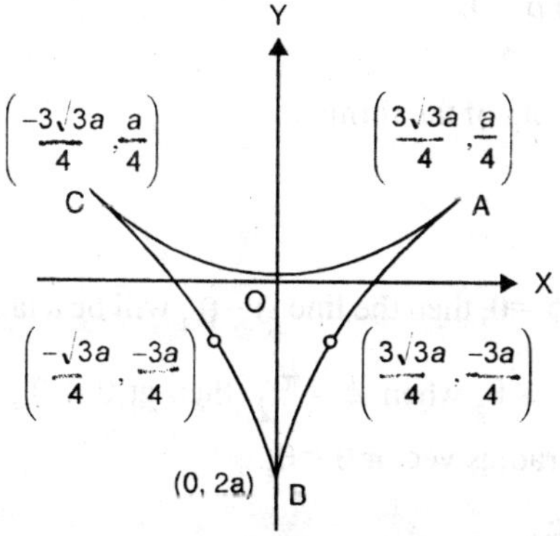

Figure 15.16

15.6. TRACING OF POLAR CURVES

The method of tracing curves whose polar equations are given is similar to that for tracing curves whose cartesian equations are given. The following points are sometimes useful

1. Symmetry:

(i) *If the equation of the curve remains unchanged when* θ *is changed into* $-\theta$, *then the curve is symmetrical about the initial line.*

For example, the curve $r = a(1 + \cos\theta)$ is symmetrical about the initial line.

(ii) *If the equation of the curve remains unchanged when* θ *is changed into* $\pi - \theta$, *then the curve is symmetrical about the initial line* $\theta = \pi/2$.

For example, the curve $r = a(1+\sin\theta)$ is symmetrical the line $\theta = \pi/2$.

(iii) *If the equation of the curve remains unchanged when r is changed into –r, then the curve is symmetrical about the pole.*

For example, $r^2 = a^2 \cos 2\theta$ is symmetrical about the pole.

2. Region:

Find the region in which the curve does not exist. If r is imaginary for some values of θ lying between θ_1 and θ_2 then there is no portion of the curve between the lines $\theta = \theta_1$ and $0 = 0_2$.

If the greatest and the least numerical values of r be a and b respectively then the curve lies entirely within the circle $r = a$ and entirely outside the circle $r = b$ where $a > 0$ and $b > 0$.

3. *Values of* θ:

Find θ with the help of the formula

$$\tan\theta = r\frac{d\theta}{dr}$$

If $\theta = \theta_1$, when $\phi = 0$, then the line $\theta = \theta_1$ will be a tangent to the curve at the point $\theta = \theta_1$ if $\theta = \theta_2$ when $\phi = \pi/2$ then at $\theta = \theta_2$, the tangent will be perpendicular to the radius vector $\theta = \theta_2$.

4. *Special Points:*

Trace the variations of r as θ varies. Also if $\frac{dr}{d\theta} < 0$, then r increasses as θ decreases and if $\frac{dr}{d\theta} < 0$, then r decreases as θ increases.

Taking the above points into consideration we can trace the requisite curve.

Example 1: Trace the curve

$$r = a(1 + \cos\theta)$$

Solution: The following points about the given curve can be noted.

(i) The curve is symmetrical about the initial line.

(ii) The curve passes through the pole, the tangent at the pole being the line $\theta = \pi$.

(iii) The curve crosses the initial line at the point $(2a, 0)$.

(iv) r is never greater than $2a$. Therefore the curve lies within a circle whose centre is at the pole and whose radius is $2a$. There are no asymptotes.

(v) The curve has no points of inflexion.

(vi) $r\frac{d\theta}{dr} = -\cot\theta/2 = \tan\left(\frac{\pi}{2} + \frac{\theta}{2}\right)$, so that $\theta = \frac{\pi}{2} + \frac{\theta}{2}$.

Therefore the tangent to the curve at the point $(2a, 0)$ is perpendicular to the initial line.

(vii) As θ increasses from 0 to $\pi/2$, r decreases from $2a$ to a. As θ increases form $\pi/2$ to π, r decreases from a to 0.

The portion of the graph of the curve above the initial line can now be drawn. The graph can be completed by taking the reflection of this portion in the initial line.

The complete graph is as shown in Fig 15.17

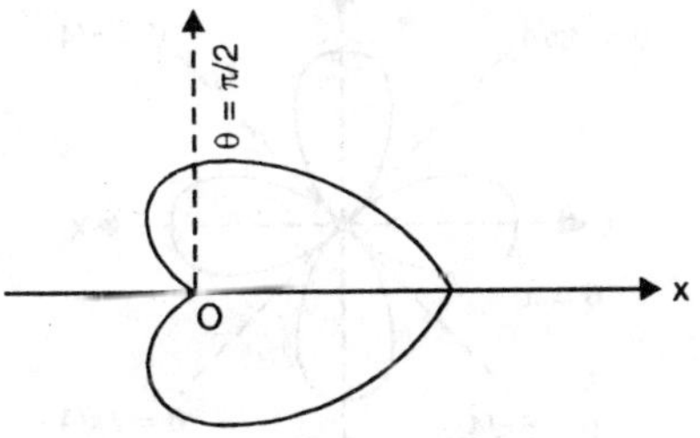

Figure 15.17

Example 2: Trace the curve

$r - a\cos 2\theta,\ a > 0$

Solution: The following points about the given curve can be noted.

(i) The curve is symmetrical about the initial line as well as about the line through the pole perpendicular to the initial line.

$r = 0$

$$\Rightarrow \theta = \frac{\pi}{4}, \frac{3\pi}{4}, \frac{5\pi}{4}, \frac{7\pi}{4}$$

Therefore the curve passes through the pole and $\theta = \frac{\pi}{4}, \frac{5\pi}{4}, \frac{5\pi}{4}, \frac{7\pi}{4}$ are the tangents at the pole.

(ii) The curve has no asymptote.

(iii) The maximum value of r is a so that the curve lies wholly within a circle of radius a.

when $\theta = 0, r = a$

(iv) As θ increases from $\frac{\pi}{4}$ to $\frac{\pi}{2}$, r is negative and decreases from 0 to $-a$.

This portion of the curve lies in the third quadrant.

(v) As θ increses from $\frac{\pi}{2}$ to $\frac{3\pi}{4}$, r increases from $-a$ to 0. and as θ increases from $\frac{3\pi}{4}$ to π, r increses from 0 to a. Since the curve is symmetrical about the initial line, the figure is as follows.

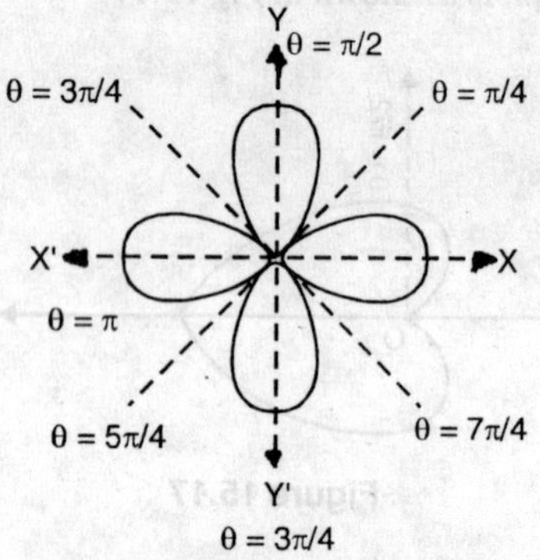

Figure 15.18

Example 3: Trace the curve

$r = a \sin 3\theta$

Solution: The following points about the given curve can be noted.

(i) The curve is symmetrical about the line through the pole prependicular to the initial line.

(ii) The curve passed through the pole and crosses the initial at the pole only.

(iii) Since, $r = 0$

$\Rightarrow 3\theta = n\pi,$

where n is any integer.

$\therefore$ The tangents at the pole are the rays $\theta = 0, \frac{\pi}{3}, \frac{2\pi}{3}, \pi, \frac{4\pi}{3}, \frac{5\pi}{3}.$

The pole is therefore, a multiple point.

(iv) r never exceeds a, so that the curve lies wholly within the circle $r = a$ and consequently are no asymptotes.

(v) When $\theta = 0$, $r = 0$.

As θ increasses from 0 to $\pi/6$, r increases from 0 to a.

As θ increases from $\pi/6$, to $\pi/3$, r decreases from a to 0.

r is negative for values of 0 lying between $\pi/3$ and $2\pi/3$. As θ increases from $\pi/3$ to $\pi/2$, r decreases from 0 to $-a$ and as θ further increases from $\pi/2$ to $2\pi/3$, r increases from $-a$ to 0 giving us the leaf between $\theta = 4\pi/3$ and $\theta = 5\pi/3$.

As θ increases from $2\pi/3$ to $5\pi/6$, r increases from 0 to a as θ further increased from $5\pi/6$ to π, r again decreases from a to 0. As θ increases further, we don't get any new points.

The curve consists of three loops and is often referred to as the three leaved rose.

The graph of the curve is as shown in Fig.15.19

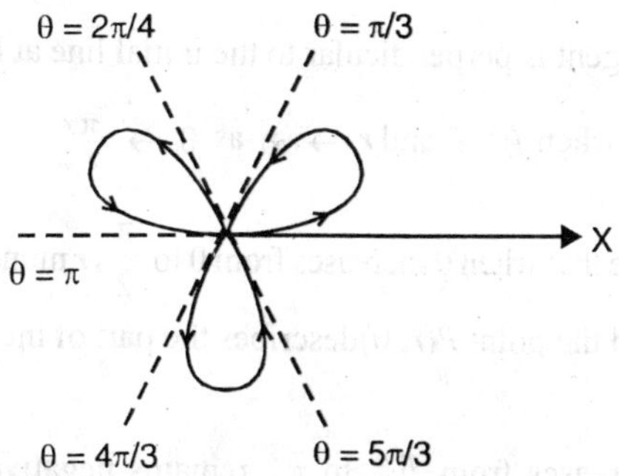

Figure 15.19

Example 4: Trace the curve:

$$r = a(\sec\theta + \cos\theta)$$

Solution: We have,

$$r = a(\sec\theta + \cos\theta)$$

$$= a\left(\frac{1}{\cos\theta} + \cos\theta\right)$$

$$= a\frac{1+\cos^2\theta}{\cos\theta}$$

The following points about the given curve can be noted

(i) The curve is symmetrical about the initial line.

(ii) $r\cos\theta = a$

$\Rightarrow x = a$ is its asymptote.

(iii) $\dfrac{dr}{d\theta} = a\dfrac{\sin^3\theta}{\cos^2\theta}$ so that

$\dfrac{dr}{d\theta}$ is positive when θ lies between 0 and $\pi/2$.

When $\theta = 0$,

$$\frac{dr}{d\theta} = 0$$

so that $\phi = \pi/2$ and,

therefore, the tangent is perpendicular to the initial line at the point $(2a, 0)$.

Also, $r = 2a$, when $\theta = 0$ and $r \to \infty$ as $\theta \to \pi/2$.

Hence we see that when θ increases from 0 to $\dfrac{\pi}{2}$, r monotonically increases from $2a$ to ∞ and the point $P(r, \theta)$ describes the part of the curve drawn in the first quadrant.

When θ increases from $\pi/2$ to π, remains negative and decreases in numerical value from ∞ to $-2a$ and so the point $P(r,\theta)$ describes the part of the curve as shown in the fourth quadrant.

As the curve is symmetrical about the initial line, no new point will be obtained when θ varies from π to 2θ.

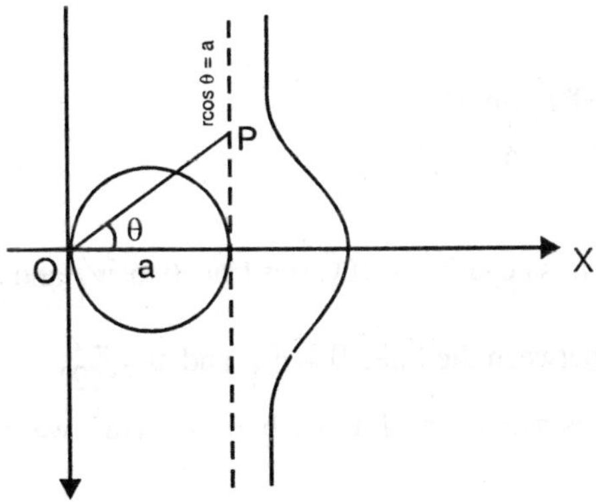

Figure 15.20

15.7. A MIXED APPROACH

In case of some curves, it is found convenient to make use of the polar as well as the cartesian from of their equations. Some facts are obtained from the cartesian and the others from the polar form.

Curves whose cartesian equations can not be solved either for y in terms of x or for x in terms of y, but whose polar equations can be solved for r in terms of θ, are best handled by this mixed approach.

Example 1: Trace the curve:

$$x^4 + y^4 = a^2(x^2 - y^2)$$

Solution: The following points about the given curve can be noted:

(i) It is symmetrical about both the axes.

(ii) Origin is a node on the curve and $y = \pm x$ are the nodal tangents.

(iii) It meets x-axis at $(0, 0)$, $(a, 0)$, but meets y-axis at $(a, 0)$ only $x = a$ and $x = -a$ are the tangents at $(a, 0)$ and $(-a, 0)$

(iv) It has no asymptotes.

(v) On changing to polar co-ordinates, the equation becomes

$$r^2 = \frac{a^2 \cos 2\theta}{\cos^4 + \sin^4 \theta}$$

We see that

$$r\frac{dr}{d\theta}=\frac{-8a^2\sin^3\theta\cos^3\theta}{\left(\cos^4\theta+\sin^4\theta\right)^2}$$

so that $dr/d\theta$ remains negative as θ varies from 0 to $\pi/4$ and threfore no point on the curve lies between the lines $\theta=\pi/4$ and $\theta=\pi/2$.

As the curve is symmetrical about both the axes, we have its shape as shown.

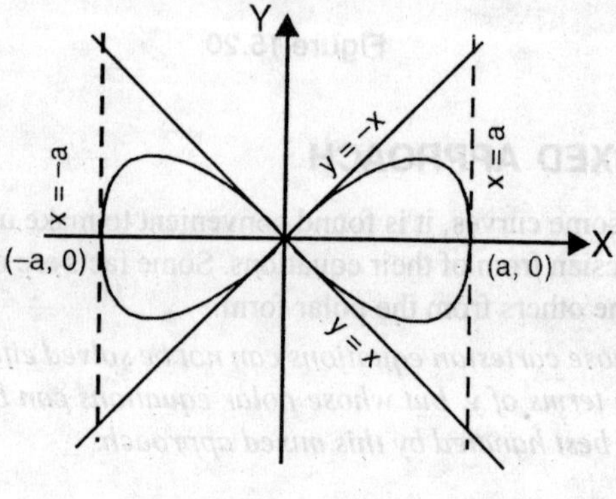

Figure 15.21

Example 2: Trace the folium of descartes:

$$x^3+y^3=3axy \quad a>0$$

Solution:

(i) The curve is symmetrical about the origin. It is also symmetrical about the line $y=x$ and meets it in the point $\left(3a/2, 3a/2\right)$

(ii) x, y cannot both be negative so that no part of the curve lies in the third quadrant.

(iii) $x+y+a=0$ is its only asymptote.

(iv) The curve passes through the origin. It does not meet the axes at any other point.

(v) The equation of the tangents at the origin is $xy = 0$, so that the origin is a node.

(vi) On transforming to polar co-ordinates, we get

$$r = \frac{3a\sin\theta\cos\theta}{\cos^3\theta + \sin^3\theta}$$

Now, $r = 0$ for $\theta = 0$ and $\theta = \frac{\pi}{2}$

$$\frac{dr}{d\theta} = \frac{3a(\cos\theta - \sin\theta)(1 + \sin\theta.\cos\theta + \sin^2\theta.\cos^2\theta)}{(\cos^2\theta + \sin^2\theta)^2}$$

which vanishes only when

$$\cos\theta - \sin\theta = 0$$

$$\Rightarrow \tan\theta = 1$$

$$\Rightarrow \theta = \frac{\pi}{4} \text{ or } \frac{5\pi}{4}$$

for $\theta = \frac{\pi}{4}, r = \frac{3a}{\sqrt{2}}$

Thus r monotonically increases from 0 to $3a/\sqrt{2}$ as θ increases from 0 to $\pi/4$ and monotonically decreases from $3a/\sqrt{2}$ to 0, as θ increases from $\pi/4$ to $\pi/2$.

Again as θ increases from $\pi/2$ to $3\pi/4$, r remains negative and numerically increased from 0 to ∞ so that the point (r, θ) describes the part of the curve shown in the fourth quadrant.

As θ increases from $3\pi/4$ to π, r remains positive and decreases from ∞ to 0 so that the point describes the part of the curve shown in the second quadrant.

It is easy to see that we do not get any new point on the curve when, θ increases from π to 2π.

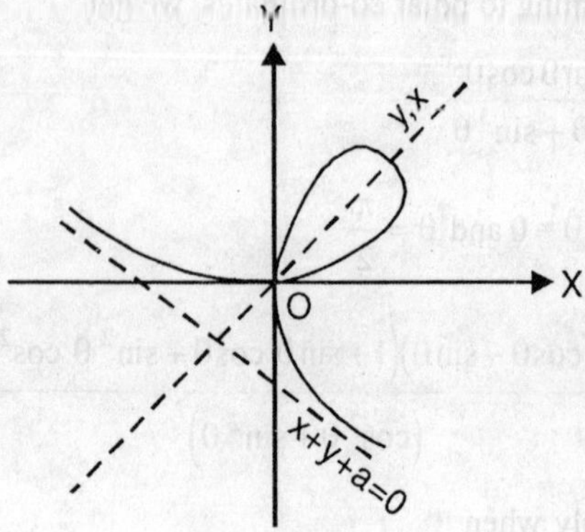

Figure 15.22

Example 3: Trace the curve

$$x^5 + y^5 = 5a^2x^2y$$

Solution:

(i) The curve has rotational symmetry about the origin. There are no other symmetries.

(ii) The curve passes through the origin. It does not meet the axes at any other point.

(iii) The tangents at the origin are $x = 0$, $y = 0$ so that the origin is a node.

(iv) $x + y = 0$ is an asymptote. There are no other asymptote.

(v) Transforming to polar co-ordinates, we have

$$r^2 = \frac{5a^2 \cos^2\theta \sin\theta}{\cos^5\theta + \sin^5\theta}$$

when $\theta = 0$, $r = 0$.

when $\theta = \pi/2$, $r = 0$.

As θ increases from $\pi/2$ to $3\pi/4$, r^2 remains positive and tends to $+\infty$ as $\theta \to 3\pi/4$. When θ lies between $3\pi/4$ and π r^2 is negative and threfore there are no points on the graph corresponding to these values of θ.

By using symmetry, the graph can be completed. as in Fig. 15.23.

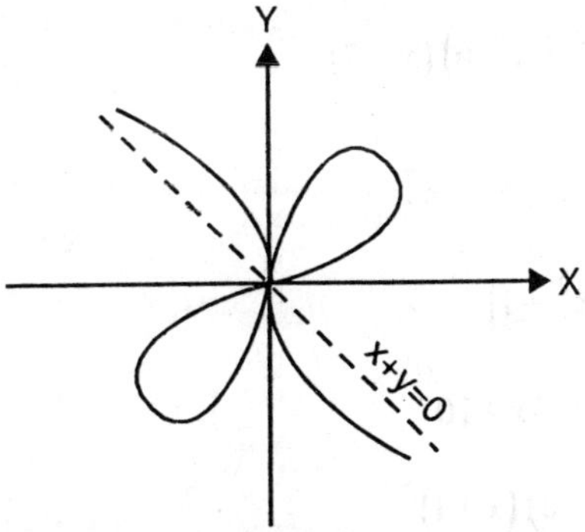

Figure 15.23

Exercise – 15

1. *Trace the following curves:*

   (i) $y = (x^2 - x - 6)(x - 7)$

   (ii) $y = \frac{1}{6}x^6 - x^5 + x^4$

   (iii) $y = x(x^2 + 1)$

   (iv) $y = x^4 - 2x + 10$

   (v) $y = (x-2)(x+1)^2$

   (vi) $y = x^4 - 24x^2 + 80$

   (vii) $y = \frac{1}{5}x^5 - \frac{1}{3}x^3 + 2$

   (viii) $y = 4 - 8x + 5x^2 - x^3$

   (ix) $y = \frac{3}{2}x^4 + \frac{1}{2}x^2 - 4$

   (x) $y = 3x^4 - 16x^3 + 24x^2$

2. *Trace the following curves:*

   (i) $y = \frac{x^2}{1+x^2}$

   (ii) $y = \frac{x^2 - 3x}{x - 1}$

   (iii) $y = \frac{x^2 - 12x + 27}{x^2 - 4x + 5}$

   (iv) $y = \frac{2x^2 - 7x + 5}{x^2 - 5x + 7}$

   (v) $y = \frac{x^3}{x-1}$

   (vi) $y = \frac{x}{x^2 - 1}$

   (vii) $y = \frac{3x+6}{(x-1)^2}$

   (viii) $y = \frac{2x-3}{3x^2 - 4x}$

3. *Trace the following curves*

(i) $y^2 = (x-1)^3$ (ii) $y^2 = x(x+1)^2$

(iii) $ay^2 = x(a^2 - x^2)$ (iv) $ay^2 = x(a^2 + x^2)$

(v) $a^2y^2 = x^2(a^2 - x^2)$ (vi) $a^2y^2 = x^2(2a-x)(x-a)$

(vii) $y^2x = a^2(a-x)$ (viii) $y^2x = x^2 + 1$

(ix) $y^2x^2 = x^2 + a^2$ (x) $y^2x^2 = x^2 - 1$

(xi) $y^2(x+2) = x+1$ (xii) $y^2(a^2 + x^2) = a^2x^2$

(xiii) $y^2 = (x-2)^2(x-3)$ (xiv) $y^2 = (x-a)^2(x-b)$

(xv) $y^2 = x^2(x-4)$ (xvi) $y^2 = x^2(3-x)$

(xvii) $x = (y-1)(y-2)(y-3)$ (xviii) $ay^2 = x^3$

(xix) $a^4y^2 = a^2x^4 - x^6$ (xx) $y^3 = a^2x - x^3$

4. *Trace the following curves:*

(i) $x = a\cos^3 t,\ \ y = a\cos^3 t$

(ii) $x = a(\theta - \sin\theta),\ y = a(1-\cos\theta)$

(iii) $x = a(\theta + \sin\theta),\ \ y = a(1+\cos\theta)$

(iv) $x = \dfrac{1-t^2}{1+t^2},\ y = \dfrac{1-t^3}{1+t^2}$

(v) $x = \dfrac{a\sin 2\theta}{1+\cos 2\theta},\ y = \dfrac{a\cos 2\theta}{1-\cos 20}$

(vi) $x = a(3\cos\theta - \cos^3\theta),\ y = a(3\sin\theta - \sin^3\theta)$

(vii) $x = a(\cos\theta + \theta\sin\theta),\ y = a(\sin\theta - \theta\cos\theta)$

(viii) $x = a\left(\sin\theta + \frac{1}{3}\sin 3\theta\right), y = a\left(\cos\theta - \frac{1}{3}\cos 3\theta\right)$

(ix) $x = a\ln|\sec\theta + \tan\theta|, y = a\ln|\operatorname{cosec}\theta + \cot\theta|$

(x) $x = \cos t, y = \cot t$

(xi) $x = 1 - e^{-t}, y = t^2 + 1$

(xii) $x = e^t + e^{-t}, y = e^t - e^{-t}$

(xiii) $x = 1 + \sin t, y = 2\cos 2t$

(xiv) $x = \frac{1}{2\cos t}, y = 1 + \cot t$

(xv) $x = a(1+\cos\theta)\cos\theta, y = a(1+\cos\theta)\sin\theta$

5. *Trace the following curves:*

(i) $r\cos\theta = a\cos 2\theta$ (ii) $r\cos\theta = 2a\sin^2\theta$

(iii) $r^2\cos^2\theta = a^2\sin^3\theta$ (iv) $r^2\cos^2\theta = a^2\sin 3\theta$

(v) $r = a\log\theta$ (vi) $r\log\theta = a$

(vii) $r = a(1+\sin\theta)$ (viii) $r = a(\theta + \sin\theta)$

(ix) $r = a + b\cos\theta$ (x) $r = ae^{m\theta}$

6. *Trace the following curves*

(i) $x^3 + y^3 = 3ax^2$ (ii) $x^3 + y^3 = a^2 x$

(iii) $x^4 + y^4 = 4a^2 xy$ (iv) $x^4 + y^4 = 4axy^2$

(v) $x^5 + y^5 = 5a^2 x^2 y$ (vi) $x^5 + y^5 = 5ax^2 y^2$

(vii) $x^6 + y^6 = 6a^2 x^2 y^2$ (viii) $y^3 - x^3 = 3ax^2$

(ix) $x^3 - y^3 = 3axy$ (x) $x^3 + y^3 = a$

Answers

Exercise – 2.1

1. (i) **R** (ii) $(0, \infty)$ (iii) $\mathbf{R} - \{-1, 1\}$

(iv) **R** (v) **R** (vi) $[-1, 1] - \{0\}$

(vii) $]1, +\infty[$ (viii) $\left\{\left[\left(2n - \frac{1}{6}\right)\pi, \left(2n + \frac{7}{6}\right)\pi\right] : n \in I\right\}$

(ix) $[-1, 1] \cup]2, +\infty[\cup]-\infty, -2[$

2. (i) Same domain (ii) $]-\infty, 1[, [-1, 1]$

(iii) $[-1, 1], \quad \mathbf{R} -]-1, 1[$ (iv) **R, R**

3. (i) $(-\infty, 0) \cup \left(\frac{3}{2}, \infty\right)$ (ii) $\left[-\frac{1}{2}, \frac{1}{2}\right]$ (iii) $(-\infty, 1)$

(iv) $\{-1, 1\}$ (v) $\{0, 4\}$ (vi) $\{0, \infty\}$

4. (i) Domain : $N \cup \{0\}$, Range : $\{(n1) : n = 0, 1, 2.........\}$

(ii) Domain : $(0, \infty)$, Range : $(1, \infty)$

(iii) Domain : **R**, Ranage : $(-\infty, 1)$

(iv) Domain : **R**, Range : $\{1\}$

(v) Domain : **R**, Range : $\frac{1}{\sqrt{7}} \le y \le 1$

Exercise – 2.2

1. $g(x)=\frac{3(x+1)}{x-2}, \quad x \ne 2$

2. $g(x)=(x-1)+|x-1|$ on $0<x\le 2$

3. $g(x)=\begin{cases} 2+x & \text{for } 0\le x\le 1 \\ 2-x & \text{for } 1<x\le 2 \\ 4-x & \text{for } 2<x\le 3 \end{cases}$

4. (i) $\sqrt{1+x}+\sqrt{1-x}$

   (ii) $\sqrt{1+x}+\sqrt{1-x}$

   (iii) $\sqrt{1-x^2}$

   (iv) $\sqrt{1+x}/\sqrt{1-x}$

5. (i) $f(x)=\sin x,\ g(u)=\sqrt{u},\ (gof)(x)=\sqrt{\sin x}$

   (ii) $f(x)=\tan x,\ (fof)(x)=\tan(\tan x)$

   (iii) $f(x)=x^2+1,\ g(u)=\sqrt{u},\ h(v)=\sin v,$

   $$(hog)of=\sin\sqrt{\left(x^2+1\right)}$$

   (iv) $f(x)=x^2-3x+2,\ g(x)=\sqrt{x},$

6. (i) $\sin x$ (ii) $\sin x$ (iii) $\sin^2\sqrt{x}$

Exercise – 3.1

6. $\lim f(x) = +\infty, -\infty, -\infty, +\infty, 1, 1$ according as x tends to $-2-0, -2+0, 2-0, 2+0, +\infty, -\infty$ respectively.

12. (i) $+\infty$ (ii) $\frac{1}{4}$ (iii) 12

(iv) $\frac{1}{2}$ (v) 2 (vi) $e^3 - 1$

(vii) $2\sqrt{a}\cos a$ (viii) 1 (ix) $9\log e^3$

(x) $\frac{1}{2}$ (xi) $\frac{m^2}{n^2}$ (xii) 1

(xiii) 32 (xiv) $-\frac{1}{\sqrt{2}}$

Exercise – 3.2

1. (i) continous (ii) continous (iii) continous
(iv) discontinous (v) continous (vi) continous
(vii) discontinous (viii) discontinous (ix) discontinous
(x) discontinous

2. Limit does not exist.

3. discontinous

4. Discontinuity of the first kind from left.

5. (i) 6 (ii) $-\frac{1}{3}$ (iii) $\frac{1}{4}$

6. $a = \pi/6,\ b = -\pi/12,$

Exercise – 4.1

2. (i) 9 (ii) $-\frac{1}{9}$ (iii) $\frac{1}{3}$

(iv) $-\frac{1}{2}$ (v) 12

Exercise – 4.2

2. Continuous and not derivable.
5. Continuous and derivable at $x = 1, 2$.
7. Continuous and derivable.
8. $f'(0) = 0$
9. f is not continuous for $x = 0$ and not derivable for $x = 0$.

   f is continuous and derivable for $x = \pi/2$.

Exercise – 4.3

1. (i) $\dfrac{ax+b}{\sqrt{ax^2+bx+c}}$ (ii) $\dfrac{-2a}{3x^{\frac{2}{3}}\left(1+ax^{\frac{1}{3}}\right)^2}$

   (iii) $9x^2 - 44x + 21$ (iv) $\dfrac{-2x^7 - 6x^4 + 6x^2 + 2x}{\left(x^2-1\right)^2\left(x^3-1\right)^2}$

   (v) $\dfrac{2(x\sin x + \cos x)}{(x+\cos x)^2}$ (vi) $\dfrac{(x+1)\sin^2 x + 1}{2\sqrt{\left(1+\sin x^2\right)}.\sqrt{x}}$

   (vii) $\left(m\cos^2 x - n\sin^2 x\right)\sin^{m-1} x\cos^{n-1} x$

   (viii) $\dfrac{2}{3}.\dfrac{\cos x}{\sqrt[3]{\sin x}}$

   (ix) $\dfrac{-1}{\sqrt{\cos 2x}(\cos x + \sin x)}$ (x) $-\dfrac{2x+3}{2\sqrt{x^2+3x+4}}$

   (xi) $\dfrac{\sqrt{\left(b^2-a^2\right)}}{b+a\cos x}$ (xii) $\dfrac{-2}{1+x^2}$

   (xiii) $\dfrac{1}{\sqrt{\left(x^2+a^2\right)}}$ (xiv) $\dfrac{\cot(\log x)}{x}$

(xv) $\frac{1}{2}$ (xvi) $\tan^{-1}\frac{a}{b} - x$

(xvii) $\frac{1}{\sqrt{1-x^2}} - \frac{1}{2\sqrt{x}\sqrt{1-x}}$ (xviii) $\frac{3a}{a^2+x^2}$

(xix) $\pi - 2\tan^{-1} x^n$ (xx) $\frac{1}{2}$

11. $\frac{15-25x}{3(1-x)^{\frac{5}{3}}} - 2x\sin(4x+2)$

15. $\frac{2}{\left(x+x^2\right)^{\frac{3}{2}}}$

Exercise – 4.4

1. (i) $x^{\left(x^2+1\right)}\log\left(ex^2\right)$

(ii) $\left[1+x\log x\ \log ex\right]x^{\left(x^2\right)+x-1}$

(iii) $\left(10^{10^x}\right).10^x.(\log 10)^2$

(iv) $\left(\frac{x}{1+x}\right)^x\left[\frac{x^2}{2}\log\left(\frac{x}{1+x}\right)-\frac{x}{2}+\frac{1}{2}\log(x+1)\right]$

(v) $(x\log x)^{\log\log x-1}\{\log(\log x)(\log x+1)+\log(x\log x)\}$

(vi) $\left[\frac{1}{2}\text{cosec}\,2x\log x+\frac{1}{x}\log(\tan x)\right](\tan x)^{\log x}$

(vii) $(\tan x)^{\cot x}\text{cosec}^2 x\log(e\cot x)-(\cot x)^{\tan x}\sec^2 x\log(e\tan x)$

(viii) $e^{x\sin x^3}\left(3x^3\cos x^3+\sin x^3\right)+(\tan x)^x\left\{\log(\tan x)+\frac{1}{2}x\text{cosec}\,2x\right\}$

(ix) $\left(x+\frac{1}{x}\right)^x\left[\frac{x^2-1}{x^2+1}+\log\left(x+\frac{1}{x}\right)\right]+x^{1+\frac{1}{x}}\left(\frac{x+1}{x^2}-\frac{\log x}{x^2}\right)$

12. $\frac{\pi}{4+\tan^{-1}x}+1$

13. $\frac{1}{2}\sin 2x$

14. $\sin\theta$

15. $\frac{x^{\sin x-1}(\sin x+x\cos x\log x)}{(\sin x)^{x-1}(x\cos x+\sin x\log\sin x)}$

16. (i) $-\text{cosec}^2 x\cot hx-\text{cosec}\,h^2 x\cot x$

(ii) $a^x\{\sin hx+x(\sin hx\log a+\cos hx)\}$

17. $\tan 2\theta$

18. $(\tan x)^{\cot x}\text{cosec}^2 x\log(e\cot x)-(\cot x)^{\tan x}\sec^2 x\log(e\tan x)$

19. (i) $-\frac{a}{b}$ (ii) $\cot\theta$ (iii) $\frac{e^t+e^{-t}}{e^t+e^{-t}}$

(iv) -1 (v) 0

Exercise – 5.1

10. $A=\frac{2}{3},\ B=\frac{-1}{3}$

20. $\frac{d^2y}{d\theta^2}+y=0$

Exercise – 5.2

1. (i) $2^{n-1}\cos(2x+n\pi/2)+2^{2n-3}\cos(4x+n\pi/2)$

(ii) $\frac{1}{2}\left[\left(\sqrt{37}\right)^n e^x \cos\left(6x + n\tan^{-1} 6\right) + \left(\sqrt{5}\right)^n e^x \cos\left(2x + n\tan^{-1} 2\right)\right]$

(iii) $\frac{1}{4}\left[\left(\sqrt{13}\right)^n e^{3x} \sin\left(2x + n\tan^{-1}(2/3)\right) +\right.$

$\left. 5^n e^{3x} \sin\left(4x + n\tan^{-1}(4/3)\right) + \left(\sqrt{39}\right)^n \sin\left(6x + n\tan^{-1} 2\right)\right]$

(iv) $2^{-4}\left[2\cos(x + n\pi/2) - 3^n \cos(3x + n\pi/2) - 5^n \cos(5x + n\pi/2)\right]$

(v) $\frac{1}{4}e^x\left[\left(a^2+1\right)^{n/2} \sin\left(x + n\tan^{-1}\frac{1}{a}\right) + \left(a^2+9\right)^{n/2}\right.$

$\left. \sin\left(3x + n\tan^{-1}\frac{3}{a}\right)\right]$

(vi) $\frac{1}{8}e^x\left[3 - 4\cdot 5^{n/2} \cos\left(2x + n\tan^{-1} 2\right) - 17^{n/2} \cos\left(4x + n\tan^{-1} 4\right)\right]$

(vii) $\frac{1}{4}\left[2^n \cos\left(2x + \frac{1}{2}n\pi\right) + 4^n \cos\left(4x + \frac{1}{2}n\pi\right) + 6^n \cos\left(6x + \frac{1}{2}n\pi\right)\right]$

(viii) $\frac{3}{4}\sin\left(x + \frac{1}{2}n\pi\right) - \left(\frac{3}{4}\right)^n \sin\left(3x + \frac{1}{2}n\pi\right)$

(ix) $2^{n-1}\cos\left(2x + \frac{n\pi}{2}\right) - 2^{3n-1}\left(8x + \frac{n\pi}{2}\right)$

(x) $(-1)^{n-1}2^{-n}\cdot 1\cdot 3 5 \ldots\ldots\ldots\ldots (2n-3)\cdot 9^n (9x+8)^{(1-2n)/2}$

(xi) $\frac{(-1)^{n-1}(n-1)!}{(x+2)^n}$

(xii) $\frac{n!b^n}{(a-bx)^{n+1}}$

2. (i) $(-1)^n n!\left[\dfrac{16}{(x-2)^{n+1}} - \dfrac{1}{(x-1)^{n+1}}\right], n > 2$

(ii) $\dfrac{(-1)^n n!}{4}\left[\dfrac{3}{(x-2)^{n+1}} + \dfrac{1}{(x+2)^{n+1}}\right]$

(iii) $(-1)^{n-1}\cdot(n!)\left[\dfrac{\sin(n+1)\theta \sin^{n+1}\theta}{2^{(n+2)/2}} - \sin(n+1)\phi \sin^{n+1}\phi\right]$

where, $x = \sqrt{2}\cot\theta = \cot\phi$

(iv) $\dfrac{(-1)^n \cdot n!}{3(x+1)^{n+1}} - \dfrac{(-1)^n n!}{3}\cdot\left(\dfrac{2}{\sqrt{3}}\right)^{n+1}$

$\left[\cos(n+1)\theta - \sqrt{3}\sin(n+1)\theta \sin^{n+1}\theta\right]$

where, $\theta = \tan^{-1}\left\{\sqrt{3}/(2x-1)\right\}$

(v) $(-1)^n n!\left[\dfrac{1}{(x-1)^{n+1}} - \dfrac{1}{(x+1)^{n+1}} + \dfrac{2(n+1)}{(x-1)^{n+2}}\right]$

(vi) $(-1)^{n-1}(n-2)!\sin^{n-1}\theta\cos\theta\cos n\theta[n\tan\theta - \tan n\theta]$

where, $\theta = \tan^{-1}(1/x)$

(vii) $(-1)^{n-1}(n-1)!\sin n\theta \sin^n\theta\,\text{cosec}^n \propto$,

where $\cot\theta = x\,\text{cosec} \propto - \cot \propto$

(viii) $\dfrac{a^2}{(a-b)(a-c)}\cdot\dfrac{(-1)^n\cdot n!}{(x-a)^{n+1}} + \dfrac{b^2}{(b-c)(b-a)}\cdot\dfrac{(-1)^n n!}{(x-b)^{n+1}} +$

$\dfrac{c^2}{(c-a)(c-b)}\cdot\dfrac{(-1)^n n!}{(x-c)^{n+1}}$

(ix) $(-1)^n n!\left[\frac{1}{(x+1)^{n+1}}-\frac{2^n}{(2x+1)^{n+1}}\right]$

(x) $\frac{(-1)^n n!}{4a^3}\left[\frac{1}{(x-a)^{n+1}}-\frac{1}{(x+a)^{n+1}}-\frac{2\sin\left[(n+1)\cot^{-1}(x/a)\right]}{\left(a^2+x^2\right)^{(n+1)/2}}\right]$

(xi) $(-1)^{n-1}(n-1)!\sin^n\theta\sin n\theta$, where $\theta=\cot^{-1}x$

(xii) $\frac{1}{2}(-1)^{n-1}(n-1)!\sin^n\theta\cdot\sin n\theta$, *where* $\theta=\cot^{-1}x$

Exercise – 5.3

1. (i) $2^{(n-2)/2}e^x\left[2x^2\cos\left(x+\frac{n\pi}{4}\right)+2^{\frac{3}{2}}\cdot nx\cdot\cos\left(x+\frac{(n-1)\pi}{4}\right)+\right.$

$\left.n(n-1)\cos\left(x+\frac{(n-2)\pi}{4}\right)\right]$

(ii) $a^{n-4}e^{ax}\left[a^4x^4+4na^3x^3+6n(n-1)a^2x^2+\right.$

$\left.4n(n-1)(n-2)ax+n(n-1)(n-2)(n-3)\right]$

(iii) $x^3\cos\left(x+\frac{1}{2}n\pi\right)+3nx^2\cos\left[x+\frac{1}{2}(n-1)\pi\right]+$

$3n(n-1)x\cos\left[x+\frac{1}{2}(n-2)\pi\right]+n(n-1)(n-2)$

$\cos\left[x+\frac{1}{2}(n-3)\pi\right]$

(iv) $x^3a^n\sin\left(ax+\frac{1}{2}n\pi\right)+3nx^2a^{n-1}\sin\left\{ax+\frac{1}{2}(n-1)\pi\right\}+3n(n-1)x$

$$a^{n-2}\sin\left\{ax+\frac{1}{2}(n-2)\pi\right\}+n(n-1)(n-2)\,a^{n-3}$$

$$\sin\left\{ax+\frac{1}{2}(n-3)\pi\right\}$$

(v) $e^x\left[\ln x+\frac{n}{x}-\frac{n(n-1)}{2x^2}+\frac{n(n-1)(n-2)}{3x^3}-\dots\dots\dots\dots\right.$

$$\left.\frac{(-1)^{n-1}(n!)}{nx^n}\right]$$

(vi) $(-1)^{n-1}(n-2)!\sin^{n-1}\theta\cos\theta\cos n\theta\,(n\tan\theta-\tan n\theta)$,

where $\theta=\tan^{-1}(1/x)$

(vii) $\sum_{r=0}^{n}\frac{(n!)^2}{(r!)^2(n-r)!}\left(a^2+b^2\right)^{r/2}x^r\,e^{ax}\,\sin\left\{bx+r\tan^{-1}(b/a)\right\}$

(viii) $a^{n+2}\cdot e^{ax}\cdot x^2$

13. If n is even, $y_n(0)=0$;

if n is odd, $y_n(0)=n!(3n-5)\frac{1}{2^{n+4}}$

14. If n is even, $y_n(0)=0$;

if n is odd, $y_n(0)=(-1)^{(n-1)/2}\,1^2\cdot3^2\cdot5^2\dots\dots\dots\dots(n-2)^2$

15. If n is even, $y_n(0)=(-1)^{(n-1)/2}\cdot2\cdot2^2\cdot4^2\cdot6^2\dots\dots\dots\dots(2n-2)^2$;

if n is odd, $y_n(0)=0$

Exercise – 6.1

1. (i) $xx'+yy'=r^2$; $xy'-x'y=0$

(ii) $4x-y=0$; $x+4y=17$

(iii) $x\tan(\theta/2) - y\cos(\theta/2) = a\theta\sin(\theta/2)$;

(iv) $b\sqrt{2}x - ay = ab$; $ax + b\sqrt{2}y = \left(a^2 + b^2\right)\sqrt{2}$

(v) $yy' = 2a(x + x')$; $xy' + 2ay = (x' + 2a)y'$

(vi) $x + p^2y = 2cp$; $p^3x - py = c\left(p^4 - 1\right)$

(vii) $x + y = 0$; $x - y = 0$

(viii) $13x - 16y = 2a$; $16x + 13y = 9a$

(ix) $x\cos^3\theta + y\sin^3\theta = c$; $x\sin^3\theta - y\cos^3\theta + 2c\cot 2\theta = 0$

(x) $31x + 8y + 9a = 0$; $8x + 31y + 42a = 0$

4. $(a\cos\theta)^{m/(m-1)} + (b\sin\theta)^{m/(m-1)} = p^{m/(m-1)}$

15. $y = x + 1$, $y = x - 1$

16. $x + y = 1$

18. $(1, 0)$

Exercise – 6.2

1. (i) $45°$ (ii) $\tan^{-1}(4/3)$

(iii) $\frac{\pi}{2}$ *and* $\tan^{-1}(3/5)$ (iv) $\tan^{-1}\sqrt{16}$

(v) $\tan^{-1} 2\sqrt{2}$

2. (i) $a - b = a' - b'$

(ii) show these cut orthogonally

(iii) show these cut orthogonally

5. (i) $a\sin^2\theta$; $a\tan^2\theta\sin^2\theta$; $a\sin^2\theta\cos\theta$; $a\sin^3\theta\tan\theta$.

(ii) $2a\sin\frac{1}{2}\theta$; $2a\sin\frac{1}{2}\theta\tan\frac{1}{2}\theta$; $a\sin\theta$; $2a\sin^3\frac{1}{2}\theta\sec\frac{1}{2}\theta$

Exercise – 6.3

1. (i) $\tan^{-1}\{(1+e\cos\theta)/e\sin\theta\}$ (ii) $\pi/2+m\theta$

(iii) $\theta/2$ (iv) $3\pi/4$

(v) $\pi/4+m\theta$ (vi) $\pi/2-\theta/2$

2. (i) $\pi/2$ (ii) $\tan^{-1}3$

(iii) $\pi/2$ (iv) $\pi/2$

(v) $\tan^{-1}4/3$; $\pi/4$ at the pole (vi) $\tan^{-1}\left[2e/\left(e^2-1\right)\right]$

(vii) $\tan^{-1}\left(a^2/b^2\right)$ (viii) $2\pi/3$

(ix) $3\sin^{-1}\left(\frac{3}{4}\right)^{1/4}$ (x) $\pi/2$

13. (i) $r^m=a^m\cos m\theta$ (ii) $p=r\sin\propto$

(iii) $2p^2a=r^3$ (iv) $r^4=\left(b^2-a^2+2ar\right)p^3$

(v) $pr=a^2$ (vi) $1/p^2=1/r^2+1/a^2$

(vii) $ar^3=4p^4$ (viii) $r^{m+1}=p\sqrt{\left(a^{2m}+b^{2m}\right)}$

Exercise – 7.1

7. $$e^{ax}\cos bx=1+ax+\frac{a^2-b^2}{2!}x^2+\frac{a\left(a^2-3b^2\right)}{3!}x^3+$$

$$\text{...........}+\frac{x^n}{n!}\left(a^2+b^2\right)^{n/2}e^{a\theta x}\cos\left(b\theta x+n\tan^{-1}\frac{b}{a}\right)$$

9. $\sin^2 x = \frac{1}{2}\left[\frac{(2x)^2}{2!} - \frac{(2x)^4}{4!} + \frac{(2x)^6}{6!} - ....\right]$

$\cos^2 x = 1 - \frac{1}{2}\left[\frac{(2x)^2}{2!} - \frac{(2x)^4}{4!} + \frac{(2x)^6}{6!} - ....\right]$

16. $e^{m\sin^{-1}x} = 1 + mx + \frac{m^2}{2!}x^2 + \frac{m(1^2 + m^2)}{3!}x^3 + \frac{m^2(2^2 + m^2)}{4!}x^4 + ...$

29. $\frac{1}{2} + \frac{x}{4} - \frac{x^3}{48}$

Exercise – 8.1

3. (i) 21, 1 (ii) 29, 28

(iii) $14, \frac{-3}{4^{4/3}}$ (iv) 19, 3

4. $a = 1, b = 0$

5. 49

9. max· – 8,

10. min· – 10, – 26

11. (i) *Min* $-1, -2/3\sqrt{6}$; *Max* $1, 2/3\sqrt{6}$

(iii) *Min*· $\left(a^{2/3} + b^{2/3}\right)^{3/2}$; *Max*· $-\left(a^{2/3} + b^{2/3}\right)^{3/2}$

(iv) *Min*· $0, -2/3\sqrt{3}$; *Max*. $0, 2/3\sqrt{3}$

12. $1/e$

13. Max, for $x = n\pi$, where n is an odd positive or even negative integer
Min, for $x = n\pi$, where n is an odd positive or odd negative integer.

14. max· for $x = \sqrt[4]{3}$; min· for $x = -\sqrt[4]{3}$.

15. max· 5; min· 50

Exercise – 8.2

1. 32, 32

2. The cylinder with radius $\left(\frac{50}{\pi}\right)^{1/3}$

3. (i) $x = L/2$ (ii) $x = L/\sqrt{3}$

6. (4, –4)

7. $\frac{r}{\sqrt{2}}$, $\sqrt{2}r$, Area $= r^2$

8. For max. project, the daily output must be 10 radio sets.

9. $\frac{3\sqrt{3}}{4}ab$

12. $\left(\frac{9c}{16}\right)^{1/3}, 2\left(\frac{9c}{16}\right)^{1/3}$ and $\left(\frac{32c}{81}\right)^{1/3}$

13. $\pi a^2\left(1+\sqrt{5}\right)$

Exercise – 9.1

1. –2	2. 1/2	3. –1/2
4. $-2/\pi$	5. 4	6. $2a/b$
7. –1	8. 1/2	9. $\ln a/\ln b$
10. a/b	11. –2/3	12. 1/12
13. 4	14. –2	

15. $p = -5/2$, $q = -3/2$

Exercise – 9.2

1. (i) 0	(ii) 0	(iii) 1
(iv) 1	(v) 0	(vi) 0

2. (i) 1 (ii) 0 (iii) $2a/\pi$

(iv) 0 (v) $2/\pi$ (vi) 0

3. (i) 1/2 (ii) $-e$ (iii) 0

(iv) 0 (v) $\frac{-2}{3}$ (vi) 0

(vii) $\pi^2/8$ (viii) ∞ (ix) 1/2

(x) 1/2 (xi) 1/2 (xii) 0

4. (i) 1 (ii) 1 (iii) 1

(iv) 1 (v) 1 (vi) 1

(vii) e (viii) 1 (ix) 1

(x) e (xi) $e^{1/3}$ (xii) $e^{-2/\pi}$

(xiii) e (xiv) $e^{-1/6}$ (xv) 0

(xvi) $\log(e/b)\log be$ (xvii) -1 (xviii) $11e/24$

(xix) -0 (xx) 1/8

6. e^a 7. e^a

Exercise – 10.1

1. (i) $f_x = \dfrac{-1}{1+(x+y)^2}$, $f_y = \dfrac{-1}{1+(x+y)^2}$

(ii) $f_x = ae^{ax}\cos by$, $f_y = ae^{ax}\sin by$

(iii) $f_x = 2xy^2\cos(x^2y^2)$, $f_y = 2x^2y\cos(x^2y^2)$

(iv) $f_x = \dfrac{2x}{(x-y)^2}$, $f_y = \dfrac{-2y}{(x-y)^2}$

2. (i) $f_{xx} = 2y(1+y^2)(1-xy)^{-3}$,

$$f_{yy} = 2x\left(1+x^2\right)(1-xy)^{-3}$$

$$f_{xy} = f_{yx} = 2(x+y)(1-xy)^{-3}$$

(ii) $$f_{xx} = -3xy\left(1+y^2\right)\left(1+x^2+y^2\right)^{-5/2},$$

$$f_{yy} = -3xy\left(1+x^2\right)\left(1+x^2+y^2\right)^{-5/2},$$

$$f_{xy} = f_{yx} = \left(1+x^2+y^2+3x^2y^2\right)\left(1+x^2+y^2\right)^{-5/2}$$

(iii) $$f_{xx} = \frac{-1}{x^2}, \qquad f_{yy} = \frac{1+2y\tan^{-1}y}{\left(1+2y^2+y^4\right)\left(\tan^{-1}y\right)^2}$$

$$f_{xy} = f_{yx} = 0$$

(iv) $$f_{xx} = a^2e^{ax}\left[\cos\left(e^{ax}+e^{by}\right) - e^{ax}\sin\left(e^{ax}+e^{by}\right)\right]$$

$$f_{yy} = a^2e^{ax}\left[\cos\left(e^{ax}+e^{by}\right) - e^{ax}\sin\left(e^{ax}+e^{by}\right)\right]$$

$$f_{xy} = f_{yx} = -abe^{by+ax}\sin\left(e^{ax}+e^{by}\right)$$

(v) $$f_{xx} = ye^{\left(x^y\right)}x^{y-2}\left[yx^y+y-1\right]$$

$$f_{xy} = f_{yx} = e^{\left(x^y\right)}x^{y-1}\left[1+\left(1+x^y\right)\log x^y\right]$$

$$f_{yy} = e^{\left(x^y\right)}x^y\left(1+x^y\right)(\log x)^2$$

5. $1/c^{2z}$

7. 0

21. $-3/2$

Exercise – 10.2

2. $4x+2y$

4. $(\cos y - xy\sin y)/x^2$

5. (i) $\dfrac{dx}{dy}=\dfrac{1-2z}{1+4xz},\ \dfrac{dz}{dy}=\dfrac{1+2x}{1+4xz}$

(ii) $\dfrac{dx}{dz}=\dfrac{1-2z}{1+2x},\ \dfrac{dy}{dz}=\dfrac{1+4xz}{1+2x}$

6. $-\dfrac{(u\cot x\cos u\sin y+z\sin v\sin x)}{\cos v\sin y}$

7. $(xt\,e^t-y)/\{t(x^2+y^2)\}$

8. $-(1+x^2)(1+y^2)(2s+r^2)/(1-xy)^2$

9. $2e^{2t}(\cos 2t-\sin 2t)$

11. $\dfrac{\partial z}{\partial x}=2y^2e^{2x}+2xe^{-2y}-2y^2/x^3,$

$\dfrac{\partial z}{\partial x}=2ye^{2x}-2x^2e^{-2y}+2y/x^2$

15. (i) $-\dfrac{4x^3-5a^2y}{4y^3-5a^2x}$

(ii) $-\dfrac{y^x\ln y+yx^{y-1}-(x+y)^{x+y}\ln(x+y)}{x^y\ln x+xy^{x-1}-(x+y)^{x+y}\ln(x+y)}$

19. $2x\{\cos(x^2+y^2)\}(1-a^2/b^2)$

23. 0

25. $\dfrac{\partial u}{\partial x}=1,\ \dfrac{\partial u}{\partial y}=1,\ \dfrac{\partial v}{\partial x}=\dfrac{-v}{u},\ \dfrac{\partial v}{\partial y}=\dfrac{1-v}{u}$

Exercise – 11

1. (i) $c \sec^2 \psi$ (ii) $a \cos \psi$

(iii) $3a \sec^3 \psi \tan \psi$ (iv) $\frac{4}{3} a \cos \frac{1}{3} \psi$

(v) $c \tan \psi$ (vi) $am\, e^{m\psi}$

(vii) $\frac{4a}{3} \sin \frac{\psi}{3}$ (viii) $2a \sec^3 \psi$

4. (i) $\cosh \frac{x}{c}$ (ii) $\frac{a^2 + x^2}{a^2 - x^2}$

9. $\left(\frac{9}{8}, \pm 3\right)$

16. $2a^2/b, \quad 2b^2/a$

25. (i) $\frac{\left(a^4 y^2 + b^4 x^2\right)^{3/2}}{a^4 b^4}$ (ii) $\frac{\left(x^{2m-2} + y^{2m-2}\right)^{3/2}}{(m-1)x^{m-2} y^{m-2}}$

26. (i) $\frac{a\left(1+\theta^2\right)^{3/2}}{\theta^4}$ (ii) $\frac{a^n}{(n+1)\, r^{n-1}}$

(iii) $\frac{a^2}{3r}$ (iv) $\sqrt{r^2 - a^3}$

35. (i) $\sqrt{2}$ (ii) $4/5$

(iii) $5\sqrt{5}/4; \quad 5\sqrt{5}/4$

37. (i) $\frac{an}{2}$

39. $2\sqrt{2}; \quad 2\sqrt{2}$

Exercise – 12

1. (i) For $x = -b/3a$ (ii) $(5, 0), \left(\frac{119}{27}, \frac{2}{3}\right)$

(iii) $(0, 0), (1, 0), (-1, 0)$ (iv) for $x = -2$ and $1 \pm \sqrt{3}$

(v) $\left(\frac{3}{2} ae^{-3/2}, ae^{3/2}\right)$ (vi) $(1, 3)$

(vii) $(0, 2)$ (viii) For $x = -2, -4 \pm \sqrt{18}$

(ix) For $x = -a$ and $a\left(2 \pm \sqrt{3}\right)$ (x) For $x = 0$ and $\pm a\sqrt{3}$

2. Concave to the x-axis in the interval $]0, \pi[$ and convex to the x-axis in the interval $]\pi, 2\pi[$.

5. (i) Concave upwards in $[-2, 0]$ and $[2, \infty[$; concave downwards in $]-\infty, -2]$ and $[0, 2]$. Inflexions at $(-2, 198), (0, -20), (2, -238)$.

(ii) Concave upwards in $]-\infty, -1]$ and $[1, \infty[$; concave downwards in $[-1, 1]$. Inflexions at $(-1, 2e)$ and $(1, 10/e)$.

6. Convex to the x-axis in the interval $]-\infty, 1[,]2, \infty[$; concave to the x-axis in the interval $]0, 1[$; (0, 1) and (1, 0) are the points of inflexion.

7. Concave upwards in $[0, \pi/4]$ and $[5\pi/4, 2\pi]$ and concave downwards in $[\pi/4, 5\pi/4]$

11. (ii) $\left[a\left(4\pi n \pm \frac{2\pi}{3} \pm \frac{\sqrt{3}}{2}\right), \frac{3a}{2}\right]$ (iii) $\left(\pm\frac{2a}{\sqrt{3}}, \frac{3a}{2}\right)$

(iv) $\left(\pm\sqrt{3}a, \pm\frac{\sqrt{3}a}{4}\right), (0, 0)$ (v) For $x = 0, \pm\frac{\sqrt{3}}{2}$

(vi) For $x = 2 \pm \sqrt{2}$ (vii) For $x = \log 2$

(viii) No inflexion (ix) For $x : \tan x = -\frac{3}{4}$

(x) For $x = \frac{1}{2} \pm \frac{1}{2}\sqrt{11}$

12. (i) $\left(3a/2, \pm\sqrt{3}\right)$ (ii)

Exercise – 13

1. $y = \pm x$ **2.** $y = \pm x$

3. $bx = \pm ay$ **4.** $y = \pm 2x$

5. $y = \pm x$ **6.** $x = 0$

8. $y = 0$ **9.** $x = 0,\ y = \frac{1}{2}x$

10. $(y-1) = \pm(x-2)$ **11.** $\pm\sqrt{3}y = \sqrt{2(x-a)}$

12. $y = x - 1$

14. $y = x - a,\ y = -x - 3a$

15. Node at (0, 0)
16. Node at (0, 0)
17. Cusp at (0, 0)
18. Node at (0, 0)
19. Node at (2, 0)

20. $(a, 0)$ is a node, if $b < a$
$(a, 0)$ is a cusp, if $b = a$
$(a, 0)$ is an isolated point, if $b > a$

21. Conjugate point at $(1, -1)$
22. Conjugate point at $(0, 0)$
23. Conjugate point at (a, o)

24. Conjugate point at $(-a, -b)$
25. $(0, -a)$ is a double point
26. No multiple point
27. Conjugate point at (a, a)
28. Cusp at $(1, -1)$
29. Cusp at $(0, -4a)$
30. $(0, \pm a)$ are triple point : $\left(-\frac{3}{2}a, \pm a\right)$ are crups; node at $(-a, 0)$
31. $y - a = \pm(2/\sqrt{3})x$; $y = \pm\sqrt{(2/\sqrt{3})}(x - a)$; $y - a = \pm(2/3)(x - 2a)$
32. Node at each of the four points $(0, \pm 1), (\pm 1\ 0)$;

    $y = \pm\sqrt{2}(x - 1),\ y = \pm\sqrt{2}(x + 1),\ \pm\sqrt{2}(y - 1) = x,\ \pm\sqrt{2}(y + 1) = x$
33. $y - 2 = \pm 2\sqrt{2}(x - 2)$
34. Node at $(-1, -1)$; $x + 1 = 0,\ y + 1 = 0$
35. $x - y + 1 = 0, \quad x + y = 3$
40. Single cusp of the first species at the origin.
41. Single cusp of the first species at the origin.
42. Single cusp of the first species at the origin.
43. Single cusp of the second species at the origin.
44. Single cusp of the first species.
45. Single cusp of the first species.
46. Isolated point
47. Single cusp of the first species at $(9, 0)$.
48. Double cusp of second species.
49. Single cusp of the second species at $(0, -1)$.
50. Oscu-inflexion
51. Oscu-inflexion
53. 0
54. $3a/2$ for each
55. $2\sqrt{2}a,\ 2\sqrt{2}\,a/3$

56. $a\sqrt{2}$

57. $\frac{85}{2}\sqrt{17},\ 5\sqrt{2}$

59. $a, -a/6, -5^{3/2}a$

60. $b^2/2a,\ b^2/2c$

62. $2\sqrt{2}$ for each

Exercise – 14.1

1. $y=0$

2. $x=\pm 1,\ y=0$

3. $x=0,\ y=\pm a$

4. $x=0$

5. $y=3, x=2, x=3$

6. $x=a$

7. $x=\pm a,\ y=\pm b$

8. $x=2, x=3$

9. $x=\pm 1,\ y=0$

10. $x=1, y=-1$

11. $x-1=1, x+y=2$

12. $x=0, y=x, y=x+1$

13. $y=x\pm a,\ x=\pm a$

14. $x-y=1$

15. $y=x-2, y=x-3$

16. $x+1=0, y=$ [illegible], $x+y=0$

17. $y=x-3, y=x+2, x+y=0$

18. $x+y=a$

19. $y=0,\ y+x=\pm 1$

20. $y=0,\ y=x\pm\sqrt{2}$

21. $x+y=2, y=x-2, x=2y-4$

22. $x+y=\pm\sqrt{2},\ x=\pm 1,\ y=\pm 1$

23. $y=0,\ y\pm x=1$

24. $y=x, 2x+y+1=0, 2y+x+1=0$

25. $x=1$

26. $x+y=2, y=x+2, y=2x-4$

28. $x=0, y=0, 2y-3x=6$

29. $x=\pm a,\ y\pm x=a$

30. $y\pm x=0$

31. $x+y\pm a=0$

32. $x=y+a, x=y+2a$

33. $2y\pm x=0,\ y=\pm(2x+1)$

34. $x=\pm a$

36. $x+y=4a$